More Prais

THE DINOSAUF

D0255054

Finalist for the Missis
of Arts and Letters' 2019 Prize in Nonfiction

"*The Dinosaur Artist* is a breathtaking feat of writing and reporting: a strange, irresistible, and beautifully written story steeped in natural history, human nature, commerce, crime, science, and politics. It's at once laugh-out-loud funny and deeply sobering. I was blown away by the depth of its characters, its vivid details, and Paige Williams's incredible command of the facts. Bottom line: this is an extraordinary debut by one of the best nonfiction writers we've got."

—Rebecca Skloot, #1 *New York Times* bestselling
author of *The Immortal Life of Henrietta Lacks*

"Paige Williams is that rare reporter who burrows into a subject until all of its dimensions, all of its darkened corners and secret chambers, are illuminated. With *The Dinosaur Artist*, she has done more than reveal a gripping true-crime story; she has cast light on everything from obsessive fossil hunters to how the earth evolved. This is a tremendous book."

—David Grann, #1 *New York Times* bestselling
author of *Killers of the Flower Moon*

"Williams's writing is often concise and evocative.... [*The Dinosaur Artist*] is gripping and cinematic." —*Wall Street Journal*

"[*The Dinosaur Artist*] pretty much does for fossils what Susan Orlean did for orchids." —*Book Riot*

"As a reader, being given entry by Williams into this underworld, privy to the secret knowledge of a black market, is a thrill.... The strange underground world Prokopi inhabits inevitably brings us in contact with some serious oddballs, each of whom is introduced by Williams with the economy and evocative precision of a haiku.... The book's most memorable character may be Mongolia itself, a r   that defies easy generalization."

"[Williams] skillfully navigates this unique nexus of paleontology and law along with its notorious black markets." —*Science Friday*

"If you love dinosaurs, paleontology, or just a rollicking good tale, you will love this book. I couldn't put it down."
—Jennifer Ackerman, *New York Times* bestselling
author of *The Genius of Birds*

"*The Dinosaur Artist* is a triumph.... Captivating, funny, and profound, it is easily one of the strongest works of nonfiction in years."
—Ed Yong, staff writer for *The Atlantic* and the *New York Times*
bestselling author of *I Contain Multitudes*

"*The Dinosaur Artist* is a tale that has everything: passion, science, politics, intrigue, and, of course, dinosaurs. Paige Williams is a wonderful storyteller."
—Elizabeth Kolbert, Pulitzer Prize–winning
author of *The Sixth Extinction*

"A cracking combination of true crime, dinosaurs, and top-notch investigative journalism."
—Steve Brusatte, bestselling author of
The Rise and Fall of the Dinosaurs

"I am in awe of Paige Williams.... Few nonfiction writers are capable of mining their characters with such a winning blend of sympathy, wonder, and rigor."
—Liza Mundy, *New York Times* bestselling
author of *Michelle* and *Code Girls*

"Paige Williams is as deft as the fossil hunters and skeleton builders she writes about.... The result is a work of art."
—Jack E. Davis, Pulitzer Prize–winning
author of *The Gulf*

THE
DINOSAUR
ARTIST

OBSESSION, SCIENCE,
AND THE
GLOBAL QUEST
FOR FOSSILS

PAIGE WILLIAMS

 hachette
BOOKS

NEW YORK   BOSTON

For my mother

Hachette Books
Hachette Book Group
1290 Avenue of the Americas, New York, NY 10104
hachettebooks.com
twitter.com/hachettebooks

Originally published in hardcover and ebook by Hachette Books in September 2018.
First trade paperback edition: September 2019

Hachette Books is an imprint of Perseus Books, LLC, a subsidiary of Hachette Book Group, Inc. The Hachette Books name and logo are trademarks of Hachette Book Group, Inc.

The publisher is not responsible for websites (or their content) that are not owned by the publisher.

The Hachette Speakers Bureau provides a wide range of authors for speaking events. To find out more, go to www.hachettespeakersbureau.com or call (866) 376-6591.

LCCN: 2018937293
ISBN: 978-0-316-38251-9

Printed in the United States of America

LSC-C

10 9 8 7 6 5 4 3 2 1

The history of life on earth has been a history of interaction between living things and their surroundings.

—*Rachel Carson*

Fossil Hunting is by far the most fascinating of all sports. The hunter never knows what his bag will be, perhaps nothing, perhaps a creature never before seen by human eyes! The fossil hunter does not kill, he resurrects. And the result of his sport is to add to the sum of human pleasure and to the treasures of human knowledge.

—*George Gaylord Simpson, 1934*

A fed crow
Returns thirteen times

—*Mongolian proverb*

CONTENTS

x • CONTENTS

ABBREVIATIONS

AAPS Association of Applied Paleontological Sciences (commercial trade group)

AMNH American Museum of Natural History (New York City); occasionally referred to here as the "American Museum"

BLM Bureau of Land Management, a division of the U.S. Department of the Interior

CBP Customs and Border Protection, a division of the U.S. Department of Homeland Security

DOJ Department of Justice

FARA Foreign Agents Registration Act

FMNH Florida Museum of Natural History (Gainesville, Florida)

ICE Immigration and Customs Enforcement, a division of Homeland Security

IRI International Republican Institute

MPP Mongolian People's Party (early 1920s)

MPRP Mongolian People's Republic Party (starting in 1924)

MUST Mongolian University of Science and Technology

NHM Natural History Museum (London)

NMNH Smithsonian National Museum of Natural History (Washington, DC)

NPS National Park Service

OSCE Organization for Security and Co-operation in Europe

PRPA Paleontological Resources Preservation Act of 2009

SDNY Southern District of New York, a federal district court of the U.S. Department of Justice

SVP Society of Vertebrate Paleontology

AUTHOR'S NOTE

This is a work of nonfiction. No names have been changed, no information invented. My reporting began in 2009, but for the purposes of the book's final form the immersive research occurred between 2012 and 2018. In the United States, I reported in Arizona, Connecticut, Florida, Georgia, Massachusetts, Montana, New Jersey, New York, Pennsylvania, South Dakota, Virginia, Washington, DC, and Wyoming. In Mongolia, I reported in the Gobi Desert, Töv Province, and Ulaanbaatar. In Canada, I reported in Edmonton, Alberta. In Europe, I reported in Munich, Germany, and in Charmouth, London, and Lyme Regis, England. The information that I gleaned from interviews with paleontologists, geologists, fossil dealers, preparators, collectors, museum curators, auctioneers, law enforcement, and various government agents may not appear in full here, yet these generous people's insights informed the work. Written source material, some of it obtained through the Freedom of Information Act, included unclassified and declassified U.S. embassy cables and State Department reports, civil lawsuits, Department of Justice criminal case files and asset forfeiture records, library collections, news archives, peer-reviewed research papers, and county court documents. I also relied upon sources' personal photos, videos, correspondence, and papers. Mongolian documents were translated by Mongolians unrelated to the Mongolian government or the *T. bataar* case.

Much of this book grew out of "Bones of Contention," a piece that I wrote for *The New Yorker* in January 2013. There, as here, I tried to convey the nuances of the debate over who owns, or should be allowed to collect and own, natural history, and how that conflict may in turn affect a range of interests, including public policy, science, museums, and geopolitics. Various scenes I observed directly. For convenience, I occasionally interchange "dinosaur," "fossil dinosaur," and "skeleton"—writing that someone "bought a dinosaur" I of course refer to the extinct animal's stony remains. Likewise, I occasionally use "bone" for "fossil," having explained that fossilization yields rock. The title *The Dinosaur Artist* is not intended to refer exclusively to a leading subject of this book, Eric Prokopi, but rather also to dinosaurs' unparalleled power to remain culturally, scientifically, and aesthetically relevant despite extinction, and to the long, crucial intersection between science and art. Some readers may also choose to infer the formal definition of the word: "a habitual practitioner, of a specified reprehensible activity." When speaking, some scientists refer to natural history museums by their acronyms ("AMNH" instead of "the AMNH"); although "the AMNH" clangs in my ear, I use that construction for clarity. I've borrowed slivers of my own language from the original *New Yorker* piece and from a *Smithsonian* article I later wrote about the endangered *takhi* horse, a creature that was a divine thrill to see in person on the Mongolian steppe.

INTRODUCTION: ORIGINS

In the summer of 2009, I came across a newspaper item about a Montana man convicted of stealing a dinosaur. The idea sounded preposterous. How was stealing a dinosaur even possible? And who would want to?

Nearly a decade earlier, this man, Nate Murphy, who led fossil-hunting tours in a geological signature in Montana called the Judith River Formation, had become well known for unveiling Leonardo, a late Cretaceous *Brachylophosaurus* and one of the best-preserved dinosaur skeletons ever found. A volunteer fossil hunter named Dan Stephenson had found the skeleton during one of Murphy's excursions on a private ranch near the small town of Malta. The remains constituted the first sub-adult of its kind on record and, remarkably, still bore traces of "skin, scales, muscle, foot pads—and even his last meal in his stomach," *National Geographic* reported. "To find one with so much external detail available, it's like going from a horse and buggy to a steam combustion engine," Murphy told the magazine. "It will advance our science a quantum leap."

"Our science" was an intriguing phrase. Murphy wasn't a trained scientist; he was an outdoorsman who had taught himself how to hunt fossils in the Cretaceous-bearing formations that run with photogenic accessibility through states like Wyoming, Utah, Montana, and South Dakota. He believed he had something to offer paleontology, and, presumably in pursuit of this idea, he had taken fossils that didn't belong to him. (Not Leonardo; another dinosaur.) What at first appeared to be little more than a bizarre true-crime story became, to me, an absorbing question of our ongoing relationship with natural history, with the remnants of a world long gone.

We know which life-forms exist because we encounter them, but what came before? Answers can be found in rock. If you've ever picked up a shark tooth or a leaf-imprinted stone, you were holding a fossil—a time portal, a clue. By definition, fossils are prehistoric organic remains preserved in the earth's crust by natural causes. If you, yourself, would like to become a fossil, a specific chain of events must occur. Your corpse must not be eaten or scattered by scavengers, or destroyed by other ruinous forces like weather and running water. You must be buried quickly in sediments or sand: metamorphic and igneous rock, which form under conditions too superheated and volatile to preserve much of anything, are no good at making fossils, but sedimentary rock—limestone, sandstone—proves an excellent tomb. Your soft tissues and organs will decompose, but unless they're obliterated by the planet's incessant chemical and tectonic motions, the hard bits—teeth and bone—will remain. These will be infiltrated by groundwater and will mineralize according to whatever elements exist in the patch of earth that has become your grave—eventually, you may become part crystal or iron. Then, to even start to be scientifically useful, you must be discovered.

Good luck with all that. It's been estimated that less than one percent of the animal species that ever lived became fossils.

While the process is rare, the product is ubiquitous, at least regarding some species. But which fossils are important to science and how should they be protected? Paleontologists have one answer, commercial fossil dealers another, and they've been fighting about it for generations.

As the only record of life on Earth, fossils hold the key to understanding the history of the planet and its potential future. Studying them, scientists can better monitor pressing issues such as mass extinction and climate change; hunting, collecting, or viewing them, anyone may feel connected to both the universe's infinite mystery and Earth's tangible past. To see the dinosaur bone beds of the Liaoning province of northeastern China is to see a landscape that 120 million years ago featured lush lakes and forests in the shadow of active volcanoes. To encounter *Glossopteris* imprints—an extinct seed fern found in South America, Africa, Australia, and Antarctica—is to witness evidence that those continents once existed as a single landmass. To hold a Kansas clam is to touch a relic of the Western Interior Seaway, which for roughly 20 million years bisected North America, overlaying what are now North Dakota, Wyoming, Colorado, Texas, Louisiana, Mississippi, and Florida, along with parts of fourteen other states and swaths of Canada and Mexico.

Fossils are found in every part of the world, and so are fossil collectors, who are legion. Collectors spend significant chunks of their lives hunting for fossils, researching fossils, buying fossils, displaying fossils, trading fossils, visiting fossils in museums, and talking—and talking and talking—about them. Fossil enthusiasts are as obsessed a segment of natural history lovers as ever existed. "I have been in people's houses where every possible inch of their home is covered in fossils," the vertebrate paleontologist Mark Norell, of the American Museum of Natural History, once said. "Even the dishwasher has trilobites in it."

This, minus the dishwasher, has been going on for millennia. As humans collected the remains of one life form after another, naturalists built an inventory of the planet's former inhabitants. That inventory today is known as the fossil record, a compendium that is postulated, debated, and revised by paleontologists through peer-reviewed research, providing a portrait of lost time. Without fossils, an understanding of the earth's formation and history would not be possible. Without fossils, we would not know Earth's age: 4.6 billion years. We would not know when certain creatures lived, when they died out, how they looked, what they ate. Without fossils, natural history museums might not exist. The geologic time scale would not exist because knowledge of the earth's stratigraphy, or layers, would not exist. We would not know that the continents were not always where they are now, and that Earth's shifting, sliding plates rearrange land and sea. We would not know the climate has warmed and cooled and is changing still. We would not know that five mass extinctions have occurred and that we're in the sixth one now. We would have no idea of any ice age. Without fossils, we would not know that birds evolved from dinosaurs; or that Earth was already billions of years old before flowering plants appeared; or that sea creatures transitioned to life on land and primates to creatures that crafted tools, grew crops, and started wars. We would not know that rhinos once lived in Florida and sharks swam around the Midwest. We would not know that stegosaurs lived millions of years before *T. rex*, an animal that, in geologic time, is closer to human beings than to the first of its kind.

The earth's layers are finite: each has a beginning, middle, and end, like tiramisu, wherein ladyfingers meet mascarpone. The most recent layers hold mammals, fishes, and birds not terribly different from those that are alive today, but the further back one goes, the more fantastical some of the creatures. The fossil record shows that life began with microscopic organisms and flourished to the unthinkably gargantuan animals of the Mesozoic, a 160-million-year era that ended some 65.5 million years ago. In the Age of Reptiles, dinosaurs crashed through forests, terrorized prey, zipped around like overstimulated roadrunners, and lub-lubbed along, looking for something leafy to eat and trying to avoid being eaten. Their remains continually surface as weather, erosion, and civilization peel the planet layer by layer.

Fossils are the single most important clue to understanding how the planet evolved, yet attitudes toward their protection vary from continent to continent, and from state to state. The United States, a particularly fossil-rich country, is unusual: policymakers have had no desire to mess with private-property laws, so it remains true that if you find fossils on your own land, or on private property where you have permission to collect, they are yours to keep or sell or ignore or destroy, no matter what or how scientifically important the specimen may be.

Three primary groups of people seek and covet fossils: paleontologists, collectors, and commercial hunters. Paleontologists hone their expertise through undergraduate, graduate, and doctoral courses that immerse them in geology, evolutionary biology, zoology, computer science, statistical analysis, ecology, chemistry, climatology, and other maths and sciences. They pursue specialties in areas like paleobotany (fossil plants), invertebrate paleontology (animals without backbones, like mussels and corals), micropaleontology (requires a microscope), and vertebrates (backbones). Paleontologists tend to work in academia and museums, publishing their research in peer-reviewed scientific journals such as *Geology* and the *Journal of Vertebrate Paleontology*. Scientists believe it crucial to protect certain types of fossils by banning their trade. Commercial dealers, on the other hand, hunt, sell, and buy fossils, at trade shows, in privately owned natural history shops, and online. It is entirely legal to sell some fossils and illegal to sell others, and it's often been hard for consumers to know the difference. Many dealers grew up hunting fossils and might have studied natural sciences in college if they'd had the chance. Most are self-taught. Many are libertarians and believe they should be able to do whatever they want as long as they're not hurting anybody. Many loathe government regulations and feel entitled to fossils, taking the view that the earth belongs to everyone. Most fossil dealers feel that by collecting and selling fossils they're rescuing materials that otherwise would erode, and that their industry provides a valuable service by supplying classrooms and collectors and, in some cases, museums, and by encouraging widespread interest in the natural world. Commercial hunters take pride in selling to museums, but they also court wealthy, private collectors. Successful dealers can make a living in fossils, though it is rarely a get-rich game, since so much of the profit folds back into the hunt. Overseas museums, especially those proliferating in China, Japan, and the Middle East, have no problem buying commercially while public museums in the United States—those supported by tax dollars— tend not to shop the market, preferring to collect their own materials under scientific conditions. While both a commercial hunter and a paleontologist may also be a collector, no reputable paleontologist is a dealer: paleontologists do not sell fossils for much the same reason hematologists don't peddle vials of blood. Fossils *are* the data, it's been said.

A fossil's contextual information is as important as the fossil itself. Extracting a fossil minus that correlating data has been compared to removing a corpse from the scene of a homicide without noting, say, the presence of shell casings or biological evidence like semen and blood. Approximate cause and time of death may be inferred, but a fossil alone cannot tell the whole story; in fact, the whole story can *never* be told, at least not without a time machine. But the story starts to come together through the analysis of details like the circumstances of fossilization (called taphonomy), the presence of other fossil animals and plants, and stratigraphy, which helps paleontologists understand when the animal lived and died. The enormous femurs found protruding from the Big Bone Lick bogs of Kentucky (as happened in the 1700s) tell one story; the large three-toed footprints found *sans* bones in the Connecticut River Valley of Massachusetts (as happened in the 1800s) tell another.

For decades the federal government debated whether and how to regulate fossil collecting, particularly regarding vertebrates, which are less common than invertebrates. The most extreme-minded paleontologists have long wanted a ban on commercial collecting, but commercial hunters organized against the idea. They defended their trade, and paleontologists defended the objects fundamental to their science.

Despite experience and field expertise, dealers who call themselves "commercial paleontologists" are not in fact paleontologists. Paleontology would not exist without them, though. The science started at the hands of natural history lovers—started long before the words *science* and *paleontology* even existed—and became perhaps the only discipline with a commercial aspect that simultaneously infuriates scientists and claims a legitimate role in the pantheon of discovery. The work of commercial hunters has allowed paleontologists some of their biggest breakthroughs and museums their most stunning displays. Museum visitors may not realize they're often looking at specimens discovered not by scientists but rather by lay people like themselves. A California boy named Harley Garbani became obsessed with fossils in the 1930s, after finding part of a camel femur while following in the tracks of his father's plow. He became a plumber but went on to find extraordinary, tiny fossils by crawling on his hands and knees in "cheaters" (jewelers' goggles), plus the first significant *Triceratops* skeleton in over half a century and a *T. rex* skeleton so good it would take years for someone to come across a better one. By the time Garbani died, in 2011, he had collected for the Natural History Museum of Los Angeles County and the University of California–Berkeley's Museum of Paleontology. Lowell Dingus, an American Museum of Natural History (AMNH) paleontologist who knew Garbani while in grad school at Berkeley, called him "among the greatest fossil collectors that ever lived and the greatest one that I have ever known and worked with."

A more recent collector was Stan Sacrison, an electrician and plumber from Harding County, South Dakota, the self-declared "*T. rex* Capital of the World." In the 1980s and '90s, Sacrison found such notable *rex* specimens that with each new discovery his twin brother, Steve, a part-time gravedigger and equally gifted fossil hunter, carved notches into the handle of his Bobcat earthmover. Discovering even one or part of one *T. rex* was a feat, given that fewer than fifteen had been unearthed. The Sacrison twins, who lived in the tiny town of Buffalo, had grown up near fossil beds and were taken with the hunt. They had learned that it was smart to search after a big storm or a spring thaw because weather and erosion unwrap gifts of bone. They had familiarized themselves with geology, knowing it's as pointless to search for mastodon in rock formations 100 million years old as it is to look for *Vulcanodon* in sediments laid down during the Pleistocene.

Another name to remember is Kathy Wankel. When the Smithsonian National Museum of Natural History (NMNH) unveils its new hall of dinosaurs in 2019, after a five-year, $48 million renovation, it will feature, for the first time, its own *Tyrannosaurus rex*, courtesy of Wankel, a Montana rancher who in 1988 found the skeleton now known as "The Nation's *T. rex*." The specimen is considered important partly because it includes the first complete *T. rex* forelimb known to science.

Despite amateurs' contributions, science and commerce developed stark opposing arguments:

Commerce: overregulation destroys the public's interest in the natural world.
Science: commodification compromises our evolving understanding of the planet.
Commerce: science doesn't *need* hundreds or even dozens of specimens of one species.
Science: multiple specimens elucidate an organism and its environment over time.
Commerce: private collectors wind up donating their stuff to museums anyway.
Science: specimens collected under nonscientific conditions are worthless to research.
Commerce: most museum fossils land in storage, never to be studied.
Science: stored fossils have generated profound advances decades after their discovery.
Commerce: scientists are stingy and elitist, with their snooty PhDs.
Science: commercial hunters are destructive and greedy.

Such were the contours of a seemingly intractable conflict. "Whether or not it's okay to sell and buy fossils is a matter of debate on scientific and ethical grounds, with analytical rigor and professional honesty squaring off against free enterprise," the paleontologists Kenshu Shimada and Philip Currie and their colleagues wrote in *Palaeontologia Electronica.* They called "the battle against heightened commercialization" of fossils "*the* greatest challenge to paleontology of the 21st century."

On both sides, the disagreement struck people as a shame, because scientists and commercial hunters at least were united in their love of one thing: fossils. If only *more* people would take a sincere interest in "rocks that can talk to you," the paleobotanist Kirk Johnson, head of the Smithsonian's NMNH, once told me. "The fact that our planet buries its dead is an amazing thing. The fact that you can read the history of the planet in fossils is profoundly cool. A smart kid can find a fossil and tell you what happened to the planet 4 billion years ago. We finally figured out how the planet works, and we did it through fossils."

If the confessed dinosaur thief Nate Murphy became an emblem of the tension between science and commerce, he didn't reign for long. In the spring of 2012, a case emerged that surpassed all others in its international scope and labyrinthine particulars, touching on collectors, smuggling, marriage, democracy, poverty, artistry, museums, mining, Hollywood, Russia, China, criminal justice, presidential politics, explorers, Mongolian culture, the auction industry, and the history of science. This book is that untold story.

PART I

CHAPTER 1

"SUPERB TYRANNOSAURUS SKELETON"

On the last day of his old life, the dinosaur hunter went to the beach. This was Florida—Atlantic side, not Gulf. An overcast Sunday morning: the twentieth day of the fifth month of the two thousand and twelfth year CE. Eric Prokopi was thirty-eight. His daughter, Rivers, was turning three. Eric and his wife, Amanda, lugged a carload of party gear from their home in Gainesville across the upper peninsula, to St. Augustine, a sixteenth-century city named for a fourth-century theologian who, as a boy, stole pears off a tree simply because it was forbidden, later writing, "Foul was the evil and I loved it."

The Prokopis drove straight onto the foreshore, as is done in that town, and set up on the sand. The previous year's party theme had been Pirates & Princesses. Eric, a tall, muscular ex-swimmer, had dressed in a frilly pink frock and tiara, accessorized with the black wraparound sunglasses of a mercenary. This year's theme was Little Mermaid. Invitations pictured the birthday girl, a brown-eyed towhead with a heart-shaped face, wearing a finned tail and a platinum wig whose synthetic waves cascaded down her back. Now she had on a shimmering green skirt and a purple plastic crown, to which Amanda had affixed a dried starfish from the burgeoning inventory of her new interior decorating business, Everything Earth. Rivers's brother, Greyson, two years older, wore a long-sleeved black swim shirt printed with the outline of a great white shark, and wraparound shades like his dad's. All the essential elements were soon in place: tent, tables, sunscreen, cupcakes. A clear-acrylic cooler dispensed cerulean Hawaiian Punch. A watermelon, cut in the shape of an open-mouthed shark, offered a gullet full of gummy fish that glowed like sunlit rubies.

Every few minutes, Eric stepped away to pace the sand with his Blackberry to his ear, increasingly anxious about the news from New York. He should have felt relaxed by the break of the surf and the opportunity to search the shoreline for washed-up treasure, as he had loved to do since childhood, but to be distracted and stressed was to miss these pleasures even while enacting them.

The competing tensions, years in accrual, were beginning to show on his body. His eyes, brown as acorns, were bracketed by deepening crow's-feet. His right eye had developed an inflamed twitch. His dark hair sprouted silver like crabgrass after a dense rain. His enormous hands—bratwurst fingers, saucer palms—were callused and nicked. Amanda told friends that Eric worked so hard, she practically had to prop him up to make him eat. The kids barely saw him anymore. Until recently he had never been a drinker, but now he needed the dulling effect of at least one vodka cocktail before he could sleep. When working on big projects he might come to bed at four in the morning or not all.

Late at night, Amanda could step to a rear window of their house, Serenola, and look out at the huge prefabricated workshop they had recently installed in the backyard—the recent shift in Eric's vocation required more space. The shop, 5,000 square feet, with four bays and a pitched roof, stood beyond the swimming pool where the elegant landscaping gave way to the wild rear acres of the property, in a part of southwest Gainesville that developers hadn't yet managed to ruin. Without close neighbors, Eric could work as noisily and late as necessary, the night vibrating with the high whine of his air scribe. Well into the hours of ambient frog song, Amanda could see the lights burning, sparks shooting off the welder.

Eric's company was called Florida Fossils. He hunted, restored, bought, and sold the remains of prehistoric creatures, mostly shark teeth and Ice Age mammals like giant ground sloths. Several times a year he hosted a sales booth at the world's largest natural history shows—Tucson, Denver, Munich—and also attended ones like Sainte-Marie-aux-Mines, France. Otherwise he sold fossils by email blast and on eBay, or to private clients. Amanda described the lifestyle to her skeptical mother as feast or famine: "Before a big piece sells, it's famine. Then you sell it, and it's feast; you pay all your people and you pay your bills, and you have money left over to reinvest. It's no different from buying property to develop."

Recently, the trade had opened up in a way that natural history dealers found promising and paleontologists found troubling. Venerable fine-art auction houses now offered fossils with the kind of upscale presentation afforded old master paintings or Chippendale desks. The auctions often attracted overseas museums as buyers, along with private collectors who realized they could own the kind of specimens found in institutions like the Smithsonian in Washington, DC, or the American Museum of Natural History in New York. A collector with enough disposable income and square footage could more or less start his or her *own* museum.

The most popular collectibles were the ones you could put on a shelf as a conversation piece, like meteorites and mammal skulls. Dinosaurs transcended everything, though. Dinosaurs fascinated children and adults alike for their size and variety alone: fleet little deadly winged carnivores capable of taking down beasts twice their size; peace-loving, plant-eating behemoths; apex predators that ravaged their way to the top of the food chain. Dinosaurs claimed universal name recognition: *Tyrannosaurus rex* is the one species whose name everyone gets right, the vertebrate paleontologist Thomas Holtz liked to say. Dinosaurs symbolized both catastrophic death and robust life: by human standards, they were *far* more successful. Dinosaurs dominated Earth for 166 million years, fell to a mass die-off, then enjoyed a cultural comeback—66 million years later. (*Almost* all of them went extinct, that is; we've still got birds.) "Dinosaurs are the gateway to science, which is the gateway to technology, which is the gateway to the future," the paleobotanist Kirk Johnson, head of the Smithsonian National Museum of Natural History, once said. A kid who spends her childhood engrossed in *Stegosaurus* might grow up to study oceans, algorithms, epidemics, algae, volcanoes, black holes, bacteria, bees, brains, sex, weather, sleep. "Seriously, if we don't start thinking about how we view this planet from a science-and-tech standpoint, we're cooked," he said. "The fact that we can't have a science debate with the presidential debate, and the fact that half the people in this country think the earth is only six thousand years old and that dinosaur bones are a hoax—it's tied to the fact that we have crap science education. And you don't *take* science education unless certain things interest you. And it just so happens that all kids care about dinosaurs."

Eric's casual interest in dinosaurs started in childhood, but lately he'd incorporated them into his business. He had sold two enormous tyrannosaurid skulls back to back, both to major movie stars, for more money than he'd ever made, at one time, in his life. The sales suggested that he had found a thread of the trade that might provide for his family for years to come. As he celebrated his daughter's birthday at the beach, he had a whole skeleton on offer in Manhattan, expected to fetch millions of dollars through Heritage Auctions, a company based in Dallas.

The skeleton was that of *Tarbosaurus bataar*, or *T. bataar* (*buh-TAR*), the Asian cousin and near twin of *Tyrannosaurus rex*, the most famous creature that ever existed. Both lived in the late Cretaceous period, over 65 million years ago. This particular animal never reached maturity yet stood 8 feet tall and 24 feet long when it died. Its bite was devastating, if it matched that of *T. rex*, whose prey's bones "likely exploded." Over millennia the remains of this particular *T. bataar* had mineralized to a distinctively lovely shade of sand. Eric had consigned the skeleton to Heritage with a business partner, Chris Moore, a veteran fossil hunter who lived on the Jurassic Coast of England. The sale had been advertised and featured in the news for weeks. "SUPERB TYRANNOSAURUS SKELETON," read the description in Heritage's catalog, referring to the species's original, sexier taxonomic classification. Prospective bidders could see the front end of the dinosaur on the back cover of the catalog, then open to page 92, unfurl the centerfold, and take in the whole skeleton, which stood elegantly lit against a deep-blue background. *T. bataar* "ruled" the "ancient floodplains that are today's Gobi Desert," read the sales text. The "incredible, complete skeleton" had been "painstakingly excavated and prepared." The teeth were a "warm, woody brown." The dinosaur was simply "delightful."

While Eric knew that paleontologists considered commercial fossil dealers like him venal and destructive—and in this case, it would be revealed, criminal—he had creation, not devastation, in mind. He was in it for the pleasure of hunting—and, yes, for the pleasure of a livelihood— yet also saw himself as a valuable part of the long tradition of preserving prehistory. Like every other commercial hunter, he felt he was salvaging materials that otherwise would weather out once they met air. The *real* crime, in his opinion, was letting fossils go to waste. His mounts from the Pleistocene epoch, the most recent ice age, stood in natural history museums from Charleston to Shanghai, and he envisioned spending the rest of his life making more finds worthy of museums. Even before he met Amanda, Eric had imagined getting married, having children, and one day taking his family on a sort of international tour of his reconstructions: standing before a looming skeleton, he would explain to his kids how the animal's remains had lain entombed for many millions of years longer than humans could fathom, and how the forces of erosion and civilization continually pushed fossils to the earth's surface. To him, self-referential talk sounded like showing off, but, if asked, he would explain how he had prepared the bones before reassembling and mounting them like a 3-D puzzle, standing the creature on its feet again for the first time since it last breathed.

Heritage estimated the *T. bataar* would fetch between $950,000 and $1.5 million, though Eric and the auctioneers suspected it would sell for more. Such a large, fine tyrannosaurid hadn't been to auction since 1997, when Tyrannosaurus Sue, a magnificent South Dakota *T. rex*, sold for $8.36 million via Sotheby's. A major sale would generate enough income to pull Eric out of a quicksand of his own making, and until roughly seventy-two hours earlier, nothing had appeared to stand in his way.

Earlier in the week, Bolortsetseg Minjin was having lunch near the American Museum of Natural History, on the Upper West Side of New York, when she overheard a news broadcast about the auction. A Mongolian paleontologist in her early forties, Bolor had spent most of her adult life in the United States. She lived in Port Washington, on the north shore of Long Island, with her American husband and their young daughter, whose name was the Mongolian word for "rainbow."

Bolor stood five feet one-ish, though who really knew; she resisted talking in many specifics about herself. Her hair was a shiny black bob. On formal occasions she wore makeup and a bright silk *deel*—the traditional Mongolian tunic, a sash at the waist—but usually preferred casual clothes like jeans or capris and Merrell trekkers, with T-shirts that acclaimed girls in science and slogans like "Life Is Good." Her pendant earrings were simple azure stones, a color that in Mongolia symbolizes the eternal blue sky. Friendly but wary, Bolor spoke in a low, measured voice, her dark eyes narrowing when she was suspicious and crinkling at the corners when she smiled. Her English, in which she had become fluent after arriving in

America, was excellent and softly accented—*Wee-oh-ming* for Wyoming, *gommint* for government.

Bolor grew up in 1970s Ulaanbaatar, aka "UB," the capital of Mongolia, in the north-central part of the country. She came of age in the '80s, in the last days of the country's long run as a Soviet satellite state. She lived with her parents and siblings in an apartment complex west of Sukhbaatar Square, the city's main plaza, home to the imposing Government House and the crypts and statues of Communist heroes. Her father, Chuluun Minjin, a geologist who had trained in Russia like his peers, taught paleontology at the Mongolian University of Science and Technology (MUST) and worked on fossil invertebrates. As a child, Bolor wasn't allowed on his Gobi Desert expeditions—going out in the field was man's work, her mother told her—but she looked forward to seeing the rocks her dad brought home in his backpack. One day he showed her a piece of coral. Corals are marine life, and it fascinated Bolor to think of her vast country, landlocked between the gargantuan nations of Russia to the north and China everywhere else, as the onetime home of a sea.

The Gobi's most stunning creatures were the dinosaurs that had lived in the area in diverse abundance and whose remains lay preserved with unusual clarity in the largely undisturbed desert. Bolor knew about them only because her father had told her. Western kids had plenty of access to information about dinosaurs: American children in particular could participate in museum-supervised digs, watch fossil preparators work in museum labs, enjoy interactive exhibits, join nature clubs, and then go home and lose themselves in dinosaur toys, books, puzzles, and TV shows. At places like Dinosaur National Monument in Utah, they could see more than a thousand Jurassic bones still embedded in the protected cliff face of a former Carnegie Museum of Natural History quarry. On the trails at Badlands National Park in South Dakota, they could see the remains of late Eocene and Oligocene mammals poking right out of the dirt. But despite Mongolia's importance as "the largest dinosaur fossil reservoir in the world," paleontology was, as Bolor put it, "no place for kids." Once you left the forlorn Natural History Museum in downtown Ulaanbaatar, there was "basically no source of information where you can learn more about these exciting, interesting animals."

On Bolor's birthday her father liked to entertain her and her friends with slideshows about dinosaurs, and in his classroom he allowed her to examine the plastic models and fossils he used to teach paleontology. Later, as Bolor pursued a geology degree at MUST, her father formally became her professor and then her graduate adviser. When she decided to work on vertebrates he told her, "Oh, it's so hard to do vertebrates—you have to know anatomy." Bolor committed to invertebrates, but found someone to tutor her in vertebrate anatomy anyway. She was still in school in 1990 when the American Museum of Natural History (AMNH) returned to the Gobi after over sixty years of banishment. Minjin, as everyone called Bolor's dad, signed on to the museum's historic joint expedition with the Mongolian Academy of Sciences. When he asked the Mongolian side's leaders for a spot for his daughter, they said she could come if she could cook. Bolor couldn't boil water; but she accepted. Then, once in the field, she spent her time hunting fossils, telling the others, "I'm a paleontologist. I'm supposed to be doing whatever it is the paleontologists are doing."

The AMNH connection led to a short-term slot as a visiting scientist in the museum's department of vertebrate paleontology. Bolor's first view of America was New York: she moved from a vast country of barely two million people to a dense city of over seven million. After finishing her master's degree in Mongolia in 1997, she returned to Manhattan to pursue a PhD in the joint AMNH program with the City University of New York, becoming the first Mongolian paleontologist educated in the West.

By now, in the late spring of 2012, Bolor had been out of school for years, working on her own projects. Her advanced training was in multituberculates, the wee mammals that lived in the cracks and shadows of the late Cretaceous and survived the fifth extinction, but her other interest was dinosaurs. It amazed her that some of paleontology's most glorious discoveries had been made in Mongolia yet even the nomads who lived within walking distance of the Gobi bone beds had no idea. Bolor could walk down the street in Ulaanbaatar and not meet five people who could name a single Mongolian dinosaur—not *Velociraptor, Oviraptor, Protoceratops, Psittacosaurus, Alioramus,* or *Therizinosaurus.* For years, she had been trying to raise awareness in her country and beyond of the importance of Gobi fossils. She wanted to draw more of Mongolia's most promising students toward paleontology instead of the usual choices, like engineering or mining. Mongolian paleontologists should be leading the way on research involving Mongolian materials, she liked to say; in her opinion, Mongolian paleontologists ceded too much authority to foreign scientists who had built their careers on Gobi fossils and given too little in return.

Lately, she had been worried about poachers, who had been hitting Gobi dinosaur sites with increasing frequency. The difference between a fossil poacher and a fossil hunter is the same as the difference between a poacher and a hunter of wildlife—one respects boundaries, and the other doesn't. A wildlife poacher may take twice as many deer as the law allows; a fossil poacher may take bones from lands where private collection is forbidden, like federal property in the United States, or countries that ban the trade. Modern thieves go after fossils using tools as basic as a shovel and as high-tech as helicopters and Google Earth. Geology and geography altogether provide favorable conditions: dinosaurs tend to be found in vast, remote, underpopulated, undeveloped places like badlands and deserts, which are difficult to police. Poachers count on the fact that a fossil's origin becomes nearly impossible to trace once it's collected—until scientists perfect matchmaking technology for trace earth elements, it's unlikely that anyone will be able to connect a fossil to a particular hole in the ground with unimpeachable certainty. Mongolian law forbids the trade: by law, fossils are the property of the state. Yet illicit Gobi Desert dinosaurs were being sold on the open market. The paleontologist Philip Currie, Canada's eminent tyrannosaur expert and then president of the Society of Vertebrate Paleontology, had started documenting the number of savaged *T. bataar* sites alone and had counted nearly a hundred since 2000.

Bolor had been trying to get Mongolian government leaders to do something about dinosaur poaching, but politicians couldn't have been less interested. Recently, though, she had made an ally, Oyungerel "Oyuna" Tsedevdamba, a women's rights activist and aide to President Tsakhia Elbegdorj, who was finishing his first term in office. On the day that Bolor overheard the news about the Heritage auction, she listened closely. "Tyrannosaur" was a North American species, but "*bataar*"—that was her language. The word meant "hero." Convinced that Heritage was selling a black-market Mongolian dinosaur, Bolor emailed Oyuna.

Thursday night in New York is Friday morning in Ulaanbaatar, a city of 1.2 million people. Oyuna was getting ready for work at her townhouse in Zaisan, an upscale district south of the city center. She had been busy recruiting women to run for public office—the election for parliament was weeks away—and was just about to check her email when her husband, Jeffrey Falt, a semiretired American lawyer, sprinted downstairs in his bathrobe to tell Oyuna about the auction he'd just read about online. Bolor's email underscored the urgency. Oyuna wrote her, *Do everything you can to stop the auction*, and suggested that Bolor and other paleontologists contact Heritage with questions: Who was the seller? How had the dinosaur been acquired? Bolor was already on it, having quickly emailed a former AMNH colleague, Dr. Mark Norell.

Norell, a vertebrate paleontologist in his mid-fifties, was busy packing for the Gobi. Compact and wiry, with distinctive round eyeglasses and wild, silver hair, he wore skinny khakis and dark linen shirts and, on his right wrist, several slender silver bracelets. He looked like he ought to be slinging a Fender Stratocaster instead of a geologic hammer.

Norell grew up in Los Angeles, where his elementary school had allowed him to substitute science class with weekend visits to the Natural History Museum of Los Angeles County. Despite career flirtations with molecular biology and law school, he ultimately pursued paleontology, though he considered the discipline a bit "lightweight," once explaining, "There's no mathematics, there's no proofs, there's no empiricism at all." Happily, that was changing. "We approach things more empirically than before," he said. "It's much more biological."

Norell didn't like dinosaurs as an animal—a lot of paleontologists don't. Yet he liked dinosaur *science*. To him, being a scientist was all about asking the right questions. He liked thinking about the *creativity* of science, about "how we come up with ideas." Hired at the American Museum of Natural History in August 1989, he was in the Gobi Desert by the summer of 1990, helping lead the first U.S. team allowed passage since the 1920s.

Now Norell was the chairman and curator of the museum's paleontology division. He had returned to Mongolia every summer except for the year his daughter was born. His crew was known to "go feral" out there, spending shower-free weeks hunting and excavating fossils all day and relaxing at night, enjoying the excellent food, beer, and wine they'd brought in bulk. Norell remembered an early tag-along journalist who imagined such an expedition would be good for him. He'd lose weight! Quit smoking! He'd spend serene lantern-lit nights reading in his tent. The experience turned out to be more akin to "touring with the Stones." Norell knew it was "important to have a good time." Such a good time, in fact, that he usually stopped in Beijing on his way home for a total-body exfoliation. Just recently, doctors had found an old beer-can tab stuck in his throat, accumulating scar tissue.

When not in the field Norell worked out of a light-infused museum office at the top of the granite turret that had overlooked Central Park since the late 1800s. The spiritual centerpiece of his lofty space was an antique desk that had once belonged to Barnum Brown, the discoverer of *T. rex*. Otherwise there were iMacs and fresh orchids, and, currently, half-packed duffel bags containing packages of quinoa earmarked for the Gobi. Curatorial cabinets held small, fine-boned Mongolian and Chinese skeletons Norell had been studying in his decades-long effort to better understand feathered dinosaurs. These were the kinds of bones that had been showing up on the commercial market, even though it was illegal to buy and sell fossils in both countries. Poachers were "clobbering" dinosaur sites in the Gobi, which straddles the Mongolia–China border. Materials often turned up online and at Tucson, the sprawling gem, mineral, and fossil show that, for two weeks every February, attracts an influx of fifty-five thousand people. Norell admired collectors' enthusiasm as long as they honored boundaries, and most of them did. "I'm not one of the people who are of the opinion that only scientists are entitled to fossils—there are a lot of legitimate, conscientious dealers out there," he once said. "My big thing is, you need to obey the laws, as far as what can be trafficked country to country." But international regulations varied widely, shifted often, and were usually unavailable in English, and certain hunters took advantage of the confusion by exploiting the legally untested gray areas; they had created "almost a Silk Road of dinosaur-dealing."

Best he could, Norell kept an eye on the trade. Auction catalogs piled up in his office, and he regularly attended the Tucson show, where dealers might sell dubious fossils in back rooms or right out in the open. Walking around, he could tell instantly which materials were hot. Dinosaur bones from Argentina were, by definition, hot. Chinese dinosaurs, hot.

Brazil, hot. A lot of Moroccan stuff, also hot. In Canada, fossil vertebrates belonged to the crown, so if you saw, say, *Gorgosaurus* bones from the Red Deer badlands of Alberta: hot.

If there was one country's fossil laws Norell knew better than most, it was Mongolia's, after having spent so much of his career there. So when he saw Heritage offering the *T. bataar* skeleton, he clocked the specimen's illegitimacy with a glance.

Individual Gobi bones and skulls had long been sold at market, to paleontologists' impotent fury, but Norell had never seen anyone with the audacity to put up a large, mounted, museum-quality skeleton. At least four other items in the Heritage lineup appeared to be Mongolian, too, including the skull of *Saichania*, a hulking late Cretaceous herbivore with a tail-club and an armored head that, in fossil form, resembled an ornamental mask. The sales descriptions included direct references to the Gobi and vague ones to "Central Asia," which, in the dodgier quarters of the fossil trade, was often code for China and Mongolia. Now, in his letter to Heritage, Norell asserted the *T. bataar* and *Saichania* specimens "were clearly excavated in Mongolia as this is the only locality in the world where these dinosaurs are known." He added, "There is no legal mechanism (nor has there been for over 50 years) to remove vertebrate fossil material from Mongolia. These specimens are the patrimony of the Mongolian people and should be in a museum in Mongolia." In her own letter, Bolor Minjin wrote, "The auctioning of such specimens fuels the illegal fossil trade and must be stopped." Politely, she asked how Heritage, a thirty-six-year-old company founded by a pair of rare-coin collectors, happened to have acquired a Mongolian dinosaur in the first place.

The species name was clue enough that the skeleton was Mongolian, but the bones themselves were the best evidence. To be sure of their facts, Norell and Bolor went to the auction preview to see the specimen in person.

The auction was scheduled for two p.m. Sunday, on West 22nd Street in Chelsea, in a converted warehouse formerly occupied by the Dia Art Foundation, an organization created in the 1970s to help artists produce projects that "might not otherwise be realized because of scale or scope." Scale and scope were what Norell and Bolor found. A space that once showcased work by modernists now held nearly two hundred objects billions of years in the making. The minerals and gemstones glowed flamingo pink, canary yellow, Prince purple. Rare quartzes and crystals resembled wayward *Lord of the Rings* props. A pair of insects in amber told a beautiful story of terrible timing: the bugs died in coitus, forever captured in fossil tree resin, a preservative so clear and golden the couple appeared to be floating in hardened honey or a nice autumnal ale. Each lot wasn't just fascinating to look at; it represented a moment in planetary history and geological process. A "boulder" of gold signified molten rock that flowed, cooled, solidified. A giant frond and stingray, imprinted in limestone, showed that palm trees and sea creatures existed some 50 million years ago in what is now Wyoming. The descriptions borrowed from the scientific lexicon, their curious names evoking a cross between hard-candy confections and bicycle parts—*rutilated quartz, fluorescent scheelite, spinel twin*. David Herskowitz, the natural history broker who had organized this, Heritage's first auction dedicated exclusively to natural history, had curated the pieces himself.

The most arresting sight in the showroom loomed behind security ropes: Lot 49135, the "crown jewel," *T. bataar*. With its arms out and its jaws open, the dinosaur appeared to be hunting the cast Komodo dragon crouching nearby, on blue velvet. Norell, who had spent two decades working in the only area known to produce significant *T. bataar* remains, confirmed the dinosaur as Mongolian.

Meanwhile in Mongolia, Oyuna contacted a friend in Connecticut who now spread the word about the *bataar* sale. News articles appeared online. A paleontologist in California started an internet petition, which quickly collected nearly two thousand signatures and comments: "Mongolian fossils are spectacular...selling them as mantelpieces is akin to using the Mona Lisa as a placemat." "Actions like this only serve to increase an air of mistrust in the scientific community, which needs to unite globally now more than ever." "This is bad for science, politics, and the world." Yet Heritage wasn't yielding. The company's attorney in New York responded to Bolor's letter, saying that in their opinion, "no impropriety exists"— Heritage had no reason to believe U.S. law had been broken, and the company was "unaware that Mongolian law would have prevented export from

Mongolia." Besides, the letter added, "Mongolia won its independence in 1921 and this specimen is quite a bit older than that."

Bolor found the response ludicrous. She told Oyuna, "We need a lawyer."

The auction went forward as planned. Forty-eight seconds of bidding yielded a hammer price of $1,052,500, an impending payout that Eric Prokopi desperately needed to pocket. On the beach in St. Augustine, he hung up the phone filled with unease, and feeling a very long way from where all of this had started.

CHAPTER 2

LAND O' LAKES

Florida is flat and rather featureless compared to other states, but what it lacks in topographical drama, it makes up for in fluids. The Everglades have been called both a river of grass and a prairie of water. Nearly eight thousand lakes have a surface area of at least ten acres. There are 11,000 miles of streams, rivers, and waterways; over 2,000 miles of tidal shoreline; and more major springs than anywhere else in the country. A Floridian who doesn't go into the drink is a Himalayan who doesn't climb.

To this waterland came a young German named Dorothea "Doris" Trappe, daughter of Cologne, a trading port on the Rhine. Her father, a mechanic, owned a "big, big, big, big, big, big" boat then lost it in an accident. In May 1940, when Doris was four, the Allied bombs of World War II began falling on the city. Tens of thousands of people fled over the next five years, including Doris's family, who returned after the war minus one: Doris's older sister died of diphtheria on the journey home.

Cologne was destroyed. The family lived in half a house. Doris's dad got a job rebuilding the city but had a heart attack and died. Doris eventually followed a boyfriend to New York and then to Florida and married him, making her home in a place that had always "represented the middle-class dream of a place" where "you could find health and warmth and leisure," as Susan Orlean put it in *The Orchid Thief.* "Florida wasn't grimy or industrial or hidebound or ingrown.... It was luscious and fruitful. It felt new and it looked new, with all its newly minted land and all the billboards pointing to new developments and the bright new sand that had been dredged up and added to the beach. Florida was to Americans what America had always been to the rest of the world—a fresh, free, unspoiled start."

Not long after Doris had a son, Gordon, her tempestuous marriage ended in divorce. But she hung on to both her child and to an important recreational relic of the relationship: *Freikörperkultur*, naturism. She remained a devoted nudist. Living in harmony with the outdoors promoted vigor and happiness, and took out all the fuss. The pastime linked Doris to Europe, but she practiced at a local colony called Lake Como. ("A Family Nudist Resort Since 1941." "Go Barefoot All Over!") A resident-owned co-op, Lake Como offered a motel, cabins, tents, and activities like tennis and volleyball tournaments. Members ate at the Bare Buns Café and took their libations in the Butt Hutt. Doris, both a Lake Como member and employee, worked in the kitchen and as a housekeeper. Slick with suntan oil, lips white with zinc cream, she mowed people's yards. Outside of the community she worked as an elementary school cook, all together saving enough money to buy a patch of property and a mobile home at the sandy edge of Lake Como, where she raised her boy alone in the bone-soaking sear of central Florida.

This was the late sixties, early seventies. Petite and slightly pigeon-toed, Doris wore big glasses and shoulder-length hair. Her skin and eyes were coffee-bean brown. When clothed she favored sandals, plaid shirts, and shorts or miniskirts. She was hardworking, genial, and extremely (and evenly) tanned. "She adds an extra thirty minutes for small talk," the *Tampa Tribune* once reported in a story referencing her charity work delivering hot meals to the homebound.

One day in the Lake Como lunchroom, she stumbled into a slender, bespectacled fellow in his early forties named Bill Prokopi, a bachelor and non-nudist who had dropped by on a whim. Bill was an elementary school music teacher, originally from Canada, whose family had moved to Winnipeg from Ukraine. A musician, Bill had come to Tallahassee for college, to study with a composer he admired, and had never left, never married, never found anyone, until Doris. He invited her paddle-boating and then to a movie. Afterward he told her, "I'm going to Canada, and when I come back I'm going to ask you to marry me. Think it over."

They married on May 17, 1973, Bill in a gray suit and crimson tie,

Doris with her hair up, wearing a white lace minidress with bell sleeves. On August 13 of the next year, Doris gave birth to their only child, Eric.

Doris's boys were nearly fifteen years apart, Gordon practically out the door while Eric was still in diapers. Bill's mother disliked that her son's family lived in a trailer "like gypsies," so when Eric was two, she helped them buy a small, brown-stucco house, with an attached garage, on four-tenths of an acre in Land O' Lakes, population 31,000, a half hour north of Tampa.

The hundreds of lakes in the aptly named place ranged in size from glorified ditch to boat-worthy, with names like Red Bug and Treasure. The Prokopis lived on Grove Lane, in the Lake Padgett neighborhood, atop what was once an orange grove. Their three-year-old house was ranch-style, with several small bedrooms, two baths, shag carpet, and sliding patio doors. They ate their meals at the dining room table, surrounded by Doris's collection of German beer steins. Bill always thanked Doris for cooking, and every time she went somewhere, he walked her to her car. Doris's devotion to Eric occasionally seemed to irritate Bill, who told her, "That's for the husband, not the son."

Doris liked to put Eric in a bicycle seat and ride him around the neighborhood. She took him with her when she cleaned houses at Lake Como or shopped for clients or drove them to the VA hospital. He never had a babysitter or went to nursery school or day care. On their daily walks, Doris and Eric collected aluminum cans together, shaking them free of snakes. Whenever they saw something interesting, they picked it up.

Eric wasn't averse to other children or afraid of them, but he never seemed to mind being alone. His talkative parents sometimes wondered what he might choose for himself in life. Bill hoped his son would like music. His own chosen instruments were the accordion and the bassoon, and he directed the choir at the Ukrainian Orthodox church where he was a member. For Eric, Bill suggested the saxophone. Eric went along, hating it; suiting up in a blue satin uniform and a plumed helmet to perform Del Borgo's "Canterbury Overture" in the school symphony did not do it for him. "I don't like it—you have to tell Dad," he told his mom. Bill was duly notified. Eric abandoned the uniform, relaxed his embouchure, and quit music altogether, but for the ambient sound of his father's classical records and opera on his Sunday radio.

Doris had heard that you weren't supposed to force children into activities, so she told Eric, "You choose." He had tried karate and a few other things, but decided to focus on swimming. Doris coached him herself, pleased to have perhaps transmitted the water gene, seeing as how Bill couldn't swim at all.

———————

Eric wasn't the top swimmer, but his dedication and focus impressed the coaches of his club teams. Land O' Lakes High had no swim team, so Doris lobbied the school system to let Eric function as a squad of one. Local newspapers ran stories about "Prokopi, the 'unknown swimmer'" whose coach was basically his mom. "Swimming is in his blood," Doris told the press. When Eric made it to the state championships, a *Tampa Bay Times* columnist wrote, "OK, so the Gators have a little trouble with relays, but this is for real." Reporters generally withheld observations about Eric's most pronounced personality trait, though one did not—

> To say Prokopi is quiet would be to suggest that Picasso was a painter. "It's no big deal," Prokopi said, possibly on his way to setting a personal record for most words spoken to a stranger. "I just want to swim. My friends know. That's enough."

When Eric did speak, his words tended to snag like a faulty zipper. He didn't stutter; rather, he put distance between words, as if measuring what he wanted to say. The pattern required patience of his listeners, and perhaps frustrated them. "Why do you swim?" people would ask. "I just like it," Eric would answer. He expressed himself in other ways, like pounding the pool in the backstroke or goofing around with a stink bomb at school. A couple of times he flew paper airplanes against the rules, and ran in class, also against the rules. "Defiance of authority," wrote the defied authorities. Doris would tootle up to the school in her VW bus or her station wagon and shrug her freckled shoulders. What trouble? This boy was just being a boy. Doris had always done the talking for him.

Eric performed at the top of his class, making almost all A's in his Advanced Placement and honors classes. Doris, an enthusiastic archivist, documented the grades, the swimming medals, the newspaper clippings, the teachers' compliments on Eric's hard work and positive attitude, the notices from the free-meals program that fed her boy breakfast and lunch.

The family photo collection showed Eric in a high chair, being spoon-fed by his unclad mom; standing as an altar boy among the religious iconography at his father's church; gazing into a picture book called *The Age of Dinosaurs*; hoisting a swim trophy; wearing a scuba suit; holding a parakeet; holding a turtle; holding a chicken; holding a puppy; holding a red-breasted macaw; kneeling beside a skeptical-looking pelican, wanting, clearly, to hold it. At home, he kept lizards, turtles, a rabbit. The first big fish he ever caught, Doris mounted as a trophy.

Bill wanted Doris to stay home, but Doris believed it impractical for a family of four to try living on a single annual income of $18,000. A wartime childhood and a frugal life had taught her to save dribbles of money. Shopping at thrift stores, she thrilled at bargains. At one point she and Eric started their own company, handing out half-fanciful business cards—

DORIS AND ERIC PROKOPI
Chimney Sweeping and Grass Cutting

On weekends, the Prokopis often drove down the Gulf, to the beach town of Venice, where Doris had an aunt who lived within walking distance of the dunes.

———————

Roughly three million to twelve thousand years ago, Florida emerged from ice and warmed beneath a shallow ocean. Over time, sea levels dropped enough for the beginnings of the landmass that we now recognize as Florida to surface. Animals of this epoch, the Pleistocene, walked back and forth between continents. In balmy, interglacial periods, extraordinary creatures lived abundantly in the waters and on land—saber-toothed cats, dire wolves, mammoth, mastodon. The megafauna included birds with twelve-foot wingspans, armadillos the size of Volkswagen Beetles. Florida had beavers the size of small bears, horses the size of dogs, dogs the size of gerbils. Their corpses wound up in rivers or drifted to the ocean floor, to be covered and preserved by sediment. "Hardly a roadcut or realignment of a ditch is made in Florida without some fossil being turned up," S. J. Olsen wrote in 1959 for the Florida Geological Survey. Fossils surfaced during quarrying and dredging, and during the paving of parking lots.

They became so popular as collectors' items that hobbyists formed societies where they could share their finds.

Natural history lovers often build collections around what they can easily hunt. In Wyoming, the gateway fossil may be Eocene fish preserved in paper-thin rock slabs; in Kansas, it may be prehistoric sea lilies. South Dakotans might collect turtles; Moroccans, trilobites; Germans, traces of dragonfly pulled from the *plattenkalk*. In Dorset, on the southwest coast of England, collectors hunt ammonites and more on the beaches, where the high bluffs slough off with each passing storm. Montana and Utah yield extraordinary dinosaurs.

Florida's gateway fossil is shark teeth. Sharks have swum the oceans for over 300 million years; a single shark may sport up to three thousand teeth at a time, losing and replacing them like a tabby sheds hair. The teeth litter the ocean floor as blue and yellow and black baubles, often still shiny with enamel. Their various shapes—witch's hat, fat comma, small awl—denote the animal's onetime habitat and behavior. The wee tooth of a bull shark may fit on a fingertip, like a thorn; the teeth of megalodon, which patrolled the early waters like a huge submarine, grew nearly 10 inches long, dwarfing those of its daintier descendant, the great white.

The broad shallows and calm waters around Venice made shark teeth so easy to find, the town long ago nicknamed itself the Shark Tooth Capital of the World and began hosting an annual festival in May. The beach was always filled with people who walked slowly, looking down, as if an entire town was out searching for a lost contact lens. Collectors sat in the surf with homemade sifting screens and waded in with box dredges. The souvenir shops sold handbooks like *Let's Find Fossils on the Beach*: "With patience, a pail, your eyes all seeing, and your back strong, you, too, can become a collector of the ages."

Eric was maybe five when he found his first shark tooth at Venice, its serrated edges like the tines of a miniature pocket comb or the rim of a new dime. The tooth wasn't just a souvenir; it once lived in the real mouth of a real animal that swam in a real ocean not unlike the one licking his bare feet. "The important thing to remember is that vertebrate fossils truly represent life," read one handbook. "They are not just dry bones but are animals that ate, drank, fought, and reproduced much in the same manner as similar animals are doing today."

Eric pocketed the tooth, hooked.

CHAPTER 3

GARCIA, KING OF THE ICE AGE

In Old West Tampa, there lived a son of New York named Frank Garcia. His parents, Cuban immigrants employed by Tampa Linen Service, were good caregivers, but Frank couldn't help feeling his father might have worked a little harder on warmth. After failing first grade, Frank decided he could forgive his old man for comments such as "You'll never be a brain surgeon," but he never forgot them, and he went on to live as if he had something to prove. As a fourth-grader he won a soap-selling contest. In high school he played the saxophone exactly once, then started a band. His many jobs included shoe shiner, lawn mower, milk shake maker, and door-to-door crab salesman. On the first date of his life, romantically overcome, he let loose a round of "That's Amore." The woman to whom he crooned "What's New Pussycat?" married him. Undeterred by divorce, he married five times more, ultimately on a covered bridge to a woman who shared his enthusiasm for open skies, big guns, and old bones.

As a young man Frank wore his hair bushy, his mustache bushier, and an earring. His sleeveless shirts showed off his ropy arms, which, tanned in perpetuity, were the color of a fresh cigar. His voice was pleasant to listen to, and he smiled a lot, a divot in his chin. The media sometimes called him Florida's best-known paleontologist, and Frank was known to refer to himself as "the most interesting man in the world," but for the better part of his early adulthood, before his license plate read ICE AGE, he worked as an insulator, wrapping industrial pipes and boilers in asbestos-laden cement.

People told Frank he had a beautiful voice, often comparing him to

Johnny Cash. At one of his jobs he liked to sneak onto the PA system and announce, "Hello, this is Spanish Cash," then anonymously serenade listeners with songs he had written. One time, the furious supervisor, unable to figure out who was haunting his airwaves, got on the loudspeaker and vowed that Spanish Cash would "never, ever work at a Florida Power and Light job again." Frank rebutted with a tune he called "Super Snake." He claimed to have once walked off a job because the boss ordered him to quit laughing so much.

One day when Frank was about ten, he went fishing with his grandfather at Lake Okeechobee, over 700 square miles of natural freshwater near Palm Beach. As he waited for strikes, Frank noticed fossil sand dollars embedded in the spoil of dredged limestone. He dug out one of the fossils, cradling in his hands what he later called "the fuel of my life."

Frank couldn't quite manage to excel in school, but he loved the library, where he came across books about dinosaurs. Dinosaurs had never been found in Florida because the Mesozoic layers are thousands of feet below the surface—the closest outcrops were said to be near Selma, Alabama, and Tupelo, Mississippi. That didn't stop Frank from fantasizing about making magnificent discoveries. As an eighth-grader, he found mastodon bones embedded in a bank of the Hillsborough River at low tide. Another time, he found a bone he couldn't identify, and took it to a University of South Florida paleontologist for identification.

A camel, the paleontologist told him, examining the specimen.

"We had *camels* in Florida?" Frank said, learning that for thousands of years Floridians have come across the fossil remains of prehistoric creatures usually associated with contemporary Africa.

Frank was getting into natural history just as gem and mineral shows came to include fossils. Not long out of the navy, he stumbled into a fossil museum housed in an RV, owned by a native Texan named Joe Larned, who wore cowboy shirts and a shark-tooth bolo tie. Larned, a former World War II air force mechanic, was a "science fanatic" who especially liked talking about meteorology, and often mentioned the coming of a new ice age owing to "man's slash-and-burn tactics." He worked in maintenance at a phosphate mine and lived in Polk County, east of Tampa, an area called Bone Valley because so many fossils have been found there. Companies like International Minerals & Chemical Corp. allowed the public to hunt fossils on their property, and when Larned led field trips, some collectors drove all day to join him. "You realize you're the first human being to ever see that," he would tell the new owner of an old bone. "When that was laid down there, we weren't even on the drawing board."

Larned started a tin-shed museum called Bone Valley Museum, later selling the collection to the town of Mulberry for $30,000. He lived the way Frank decided he wanted to live. The first time they went hunting together, Frank realized he'd found something that he could do happily for the rest of his life.

Frank also wanted to be known for having contributed to science. When his cousin found a strange fossil horn, Frank drove him to Gainesville to have it identified at the Florida Museum of Natural History (FMNH). The museum had seized on the public's growing interest in fossils by creating the Florida Paleontological Society, with a newsletter called *The Plaster Jacket*, named for the protective casing used to transport specimens from excavation sites to the lab. The society educated and inspired the public while unofficially monitoring who was out there hunting, and what they were finding.

Frank arranged for his cousin to show the horn to Dr. David Webb, the museum's curator of vertebrate paleontology, who worked on Pleistocene animals. He expected to meet "the typical scientist-type fellow," someone who "looked intelligent and dignified—maybe reserved" yet was delighted to find Webb dressed not unlike his mentor, Joe Larned, in a cowboy shirt and a white hat. The fossil horn so impressed Webb that he asked Frank's cousin to donate it to the museum, later determining it to be that of an extinct deer. Frank had a similar fossil in his collection, so he sent it to Gainesville; Webb thanked him with a cast replica. "I think what impressed me most about Frank—and it's a feeling that I think anybody involved in paleontology shares—was that really deep sense of mystery," Webb later said. "You're touching the past, you're somehow getting a new understanding of ancient life."

Expanding his search area, Frank taught himself how to scuba dive. In the rivers he found Pleistocene mammals as well as spear points and scrapers and harpoons, some made of ivory. On land, he befriended mining executives who let him go deep into the cuts. He found "the world's largest dolphin skull," a rare *Amebelodon* site, the "only known prehistoric giraffe skull" of its type, a sperm whale jaw, the "biggest giant sloth claw ever," and dugongs—sea cows. The dugongs included the first *in situ* example of a new pygmy species, which a Smithsonian scientist later named after Frank: *Nanosiren garciae*.

When you see a skeleton on display in a natural history museum, you're seeing the end of a process. Fossils are never found clean and neat like in the movies, ready to lift whole after a few swipes with a hand broom. Some skeletons are articulated, which means connected or intact, but most are either strewn about or crushed or jumbled, like a box of dropped toothpicks. A fossil embedded in rock matrix often must be lifted as a whole. Extracting it can entail hardening degraded bones with a cyanoacrylate adhesive (the active ingredient in Super Glue), and digging a trench around the base until the block rests on an earthen pedestal. The block is then layered with a protective material (like aluminum foil), covered in plaster-soaked burlap or paper (which hardens like a cast on a broken arm), and carefully flipped and fully jacketed. The plaster dries white, resembling a blob of chalk or a misshapen sarcophagus or, from a distance, when still in the field, a patch of rogue snow. When it is sawed open in the lab, the contents may look like nothing but dirt and bones cupped in half a giant egg. Preparators spend long, arthritic hours scraping and blowing, chipping and blowing, chiseling and brushing, millimeter by millimeter, until the fossil is clean. Frank still knew hardly any of this when he came across the extraordinary dolphin skull. He had never jacketed a fossil in his life, but he had seen it done. He drove to Bartow to buy plaster. Once back at the site he realized he'd forgotten to get burlap, so he took off his jeans, cut them into strips with his pocketknife, and jacketed the skull with the britches right off his backside.

Construction, road building, and mining were churning earth nationwide. Natural history clubs were popping up in every state. Frank began writing a column called "Fossil Facts and Philosophy" for the newsletter of the Tampa Bay Mineral and Science Club. Organizations and teachers invited him to give presentations and lectures. "Did you learn about fossils in college?" a student sometimes asked. Frank had gone to the navy, not college, but he came up with an answer that satisfied him: he had learned about fossils as a student of the natural world.

Frank was made a Smithsonian "field associate" after finding the extraordinary skull of a baleen whale and donating it to the museum. The honorary title recognized what one scientist called his "heroic" volume of collecting. Daryl Domning, a Howard University anatomist and Smithsonian affiliate, shared one of his National Science Foundation grants with Frank, to study dugongs. By now Frank had self-published a self-illustrated book about Florida fossils, *Illustrated Guide to Fossil Vertebrates*, but the Smithsonian's recognition was what floored him. He had never been so proud of anything in his life. When he received the official letter, he sat on his doorstep and cried at the thought of being a thirty-three-year-old pipe insulator with a high school diploma, acknowledged by one of the world's most prestigious scientific institutions.

On June 27, 1983, a rainy Monday, Frank drove to Ruskin, to one of his favorite hunting spots, a former tomato field now being quarried for crushed seashells. He often prospected there with the permission of the landowners, a family company called the Leisey Shell Corporation. He had once found an elephant skull on the property by following the fossil signature of a stream bed, and ever since then he had been checking the draglines, expecting more bones to surface.

A new cut had opened up several days earlier and the dragline operator had immediately gone on vacation. The whole mine had knocked off for the day by the time Frank wandered into the pit. Surrounded by high banks, he looked up to see fossils the likes of which he would not have thought possible. They jutted like wayward tree roots from a bed 2 feet thick and 60 feet long. The earthen wall so bristled with bone it was like looking at an earth sandwich, overstuffed with body parts.

After some joyful sobbing he phoned David Webb at the FMNH, only to learn that the museum wasn't interested in the Leisey shell pit. No problemo, Frank said. He rounded up a bunch of friends from the insulation business and started an independent excavation.

Every day, the crew took photos and measurements as they worked, documenting the dig. Frank sketched the finds to scale. His brother-in-law, Mickey, a long-haired, big-bearded insulation installer, brought in an RV so they could guard the pit 24/7, and tracked each day's assignments, often wearing little more than a Speedo. The shell pit's surface temperature could reach 140 degrees and the mosquitoes were the size of wasps, but nearly two hundred volunteers, many from local fossil clubs, showed up to help. Frank already knew he was onto something important. As the bootleg crew broke for Fourth of July picnicking, he climbed a spoil and planted an American flag.

The Leisey pocket produced bones one million to two million years old, representing 140 species, dozens of them previously unknown to science. The site yielded a hodgepodge of legs, ribs, skulls, vertebrae, tusks, arms, teeth, and claws, as if carcasses had piled up in a flood—bits of cheetah, jaguar, llama, bear, monk seal, shark, wrasse, snook, turtle, flamingo, and toad. The extinct creatures included the cotton rat, a condor with a wingspan of over 11 feet, and a rare gomphothere. A giant armadillo stood 4 feet tall, a beaver nearly 8, a ground sloth nearly 20. The temporary population of Ruskin seemed to triple in the weeks that followed. The FMNH eventually launched a formal dig, becoming the fossils' repository. At that point, Webb was calling the discovery "extremely significant," saying, "This is like finding a new chapter in the history of life."

Frank had discovered the richest Pleistocene fossil bed in all of North America, and volunteers had helped get the bones out. Euphoric, he went live on NBC's *Today* show, beaming in from a Florida studio to chat with the host, Bryant Gumbel. Wearing white jeans, big hair, and an orange patterned shirt, he sat next to a small table that held a sloth claw and other fossils he'd brought along as props. "You've been offered a great deal of money, I'm told, for your fossils, but you've chosen to donate them to science rather than sell them," Gumbel said. "Why?"

"Because they belong to the people of Florida," Frank said, smiling nervously.

"You have no formal training in paleontology," Gumbel said. "Why have you pursued it?"

Frank explained that he'd rather not spend his life in an air-conditioned building. "I like going out to the field and discovering. That's exciting."

"What about scientists who've now taken a look at what you've come up with and have marveled at it?" Gumbel said. "Are they—this may sound kind of strange—are they accepting you as an equal or are they just looking at you as some kind of an amateur who got lucky?"

It was an interesting question. Frank had donated hundreds of finds to museums. The Smithsonian had awarded him a grant. Scientists had named species after him. Frank had added at least five type specimens to the fossil record—meaning he'd found the holotype, or the first of a creature's kind—including a bizarre, distant relative of camels that David Webb named *Kyptoceras amatorum*, honoring all amateur fossil hunters. The Smithsonian's Clayton Ray thought of Frank as "an amateur in the best sense of the word. He's a full pro when it comes to field work, but he just doesn't make any money at it." As Webb put it, "In twenty years he's produced more than any other single human being active in Florida in the way of exciting new finds."

Frank told Bryant Gumbel, "No, I think they have respected me now."

"Huh," Gumbel said. He wanted to know what Frank planned for an encore. "I mean, having done this you certainly can't hope to top it."

Frank grinned and said, "Oh, yes, I probably can."

That summer, Frank was asked to help persuade state lawmakers to toughen the law concerning fossil collecting in Florida. The federal government was struggling with the question of how to protect the nation's fossils, and various states were taking up the matter on their own. There are many more legitimate dealers out there than not, and as long as the law allows them to collect, they want to be taken seriously as professionals. In the late 1970s commercial hunters had formed a trade group, the Association of Applied Paleontological Sciences (AAPS), eventually modeling their publication, the *Journal of Paleontological Sciences* ("Paleontology in the spirit of cooperation"), on esteemed scientific journals. Dealers gave their businesses scientific-sounding names. They established a code of ethics, which required AAPS members to obey all regulations, cooperate with agencies and institutions, and at least attempt to sell items of "unique scientific" interest to responsible buyers "for study, research, and preservation." Collectors and hobbyists strolled the aisles of the organization's fantastical fossil fairs and shopped online, many "unaware that the commercialization of fossils is even a problem," Shimada, Currie, and their colleagues wrote. In the paleontologists' view, the United States had a regulation problem, certain countries had a black-market problem, and paleontology overall had a PR problem. The public's "misguided perceptions" about fossils not only hindered scientists' ability to conduct academic research but also imperiled their efforts to secure funding and keep their jobs. And the rising commercial value of fossils led to poaching. Paleontologists were increasingly lobbying the U.S. government to make fossils altogether off-limits to commercial hunters. Other fossil-rich nations already outright banned the private ownership and sale of paleontological resources, but the United States struggled to decide how to fairly handle objects that served as the foundation of a science, a trade, and a hobby, the latter of which connected millions of people to the natural world in an increasingly sedentary age of screens.

Frank saw both sides of the tension between paleontologists and commercial hunters, but he testified before the Florida Senate in favor of requiring amateurs to purchase an annual $5 collection permit and report any unusual or potentially significant finds; the state would then decide who kept them. The proposition vaguely echoed the law in England, except that in England the crown either bought hunters' fossils or allowed them to be sold elsewhere if scientists didn't want them. The Florida legislation provided no such provision. To hunters it didn't seem fair that they'd spend time and money finding something great and not be reimbursed for their efforts. Many considered the regulatory attempts a threat to their favorite pastime and livelihood, and were so furious that Frank claimed he got death threats.

The legislation passed. Frank carried on, ablaze with new purpose. He won an award once given to the famous shipwreck explorer Mel Fisher, drank Singapore slings with the pro wrestler Dusty Rhodes, and founded the Tampa Bay Fossil Club. When invited to speak at colleges he turned what was supposed to be a paleontology talk into a dazzling, if baffling, motivational lecture. "Attitude—it's all about attitude!" he told students at schools like Yale.

The insulation business was not big enough to hold him. He announced his plans to quit. Wait a month, and you'll get a gold pin, the union told him.

Gold or gold-*plated*? Frank asked.

Plated, said the union.

See ya, said Frank.

Now pursuing fossils full time, he called himself a freelance paleontologist. He self-published a memoir, *Sunrise at Bone Valley*, with his fossil sketches in the back pages and himself on the soft cover, in an "I Dig

Fossils" T-shirt, posing beside the semi-interred rib cage of a huge dugong. The book began, "Mystical, magical, wonderful, and exciting—these are just a few of the words I use to describe my life." He taped himself singing "Besame Mucho," one of his favorite songs, and eventually put it on YouTube. He won karaoke contests. He wrote a tune called "Corazon de Tampa" and lobbied politicians to declare it the city's official song, at one point vowing to turn it into a Broadway musical. Instead, he worked on his second of three memoirs, *I Don't Have Time to Be Sane: The Life Story of One of the Most Notorious Fossil Hunters in America.* The cover showed Frank in a white-tie tuxedo and pearl earring, playing, like a piano, the ribs of that same dugong. The preface, written by the president of the Delaware Paleontological Society, read, "Imagine a character composed of (Forrest) Gump, Pied Piper, Rocky Balboa, Indiana Jones, and Norman Vincent Peale—add some hot sauce and you've got Frank."

Some admired Frank's confidence; others thought Frank was way too full of Frank. Frank didn't care. "His lectures are always educational and entertaining, and generally result in 'Standing Room Only,' " read Frank's bio, written by Frank. His goal was to convey one exuberant message: "You are never too old, too weak, too sick, too poor, too tied down, too far away—too anything—to discover and learn."

Collectors continually offload their fossils, whether to make bank or to make room for more. And so it was that Frank announced that he'd sold part of his personal stash to a phosphate company, which in turn donated it to Tampa's Museum of Science and Industry. Other hunters from other states were well known in the trade—the Larson brothers in South Dakota, Mike Triebold in Colorado, Henry Galiano in New York, Tom Lindgren in Utah, the Ulrich family in Wyoming—but Frank was the face of Florida.

Among his protégés was the kid from Land O' Lakes, Eric Prokopi.

CHAPTER 4

DIVE

THE LEISEY SHELL PIT FLANKED A SUN-BEATEN TWO-LANER called Agricola. Parked cars lined the road almost from the moment Frank Garcia brought in his band of volunteers. Shovels and sifting screens lay scattered among the spoils. Hunters planted golf umbrellas for shade. They drank out of coolers. Women worked in bikinis. A radio was always going. Brief, hard rains bathed the bone, raising a steamy mist. A medical professional stood by because the heat could drop you like a sniper, boy. Could anything be more irresistible to a kid?

Every possible weekend, Eric and his mom drove to the Ice Age grave-yard near Cockroach Bay, ninety minutes round trip. They joined the crowds, searching on foot and knee like everyone else, like every fossil hunter ever. Paleontologists are always asked how they find what they find, and they're always saying, "Just look down." Because for all of pale-ontology's technological advances—3-D, digital data-sharing, photo-grammetry, synchrotron-radiation tomography—fossil hunting remains an analog activity. It starts by walking around and paying attention, by looking for bits of earth that don't quite match their surroundings. As the volunteers worked, Frank answered questions, a dagger hanging from his belt. "Is this anything?" they asked. "Is this anything?" is the fossil hunter's refrain. The *this* might be the ear bone of a whale or a concretion, round as buckshot with a secret center: split neatly with the whack of a geological hammer, it may yield a nicely preserved prehistoric bug or other fantastical remains, depending on where in the world it was found. Then again, rock may simply be rock: Scraposaurus, junk, float, frag.

Hunters like Frank seemed born with the ability to "see" fossils; others had to learn it, and some never learned it. Doris and Eric would take Bill hunting, and as their buckets overflowed, he'd find nothing. Doris had a good eye, but Eric's was better. "He *smells* them," she said. In fact, Doris thought Eric might have the Garcia gift.

The remains of megafauna routinely appeared in the toothed scoops of construction backhoes, but also in water. The rivers beautifully preserved the state's fossil record, which extends to 50 million years ago. Hunters reported wading into streams and discovering near-complete skeletons of giant sloths. Fossils were found in sinkholes, rivers, and quarries, and in the limestone caves that honeycombed the Panhandle. After high winds and heavy seas, hunters spotted bones among beached bunches of seaweed. Joe Larned, Frank Garcia's mentor, once said he used to see so many fossils while canoeing down rivers that he and his friends would "just pick out whatever we wanted with the oars." He said, "We wouldn't even stop. If we saw a big tooth, we'd just reach out with the paddle like a baker with bread."

By now, Eric was diving. When he was ten, he had asked his mom to stop at the Land O' Lakes Scuba Center, a cinder-block shop on the highway into town, with a dive flag hanging from a pole out front. The shop stocked fins, tanks, hoses, masks, floats, ropes, belts, lights, knives, all the tools a human needed in order to move between worlds.

Doris had given birth to Eric in a Tampa hospital, but it was to the waters that he seemed born. He had been swimming since he was a small child, and swimming competitively since age seven. Diving interested him because he wanted to look at fish, but then a display cabinet drew his nose to the glass; inside lay objects larger and curiouser than shark teeth, such as enormous molars the color of fudge.

You want those kinds of fossils, you'll have to dive for them, the shop-keeper had told Eric.

Their lessons had started with clear-water diving, Eric learning by repetition. Into the drink, backward or split-legged over the side of a boat. Breathe calmly. Come up slowly. Always dive with a partner. Don't forget your dive float—you wanna get run over by a Bayliner? Then came river diving. You will see fish, snakes, turtles, antique bottles, arrowheads, tires, and train wheels. If you see an alligator, stay calm, but get out of there. You may see "sinker wood"—thousands of lumberjacked logs sank in the 1800s and early 1900s while being floated downriver to mills. Jam your hand blindly into mud and you might get cut, so don't do that. To unearth a fossil, try not to touch the sediment at all: fan the water, letting physics do the work.

Eric advanced to dark-water diving within two years, testing himself in rivers so murky that the observable world shrank to the primordial soup that swirled in the yellow beam of his headlamp. It was spooky and lonely down there, and took some getting used to. Different waters spoke different languages. Freshwater was largely silent but for the muffled drone of a motorboat. Salt water was symphonic with the crackle of shrimp. When Eric started venturing out on his own, Doris sat in the canoe, enjoying the sun. As he worked out of sight at the end of a tether, she waited between the gunwales, ready to tug the line at the first sight of an alligator.

Fossils began piling up at the house on Grove Lane. Bill often showed Eric his stamp collection, still hoping to interest him in one of his hobbies. Eric had watched his dad spend hours at his desk by the living room window, cutting the stamps off old envelopes and organizing them into books. An interior life didn't bother him, but an indoor life did.

Yet Eric enjoyed his father's fondness for maps. Every summer, before driving to Winnipeg to see the Canadian side of the family, Bill would get out his atlas and choose a new route north, through counties they hadn't visited yet, highlighting each area in yellow as they went. Sometimes they stopped at a river so Eric could snorkel near a boat ramp and see what the currents had washed up.

Eric spent much of his childhood submerged and searching. Older divers taught him to target bends in the river, where millions of years ago carcasses had likely snagged as they washed away. One day at the Withlacoochee River, he found his first mastodon tooth, a cusped hunk of blackened ivory as big as his hand. At home, he soaked the tooth in a mixture of Elmer's glue and hot water, cleaning and preserving the fossil as he'd seen older hunters do. Dry, it resembled a wood sculpture the color of caramel. Diving for fossils wasn't easier than hunting them on land—Eric was surprised to discover that it wasn't even less sweaty—but river fossils were easier to clean. Still, jewelers spent less time polishing their gems than Eric spent prepping his finds.

The first best recorded incidents of someone picking up an odd rock and wondering, *What the heck is this thing?* occurred in ancient Greece. People asked themselves what forces of heaven and earth possibly could have *coiled* a *stone*, or imprinted a fern frond in hard rock like a wayward spray of hair. Were these objects organic? Inorganic? Natural or supernatural?

The Greeks did not yet have a word for what they were doing, which was science. All of science started in the physical universe, with humans noticing and questioning the natural world—thunder, lava, larvae, fire, mountains, beaks, bark. They sought explanations for the flight patterns of birds, the greening of leaves, the angle of shadows, the color of soil, the nursing of newborns, the shape of the moon, the bitterness of salt, the ripening of fruit. Their best term for the search for meaning in nature was *episteme*, or "knowledge." Scientists of the time were "priests, government officials, kings, emperors, slaves, merchants, farmers, and aristocrats...," as Russell M. Lawson wrote in *Science in the Ancient World*. "They were artists, explorers, poets, musicians, abstract thinkers, and sensualists."

For the longest time natural phenomena were explained through magic and superstition. Then, starting around 600 BCE, philosophers in the city of Miletus, Ionia, what is now western Turkey, applied rationality. Thales, one of Greece's earliest philosophers, believed the universe originated with water and that natural objects had souls. His student Anaximander, whose poetry described humans descending from fish, argued that infinity rules: creation and destruction work as a mysterious, impersonal, unstoppable cycle governing all of existence. Anaximander's student Xenophanes of Colophon, a traveling poet who lived in the mid-400s BCE, used seashells found on mountains to theorize (somewhat rightly) that the earth had experienced alternate periods of wet and dry; he conjectured (super wrongly) that Earth, which surely was flat, was lit by a sort of conveyor belt of suns, and that with each devastating wet/dry cycle, humankind descended into, and later emerged from, mud. The idea that Earth once swam in muck wasn't original, but Xenophanes may have been the first to draw such conclusions using what we now know to be fossils. He influenced Socrates, who taught Plato, who taught Aristotle, the ancient world's greatest scientist. Aristotle in turn influenced the Roman author, soldier, and natural philosopher Pliny the Elder, whose classic ten-volume encyclopedia, *Naturalis historia*—finished in 77 CE, two years before Pliny died in the volcanic eruption that decimated Pompeii—became one of the few complete works to survive the Roman Empire. *Natural History* influenced scientists for centuries to come.

But what were fossils? The term, which first appeared in the sixteenth century, derived from the broad Latin term *fossilis*, which meant "dug up." Fossils were probably the products of ocean vapors, lightning, or the waning of the moon, naturalists decided; or maybe they were the remains of giants. Or they fell from a stormy sky, like rain. Or they grew in the ground, like plants. The triangular stones found scattered about the land were surely serpent tongues that Saint Paul, as payback for a snakebite, had turned to rock in a curse upon the vipers of Malta. In China, people came across huge skeletons embedded in the earth, some curled up as if caught in a nap. A single thigh bone might stand 6 feet taller than a full-grown man. *Mei long*, the Chinese called these creatures—"sleeping dragon." Believing that *long gu*—"dragon bone"—had healing powers, they ground the bones and ingested them in an attempt to cure everything from heart problems to insomnia. A recipe, courtesy the eleventh-century scholar Lei Xiao:

> For using dragon's bones, first cook odorous plants; bathe the bones twice in hot water, pound them to powder and put this in bags of gauze. Take a couple of young swallows and, after taking out their intestines and stomach, put the bags in the swallows and hang them over a well. After one night take the bags out of the swallows, rub the powder and mix it into medicines for strengthening the kidneys. The efficacy of such a medicine is as if it were divine!

To this day acupuncturists prescribe dragon bone. Not so long ago, a *National Geographic* writer asked the well-known paleontologist Xu Xing, How can Chinese people living in the twenty-first century still believe in mythological beasts? Xu Xing replied, "How can so many Americans still disbelieve in evolution?"

In October 1666, fishermen near Livorno, Italy, caught a large white shark. The Grand Duke of Tuscany, Ferdinando II de' Medici, ordered the severed head of the shark sent to his medical anatomist and physician,

Nicolas Steno, in Florence. Naturalists had already deduced that "corpuscles" (molecules) comprised all matter; and Leonardo da Vinci and Robert Hooke, a seventeenth-century polymath, already suspected fossils to be the remains of animals, Hooke lamenting that people too often tended to "pass over without regard these Records of Antiquity…" Steno, a dark-haired, sad-eyed Dane, was among those who rejected the idea that fossils fell from the sky or grew underfoot. Dissecting the duke's rotting shark head, he recognized the creature's teeth as identical to the objects known as tongue stones. Expounding on the notion that fossils were the remnants of animals, he conjectured that fossils became fossils after bone absorbed mineral water and turned to stone. Yet the mineralization theory seemed to explain only part of a larger puzzle. What about the weirdness of finding a rock within a rock—or "a solid naturally enclosed in a solid," as he put it? What could explain the existence of mountaintop seashells?

Scrutinizing the Italian landscape, Steno theorized that fossils found deeply embedded in the earth were once part of the planet's fluid surface. Over time, those surfaces must have settled into successive horizontal layers, with the oldest layers at the bottom and the youngest on top. Strata either perpendicular to the horizon or "inclined to the horizon" once lay parallel with the horizon, he reasoned. Strata surely existed "continuous over the surface of the Earth unless some other solid bodies stood in the way"—if molten rock cut through a stratum, it made sense that the molten rock was younger than the stratum it interrupted. Each layer, therefore, provides a "snapshot" of life on Earth at a certain time. Looking at a clean, uncomplicated cross-section of the planet, you can expect to see stegosaurs buried well below tyrannosaurs, and mastodons many layers above them, because they existed millions of years apart. Yet the oldest layers aren't necessarily the deepest layers; the planet's continuous shift and heave rearranges terrain, sometimes leaving extremely old layers on top.

Steno's strata theory survived as the principle known as the law of superposition, a tentpole of modern paleontology, geology, and evolutionary theory. His pen-and-ink drawing of the severed shark head—pointynosed, open-mouthed, simultaneously cartoonish and grotesque—lived on in the scientific sketch canon.

Around the same time Steno was devising his theories, Agostino Scilla,

a star painter of the Sicilian *sciento*, published an important paleontological treatise based on his own observations. As his frontispiece he
used a sketch of a man clutching a shark tooth and sea urchin in one
hand and "pointing to a hillside draped in fossils with the other," wrote
the paleontologist and science historian Stephen Jay Gould. By comparing fossils with other objects, Scilla deduced that they were organic. He
stressed the value of empirical evidence in science—the importance of
noticing.

Empiricism was a profound concept. Because early scientists were
also men of the church, the divine infused everything. People generally believed Earth to be six thousand years old, a number that entered
the public consciousness around 1650 via writings published by an Irish
archbishop named James Ussher: after analyzing the Holy Bible, Ussher
declared the planet's moment of birth as eight p.m., October 22, 4004
BCE, and said that whatsoever fossils were found in the earth belonged to
creatures that died in the Great Flood. "Parson-naturalists" such as John
Ray sought to support both the religious *and* the scientific but increasingly favored "observation over speculation as the watchword of a new
scientific era." Even as empiricism took off, Scilla, for one, confessed deepest interest in the former *life* that fossils represented. "I do not know how
the sea was able to reach so far inland...nor indeed do I care," he wrote. "I
know full well that the corals, shells, teeth of sharks, dogfish, sea urchins
etc. are *real* corals, *real* shells, *real* teeth..."

Someone had to organize all this knowledge, so a Swedish botanist and
zoologist named Carl Linnaeus did it. In 1735, he published *Systema Natu-
rae*, outlining kingdoms of plants, animals, and minerals, thereby laying
the groundwork for modern taxonomy, the principles by which living
things are classified. Earlier men had made attempts at classification, but
Linnaeus's binomial system was the one that stuck, in which the generic
name (genera) came first, followed by the specific (species). "Here's what
makes him a hero for our time," David Quammen once wrote in *National
Geographic*: "He treasured the diversity of nature for its own sake, not
just for its theological edification, and he hungered to embrace every possible bit of it within his own mind. He believed that humankind should
discover, name, count, understand, and appreciate every kind of creature
on Earth." When scientifically describing *Homo sapiens*, the specimen

Linnaeus used was himself: though his body is entombed at Uppsala Cathedral, in the rarified company of kings and saints, it is the holotype used to scientifically represent all of humanity.

As curiosity about the planet's history spread in the late 1700s, a Scottish farmer and naturalist named James Hutton expanded on the notion that the earth was a dynamic work in progress. He argued that the planet endlessly morphs via geological cycles of superheating, settling, cooling, and weathering. Hutton, whose Edinburgh study was once described as "so full of fossils and chemical apparatus that there is hardly room to sit down," argued against Ussher's theory of a young Earth, saying the buildup of layers showed that the planet's structural processes moved far slower than what could have taken place in mere thousands of years.

Hutton's conclusion started with a boat ride. In 1788, at age sixty-two, he sailed with friends along the coast of Scotland, coming to Siccar Point, an odd rock promontory on the North Sea. Siccar was a geometric jumble of red sandstone juxtaposed with sloping layers of graywacke sandstone 380 million years older, a bizarre arrangement that illustrated Earth's powerful, continuous forces of refiguring heat and pressurization. A travel companion of Hutton's, John Playfair, later remarked on the momentous sight: "The mind seemed to grow giddy by looking so far into the abyss of time."

Hutton presented his findings—now called the Great Unconformity, or Hutton's Unconformity—in a paper at the newly created Royal Society of Edinburgh. To his colleagues' astonishment he described a "universe . . . formed by a continuous cycle in which rocks and soil are washed into the sea, compacted into bedrock, forced up to the surface by volcanic processes, and eventually worn away into sediment once again," as Edmond Mathez put it in *Earth: Inside and Out*. Such cycles made mountains, cut rivers, moved oceans. Hutton then delivered a line that would help cement his reputation as the father of geology: "The result, therefore, of this physical enquiry is that we find no vestige of a beginning, no prospect of an end."

Such a distant past is so imponderable that Hutton, after hearing his friend Playfair's "abyss" remark, shrunk the concept to an evocative phrase: deep time. A mere two hundred years later, the author John

McPhee explained deep time with perhaps the best science metaphor ever written:

> Consider the Earth's history as the old measure of the English yard, the distance from the King's nose to the tip of his outstretched hand. One stroke of a nail file on his middle finger erases human history.

Geologists used information like Hutton's to chart Earth's structure as the geologic scale, parceling the chronological history into eras, periods, epochs, and ages. Whereas the periodic table of elements presents both horizontally and vertically, in tidy cubes, the geologic chart visually represents a vertical coring of the planet. It's something of a fluid document because we will never finish learning all there is to know about Earth, and because humans are forever finding fossils that help tell that story.

Eric and his parents belonged to the Bone Valley Fossil Society ("Digging into Florida's Past") and the Tampa Bay Fossil Club. The organizations thrived as extensions of the natural history "field clubs" that had existed in the United Kingdom as far back as the 1830s and in the States since just after the Civil War. Their thousands of members were teachers, nurses, firefighters, jet-engine technicians, insurance salesmen, physicians, and sawmill workers. Some had collected fossils since childhood; others had started only recently. Fossil hunting had led one member to quit "riding the couch all day on Sunday, watching NASCAR."

For fifteen dollars a year, members could participate in monthly meetings and regular field trips as well as fossil shows. Mines were starting to limit or ban individual prospectors for liability reasons, but some clubs still got group access because their sponsors included universities and museums. Membership came with subscriptions to the clubs' monthly newsletters or bulletins, which were staffed by volunteers and often edited by scientists. Members shared book recommendations and bought and sold fossils through the classifieds. An astute kid could learn a lot simply by reading the newsletters, which often carried illustrations that helped identify their finds. It wasn't unusual to see a geologist's piece like "Why

Janey and Johnny Don't Understand Evolution" alongside an account of a hobbyist's field trip, or an editor's note apologizing for a publishing delay—"Karen's back went out again."

Eric kept every newsletter. His burgeoning book collection included volumes by Roy Chapman Andrews, the once-famous 1920s explorer who spent his entire career at the American Museum of Natural History. Two of Andrews's works from the 1950s were *All About Dinosaurs* and *In the Days of the Dinosaurs*. "The word dinosaur (DIE NO SAWR) means 'terrible lizard.' Dinosaurs belonged to the family known as reptiles," Andrews wrote, the information unusually scientific for popular children's books of the day. "Baby dinosaurs were hatched from eggs." "The first animals could walk from one continent to another. That is why dinosaur bones are found in many parts of the world." "The best place to look for fossils is desert country." "In 1922, I organized an exploring party to…the great plateau of Mongolia because it contains a desert. This is the Gobi, the greatest desert in Asia."

CHAPTER 5

DEAL

THE FIRST DOCUMENTED IMAGE OF A NATURAL HISTORY collection appeared in 1599 in Naples, Italy, courtesy of an apothecary named Ferrante Imperato. In his book *Dell'historia naturale* he published a sketch of his "cabinet," which meant "room," which basically meant museum. This cabinet had a vaulted ceiling and architecturally intricate built-in bookcases with nesting cubbies and magical drawers. Every square inch of the room held a sample of the captive outdoors, seahorses to birds to a large crocodile, which centerpieced the crowded ceiling. It was as if Imperato had followed the procurement advice of the German artist Gabriel Kaltemarckt, who told Cristian I of Saxony that a fine personal collection must include not only sculptures and paintings but also "antlers, horns, claws, feathers and other things belonging to strange and curious animals."

Such a cabinet was also called a *Wunderkammer*, or room of wonders, an encyclopedic inventory of all things paleontologic, geologic, religious, ethnographic, and archaeologic. One prolific Wunderkammerer was the polymath Peter the Great, who, around the turn of the seventeenth century, built one of the most dazzling menageries in Russia. His collection—the foundation of what is now the Russian Academy of Sciences—included a four-headed rooster, a two-headed sheep, and the pickled noggins of infants who, pink-cheeked and puffy-lipped in their large jars, looked as if they might simply be sleeping. The czar also collected human teeth, whose impromptu extractions he ordered from a "fast-walking messenger," a "person who made tablecloths," and other random people met on the street. From a renowned anatomist he purchased fly eggs sourced from the anus of a "distinguished gentleman who sat too long in the privy," wrote Stephen Jay Gould, adding that when it came to stocking their cabinets, collectors "vied for the biggest, the most beautiful, the weirdest, and the most unusual."

Natural history museums grew out of these rooms of wonder. Paris founded its Muséum national d'Histoire naturelle in 1635. The Ashmolean Museum, at Oxford University, opened in 1667. In 1753, Sir Hans Sloane parlayed his enormous personal collection into the British Museum, which in 1881 birthed the standalone Natural History Museum at the urging of Sir Richard Owen, coiner, in 1841, of the word *dinosaur*. By 1900, many major cities had a natural history museum—Berlin, Vienna, Madrid, Philadelphia, New York, Pittsburgh. The *Wunderkammer* eventually shrank to a single piece of furniture, the curio cabinet, and to the Riker case, the bookshelf, the shadowbox.

Eric's collection started in beer flats and cigar boxes.

Doris's archive of papers and keepsakes formed buttes and levees in the family den. Bill's album collection ran wall to wall, ceiling to floor, in the room where he tended his stamps and listened to music. Eric's thousands of fossils overtook his bedroom and spilled into the communal spaces and garage, as if parades of creatures passed through the Prokopi house during the night dropping teeth and claws.

There wasn't anything about fossils that he disliked. He enjoyed hunting them, cleaning them, identifying them, sorting them, and learning about them. He wrote newsletter articles describing certain creatures: "Order Proboscidea—the mastodons, stegodonts, gomphotheres, and elephants—was found on all continents except Australia and Antarctica…" As his collection threatened to take over the house, Eric rented a booth at the Bone Valley fossil fair in Lakeland and, with his parents' help, unloaded hundreds of shark teeth and Ice Age bones, banking eight hundred dollars. The Prokopis started driving to trade shows as far away as Illinois, sometimes pocketing thousands of dollars at each event.

Eric had started collecting shark jaws in order to learn about the different species and their teeth. One day, his mom heard about a shark-fishing tournament in Tarpon Springs and said, *You shoult go get toes shocks and cut out toes jaws!* Some collectors would pay $15,000 for a single mouth of a great white. Eric borrowed a pickup and he and Doris drove over to the tournament pier and found a dozen or so lemon, tiger, and nurse sharks strung up by their tails. The fishermen said Eric could have the carcasses provided he hauled them away, so he loaded the big fish into the bed of the truck and drove them to Land O' Lakes. In the front yard, he cut them open, carving out the jaws. Each set of jaws went into an enormous Ziploc. Each Ziploc went into a garbage bag. Each garbage bag went into another garbage bag, then another. The bags went into Doris's deep-freeze. Whenever he was ready to work with a set of jaws, Eric would scrape off the flesh, soak the jaws in peroxide, then insert a cylindrical object like a coffee can or bucket into the mouth to stretch the cartilage into an "O." The average buyer prefers a pose an ocean swimmer might see just before he is eaten.

In the classifieds he found a used fishing boat. His father had given him his old Mercury Sable. Now that Eric had his own transportation he could hunt whenever he wanted when he wasn't in school or the swimming pool. Working several trade shows per year, he sometimes grossed more money than his father. His new business cards featured a sketch of a shark tooth and read—

Florida Fossils
Collector of Shark Teeth, Fossil Vertebrates,
and all Florida Fossils
Buy—Sell—Trade

He stocked up on prep tools, eventually owning not just paintbrushes and toothbrushes and dental picks but also an electric scribe. He built a blast cabinet to clean denser materials without shrapneling himself with sprayed matrix. Some days he'd sit in the driveway in a lawn chair, working a fossil the way old men patiently whittled wood.

In the fall of 1992, he started school at the University of Florida, having graduated third in his high school class. The university had awarded him grants and an academic scholarship based on his high SAT and ACT scores, and he paid his living expenses with fossil money. He had been swimming competitively for so long, he wanted to keep going, so he walked onto the swim team. One of the coaches told the press that Eric would need to work hard to earn a berth on such a competitive, successful squad, but said that what he lacked in natural talent he made up for with determination. "Eric is a very conscientious guy," coach Peter Banks told the *Tampa Tribune*. "He may not seem all that outgoing in the pool—and that has to do with swimming alone—but he has a tremendous work ethic, with swimming and everything else."

The university had a strong paleontology program affiliated with the Florida Museum of Natural History there in Gainesville. Eric volunteered at the museum as part of his work-study obligation, but he didn't enroll in any paleontology courses because he still largely thought of fossils as a fun hobby. From what he could tell, paleontologists, who worked for universities, museums, or the government, spent a lot of their time fighting bureaucracy and groveling for funding, and too little of it hunting. Eric majored in engineering science instead, to get a general knowledge of how things worked. Yet even a high-paying job in his concentration, coastal and oceanographic engineering, wouldn't be worth the price of being bound to an office or kowtowing to a boss, he decided. He wanted a life as a hunter, no matter the risks.

The risks were growing, but so were the potential commercial rewards, thanks to the case of Tyrannosaurus Sue.

People who live near badlands live with the prospect of dinosaurs in their backyard. *Mako sica*, the Lakota Indians called such areas—"land bad." The term used by Early French Canadian trappers was *les mauvaises terres pour traverser*—"bad lands to travel through."

Badlands are nothing more than the ruins of a former sea bed, but they are among nature's finest art—a "very skeleton of nature, or the wreck of an embryonic world," as scouts put it when surveying western U.S. territories for Congress in the mid-1850s. In searching for the best route to the Pacific, the government-dispatched surveyors, scientists, and artists described soil, climate, lumber, tunnels, fuel, snowfall, wildlife, harbors, good water, bad water, trading posts, trees, mosses, war parties, and "Women of great age." One author might have been describing areas of either Montana or the Gobi Desert when writing, "From the uniform, monotonous open prairie, the traveller suddenly descends...into a valley that looks as if it sunk away from the surrounding world, leaving standing all over it thousands of abrupt, irregular, prismatic, and columnar masses, frequently capped with irregular pyramids…" The description went on—

> …[T]he traveller treads his way through deep, confined labyrinthine passages, not unlike the narrow, irregular streets and lanes of some quaint old town of the European continent. Viewed in the distance, indeed, these rocky piles, in their endless succession, assume the appearance of massive artificial structures, decked out with all the accessories of buttress and turret, arched doorway and clustered shaft, pinnacle, and finial and tapering spire. One might almost imagine oneself approaching some magnificent city of the dead, where the labor and genius of forgotten nations had left behind them a multitude of monuments of art and skill.

In fact, one *was* approaching some magnificent city of the dead, for fossils were everywhere, remnants of the Western Interior Seaway, that vast body of water that bisected the continent for tens of millions of years. Fossils are more easily discovered in and around the old ghost sea because the sedimentary layers aren't covered by kudzu or neighborhoods or malls.

The Larson brothers, Peter and Neal, grew up in one such area of South Dakota in the 1950s and '60s. By the time Pete was ten, the brothers, enthusiastic fossil collectors, had scrawled MUSEUM on a scrap of wood and mounted the sign on a pole, crowning it with the sun-bleached skull of a cow. Pete went on to earn a geology degree at the South Dakota School of Mines and Technology, and to cofound a commercial hunting and prep company called the Black Hills Institute of Geological Research. By 1990, Black Hills, located in a small but stunning and biologically complex mountain range by the same name, was known as one of the world's largest hunters, preppers, mounters, and casters of museum-grade fossils; the company had donated or supplied materials to the Smithsonian, the American Museum of Natural History, the Denver Museum of Nature and Science, and the Yale Peabody Museum of Natural History.

The Larson brothers hoped to build a standalone dinosaur museum one day, and in the late summer of 1990, Pete believed he'd found his centerpiece. One strangely foggy morning at the end of the field season, he and his crew were working near Faith, a blip on the lonesome highway between Mud Butte and Red Elm, when his girlfriend, Sue Hendrickson, a successful amber hunter by trade, walked out to a sandstone cliff she'd been wanting to explore. There, she found a nearly complete *T. rex* weathering right out of the bluff. "I don't know if it's a sixth sense or luck, but there's something going on," she later wrote. "When I'm attuned to something I'm looking for, I find it."

The skeleton turned out to be 42 feet long and 13 feet tall at the haunches, with a skull the size of a bathtub and teeth that measured over 6 inches each—the largest and most complete *Tyrannosaurus rex* of the twelve or so on record. Pete later said Tyrannosaurus Sue, as he called the specimen, could have "swallowed my kids as if they were multivitamins."

What seemed like the best possible luck quickly turned into a mess. The Black Hills Institute bought the skeleton for $5,000 from Maurice Williams, the Sioux rancher on whose land the bones were located. Williams later disputed the transaction, saying he was the skeleton's rightful owner. The Cheyenne River Sioux got involved because Williams's ranch was on the reservation. Then the Department of the Interior got involved because the federal government was holding the Williams ranch in trust, in lieu of property taxes. Then federal law enforcement got involved because some officers had been wanting to get tougher on wayward fossil collectors, and here was an opportunity to delve into a commercial company's business practices. The FBI raided Black Hills and confiscated Tyrannosaurus Sue on the grounds that Larson and company had taken the bones from federal property. The news cycle filled with images of Hill City townspeople protesting with FREE SUE! signs. Black Hills' chief fossil preparator transformed his vintage Checker cab into a protest vehicle, the side panels hand-lettered with SUE HAS ALREADY SERVED 65,000,000 YEARS. Video crews captured the National Guard carting the crated dinosaur away in trucks. A real estate agent named Marv supposedly lay down in the road in an attempt to block their departure. Children chased after the dinosaur, crying.

The ensuing criminal charges echoed the Society of Vertebrate Paleontology's position that the commercial fossil trade posed a clear and significant threat to science. Robert Hunt, a University of Nebraska geology professor, later told *American Lawyer* magazine that he and other SVP members spoke out against the Black Hills Institute partly because the media incorrectly characterized paleontologists as "divided on this issue" when in fact they were "unanimous in condemnation of this supposed institute." The claim wasn't true. The AMNH's Mark Norell, for one, thought the Larsons were getting a bad deal owing to overzealous law enforcement and a careerist prosecutor, and at least two other scientists, including the Smithsonian's Clayton Ray, resigned their SVP membership over the issue. The well-known paleontologist Robert Bakker, who completed his undergraduate degree at Yale and his PhD at Harvard, spoke out against science "extremists," saying, "Because these people have their PhDs they think they have some God-given duty to protect antiquities and fossils. They're like self-appointed guardians of the faith; they want to make fossils off-limits to anyone without a doctorate. It's especially tragic because it threatens good amateurs—who've done more for the science than anyone."

Bakker resembled a time-traveling miner, with his grizzled beard and a battered straw hat that looked, as Pete Larson once put it, as if it "fell from his head during a buffalo stampede." Bakker and Larson were friends, and they often discussed why the relationship between paleontologists and commercial fossil hunters had soured so badly. Natural history museums owed their existence to independent collectors, yet scientists seemed to view them all with resentment and suspicion. Bakker blamed *Sputnik*. After Russia launched the satellite in 1957 the United States increasingly invested in science—"'the *real* sciences,' Bakker hastened to add, because everyone knows putting old bones together is a very different process than serious endeavors such as finding a cure for cancer or landing on the moon," Larson once wrote, adding that according to Bakker's theory, the space race affected all of science:

> Americans wanted to see their scientists finding things and digging things up and building things, and doing all of this before anyone in Russia did. It didn't take long for paleontologists to realize that if they wanted to get funded like other scientists, they had to start looking like other scientists. "They had to wear white lab coats," Bakker explains, "they had to make proposals and marketing plans, and move around boardrooms as if they knew what they were doing.... Competition for funding changed (the dynamic).

Defining more than how paleontologists looked, this fever sought to change who we were.

After a six-week trial, Larson was sentenced to two years in prison on charges that had nothing to do with Tyrannosaurus Sue. A jury found him guilty of misdeclaring cash and travelers' checks at customs, buying fossils that had been taken illegally from Custer Gallatin National Forest, and removing fossils valued commercially at less than a hundred dollars from Buffalo Gap National Grassland. He served eighteen months at the federal penitentiary in Florence, Colorado.

Meanwhile, Maurice Williams, the rancher, got Tyrannosaurus Sue. Sotheby's auctioned the skeleton in New York City on October 4, 1997, in a deal brokered by David Redden, the auction house's executive vice president. Redden had sold (or would sell) the Duchess of Windsor's jewels; the papers of Martin Luther King Jr.; the estates of Jacqueline Kennedy Onassis and Andy Warhol; and an "unspeakably fresh" first-run copy of the Declaration of Independence, which a collector had found at a Pennsylvania flea market, framed behind a four-dollar painting. "The Declaration of Independence is just a piece of paper. In determining its value, you can explain it to other people, and it becomes precious to them, too, and you've done more than just make the document worth $4 million or $5 million, you've actually done something significant," Redden once told the *New York Times*. "That's the irony in a capitalistic society: by conferring value on an object, you save it." Redden had approached Williams about auctioning the "world treasure" *T. rex*, later telling the media, "We have never sold anything of this importance, and nobody else has either."

The McDonald's corporation and Walt Disney Company teamed up to buy Tyrannosaurus Sue for the Field Museum in Chicago. One newspaper had wondered just how much Chicago would have paid to ensure that "'Da Bears' and 'Da Bulls' were joined by 'Da Bones,'" but now everyone knew the answer. The price, with the buyer's premium, came to a staggering, unprecedented $8.36 million. Williams walked away with $7.6 million, tax free. The Smithsonian's Kirk Johnson later said, "The day they sold Sue is the day fossils became money."

The sale proved dinosaur bones could be bought and sold in America like a Brancusi bronze. One good skeleton could put a family's kids through college or pay off the mortgage on a ranch. A single spectacular *rex* tooth alone might go for $10,000. Ranchers turned to fossil prospecting to make up for climate-related losses of livestock and crop, leasing their land to commercial crews and taking a cut of any big sale. Cretaceous formations crawled with a new generation of hunters, some with enough money to muscle underfunded scientists out of their longtime research sites. "I don't want to offend any collectors, and I know I have no right to say, 'Don't buy fossils'—it's a free country, it's a free market," the DePaul University paleobiologist Kenshu Shimada later told *Collectors Weekly*. "But if people decide to buy a fossil to display it in their home, I want them to understand the ramifications." Buying fossils removed potentially important specimens from proper study and stoked the market, which drove demand, which inspired poaching, all of which threatened to push paleontologists out of the field and fossils beyond the reach of scientific research.

At one point, Frank Garcia proposed to his Smithsonian contact Clayton Ray that scholars and dealers come together for a "town hall" meeting to work it out. Ray responded with a note that Frank would keep for the rest of his life, calling it "the smoking gun." Ray said science and the public interest had always depended on the amateur community and that "an ever larger and more sophisticated public is what will save more fossils and advance paleontology, not heavy handed (and unenforceable) restrictions." Yet "extremists" within the Society of Vertebrate Paleontology now dominated the discussion, Ray told Frank; he had "given up."

———

Eric went on building Florida Fossils as he worked his way through college. His courses were so challenging he realized he needed to drop either the swimming or the hunting. Once Doris heard the news, she said, "Okay, if you not doing the swimming anymore, I do the swimming for you." She had not competed since she was a girl in Germany, but into the pool she went, winning ten gold medals in the Florida Senior Games the year she turned fifty-nine.

At one point Eric heard there might be fossils at a quarry near the town of Brooksville, where hobbyists and commercial hunters were no longer allowed to hunt. Fossil clubs' codes of conduct now prohibited trespassing, though some hunters ignored the rules, feeling comfortable and entitled after so many years of access. Mines sprawled like moonscapes and were hard to police; guards had been known to simply ask, "Find anything today?" and go on their way. Hunters figured that if they got busted, they would simply be asked to leave.

Eric entered the Brooksville locality and found fossil birds, reptiles, amphibians, and the extraordinary remains of extremely small mammals. The average hunter wouldn't necessarily have recognized the materials as important—"Once you're dealing with the tiny little teeth of mice and bats, if you're not a trained scientist you may not be able to tell they're millions of years old, and very rare," Richard Hulbert, the FMNH vertebrate collections manager, later said—but Eric knew the fossils' significance and said he told the museum about them.

Since the fall of his freshman year, he had volunteered at the natural history museum as a laboratory assistant in the vertebrate paleontology department. David Webb, the paleontologist who had received so many of Frank Garcia's finds, had put him to work cataloging and prepping fossils, particularly shark teeth. The museum displayed a dugong vertebra that Eric had donated. The Brooksville site promised to be another contribution: it turned out to be an important deposit of Florida's earliest mammals. The locality, 25 million years old, captured a moment in the planet's history when mammals were about to grow large and flourish.

Eric loaned the museum some of the Brooksville fossils he had dug, and when he went to pick them up a year later they had accession numbers. One of the specimens that he had found had been named for another commercial hunter. Eric never found out why but he figured the FMNH scientists saw his presence at Brooksville as a serious breach. They didn't acknowledge him as they began publishing papers about the site, and the lack of recognition burned. Once the museum appeared to have finished working at Brooksville, Eric sneaked back inside, to see what was left. When a museum crew found him hunting there, he turned over the buckets of matrix he'd dug, and left. When Eric received a letter from the museum telling him to stay out of quarries where he didn't belong, he abandoned any allegiance he might've felt to science and devoted himself entirely to the hunt. Much later, he would wonder aloud why paleontology was "even important." After all, fossils are "just basically rocks," He said, "It's not like antiquities, where it's somebody's heritage and culture and all that."

———————

Bill cautioned Eric against pursuing fossils as a hobby after college, but hunting was all Eric wanted to do. To him, no other career option offered such a compelling combination of freedom and opportunity. He stuck to rivers, hacking his way into forbidding thickets and waters. An aquatic journey of ten feet could transport him millions of years back in time. The larger the animal, the more complex the prep work: cleaning a shark tooth was one thing, but mounting the three-dimensional skeleton of a giant armadillo was another, requiring an understanding of anatomy, and of balance and dimension. As the Florida Museum of Natural History once explained, "The process of building a supporting and hidden metal frame for the skeleton and articulating each bone in its proper place, as well as posing the skeleton in a lifelike stance, requires an extraordinarily high level of craftsmanship and ingenuity."

Preparators, or preppers, were often independent contractors who worked for commercial dealers as well as museums. Eric had been hiring out a lot of his prep but started doing it himself to avoid backlogs of a year or more. To understand anatomy, he read scientific papers and studied photos and sketches. To make his own mounts, he taught himself how to weld. As his inventory grew, he took his stone zoo to market.

CHAPTER 6

TUCSON

THE DIFFERENCE BETWEEN A ROCK AND A MINERAL IS LIKE THE difference between a fruitcake and the fruit that's in that cake. Minerals make up rocks. Rocks make up landforms. If rocks are the wallflower of geology, minerals are Beyoncé. They're found pocket and seam throughout the earth in wondrous forms—cylinders, needles, clusters, nuggets, spheres, dipyramids, cubes. Uvarovite looks like blocks of green Jell-O, rhodochrosite like wads of pink bubblegum, cyanotrichite like a bright blue pompom. A stone, on the other hand, is a rock fragment. Precious or semiprecious stones can be cut into gems. Geology's marvelous lexicon includes *batholith, Jolly balance, polymorph, travertine, magma, alluvium, anticline, talus, luster, topaz*. The chemistry itself is art: tourmalines are complex boron silicates, or $XY_3Z_6(T_6O_{18})(BO_3)_3V_3W$. The formula ultimately produces an object breathtaking to behold: tourmaline can be so brilliantly pink and green, there's a variety called watermelon.

Just as Floridians are surrounded by water, Arizonans are surrounded by minerals, more than eight hundred varieties of them. Collectors spend their lives haunting abandoned mines and the unexplored dump piles of dug tunnels. On the night of December 3, 1946, twenty-eight rockhounds met in the Pima County Courthouse, in Tucson. As the country emerged from World War II, many returning servicemen were looking for outdoor hobbies they could pursue with their families. Lapidary equipment—the grinders, carvers, polishers, and saws used in stonecutting and jewel faceting—was becoming affordable, and hobbyists were getting into it. The courthouse group created what is today known as the Tucson Gem & Mineral Society, which hosted field trips, started a library, and recruited University of Arizona faculty and staff as participants and scientists and authors as speakers on topics like "Perambulations through Crystallography."

In the spring of 1955, the group put on a show. Nine vendors displayed their collections in an elementary school auditorium, charging no admission. The weather was terrible, but fifteen hundred people showed up. The show then relocated to a rusty Quonset hut at the county fairgrounds, with bad toilets and a leaky roof; attendance doubled, each person paying a quarter to get in. When the show expanded to three days in 1958, so many people signed up for the field trip, the caravan consisted of seventy-five vehicles and a Tucson Police Department escort.

As the event grew, elite dealers developed a reputation beyond that of rockhound. "After all, one does not call Escoffier a chowhound, or Baryshnikov a toe dancer," the *New York Times* reported. "A rockhound might pick up a piece of petrified wood and carve it into an ashtray. The mineral collector who has a chunk of some obscure rock like Quetzalcoatlite prefers to keep it in the same pristine clothing it has worn for 50 million years."

In 1960, the show's organizers invited top museums to exhibit. The only curator who responded was Paul Desautels, the new curator of gems and minerals at the Smithsonian National Museum of Natural History. "Museum curators have traditionally been regarded as a rather phlegmatic, even laid-back group, primarily concerned with the stewardship of their collections. This was not Paul's style," wrote Daniel Appleman, the Smithsonian's associate director for science. Desautels was a former chemistry professor who had founded the Baltimore Mineral Society. "His unique contribution was to bring to museum curation all of the passionate urgency to build the world's best collection that characterizes the most successful private collectors." Desautels set up two Smithsonian exhibits at the Tucson event, showing some of the institution's most interesting specimens. He delivered talks in a "curtained off portion of the back room infamous for its hard chairs and limited ventilation," Bob Jones, a lifelong mineral collector and Scottsdale science teacher, wrote in a history of the show.

Having Desautels on hand was like having Eli Manning at summer football camp. It changed the way organizers obtained exhibits and "meant that people living too far from any major museum could still see at least a small portion of a museum's collection," Jones wrote. In exchange, curators had a chance to scope out the excellent stones found by amateurs. Some of the private collections Desautels discovered—Al Haag's wulfenites; Susie Davis's rare linarite—eventually became part of the Smithsonian's permanent collection. And as the show began turning a profit, the local community benefited in the form of university scholarships and free admission for disadvantaged children.

Before long, the show became as much about selling and buying as looking: a wholesale section was introduced. After it was temporarily eliminated, dealers took their business outside, to an abandoned gas station, an old barn, and the Holiday Inn South, where Desautels stayed when he was in town. A "quiet closed-door" side market developed at the hotel.

The Society contractually banned "pre-show selling" because they wanted to provide paying visitors with fresh selections, but when warned about infringing on free trade, show organizers lifted the rule and the side market grew. "American entrepreneurial spirit (particularly strong among independent-minded mineral dealers) meant that dealers had an irresistible urge to do business whenever and wherever the opportunity might arise," Jones wrote. Dealers started renting a "sleeping room," a "selling room," *and* a booth at the show, and also sold items out of the trunks of their cars. The hotel sales culture spread to the Desert Inn, the Vagabond, the Pueblo, the Ramada.

An old town, Tucson was already popular with tourists, filmmakers, and the mob. ("It was pretty much agreed that any mafioso who wanted to take a vacation could go to Tucson and be okay," Jones said.) Wintertime Tucson was not a hard sell: the show happens in late January and early February, as the palo verde trees and saguaro cactuses bloom. But really it's "good *rocks* that bring folks in," Jones said. So much so that in the 1970s, the event moved to the city's new convention center. The show went international after the British Museum mounted an exhibit. In came Colombian emeralds and Russian malachites. The Sorbonne brought cumengite crystals found in Baja. There were pegamites from Afghanistan. Magnesite from Brazil. Vanadinite from Morocco. Swiss gwindels. Indian zeolites. Tasmanian crocoite was "lap-carried on the plane all the way from Australia," eventually to be displayed at the Harvard Mineral-ogical & Geological Museum. An American Museum of Natural History exhibit featured J. P. Morgan gold. "Museum curators would spread the word: Oh, you've got to go to this show!" Jones said. The city's most important event used to be the rodeo, then a golf tournament; now it was gems and minerals.

For a collector, Tucson was Earth, sized to buy. It's hard to estimate the annual monetary value of the current global trade in natural history, but it's easy to say the industry would not exist to the extent it does today without Tucson. The city name itself has become synonymous with the marketplace: you don't say, "They saw it in Tucson," but rather, "They saw it *at* Tucson." Dealers, collectors, scientists, preppers, artists, curiosity seekers—and, lately, undercover federal agents—descend upon the city for what has been described as the "New York Stock Exchange of the mineral world."

If not for Tucson, fossils might never have found such an easy home in the natural history trade. They became a regular presence in the 1970s, and in 1986, commercial fossil hunters were formally invited to show. By the 1990s, a Colorado collector and promoter named Martin "Marty" Zinn III had developed fossils into their own event.

For paleontologists, Tucson was both a thrill and an abomination. "Some of the things I saw made me sick," a UC-Berkeley paleontologist, William Clemens, once said. "I saw a rare fossil amphibian from Russia on sale, accompanied by a certificate from Russia's Paleontological Institute allowing export of this treasure. The Russians must certainly be hard up to let things like that go." Others thought some of their fellow scientists were getting a little hysterical. "This whole campaign by the federal cops against the fossil business is ridiculous, considering all the murderers out there that remain to be caught," John Maisey, an AMNH ichthyologist, said at one point.

There was no real oversight at the show. If you wanted a booth, all you had to do was fill out a form and describe your inventory, and maybe send a photo or two, remembered Jones, a longtime member of the organizing committee. The vendor-vetting process happened largely by word of mouth, a "sort of self-policing thing," he said. "People on the committee would say, 'Oh yeah, I know that dealer, he's good,' or, 'I don't know about that guy.' It was an informal monitoring system. Later on, when the show got so big, it didn't matter—anybody who wanted a space would get it, if there *was* space."

For a fee, the hotels moved furniture out of the rooms, clearing the way for display cabinets and sales tables, the spaces so transformed they resembled small shops instead of cookie-cutter suites. Dealers sold fossils from right off the tops of their bedspreads and televisions. Jones, who cohosted a TV program about the show, liked walking into hotel rooms and saying, "Okay, what you got in the bathroom?" He later explained, "These are the minerals that are quietly sold to special customers who've been notified usually ahead of time: 'I've got such-and-such for you, come see it.' Those items never or rarely go on public display. There's nothing underhanded about it. They're not being secretive about it. It's simply a way of doing business." Materials that a dealer knew to be illicit were generally kept out of sight, not that law enforcement would've known the difference—they were more likely to check for work visas, though that rarely happened, either. The City of Tucson wasn't about to step in to make sure every last bone was legit: by 2014, the shows would boost the annual local economy by $120 million and generate over $10 million in city taxes. Policing everything wasn't even possible at an event with so many strands, especially when so much of the inventory was impossible to trace. Buyers learned about authenticity through experience, and by reading websites and talking to dealers who enjoyed teaching.

Forty-three shows eventually took place simultaneously across town, in ballrooms, warehouses, parking lots, hotel rooms, and tents. Foreign vendors often stowed their materials in self-serve storage units year to year, and kept U.S. bank accounts. An event that had started with nine mineral enthusiasts attracted four thousand dealers in 2017. Even if you stayed the whole two weeks, it would be impossible to fully absorb the Tucson experience, though with the right questions and contacts it was possible to learn who had the nice lapis lazuli, and which guy to see about a dinosaur.

Eric graduated from college in December 1996 and moved back to Land O' Lakes in the spring, back into his childhood bedroom, running Florida

Fossils out of his parents' house. As a student, his life had rotated around semesters; now it revolved around Tucson in February, and the Denver show in September. Most days, he left the house in the morning and stayed gone until dark, returning with a pickup filled with river bones. After supper he worked in the garage or the driveway late into the night, when the stuccoed nooks of the front stoop attracted live décor in the form of jumpy green frogs. The newspapers that once covered his swimming now wrote about his fossils: "Love of paleontology evolves into business." One photo showed Eric at the Prokopi kitchen table with a saber-tooth cat skull and a megalodon tooth half a foot long. Asked how he had made such interesting finds, Eric said, elaborately: "I just dove, collected, and read up on it."

At twenty-two, he had a buzz cut and a wide, shy smile. His hands were morphing into huge parodies of hands. "Today, in spite of his young age, Prokopi has gained a reputation in the field as a veteran fossil hunter," one newspaper noted. Doris and Bill remained unconvinced, but the paper declared that Eric had found a promising career, in which he preferred hunting to retail. "I don't like setting up shops," he said. "I'd rather be out collecting."

Fossil dealers typically stockpiled inventory, saving up for the big shows, but the advent of the internet allowed them to sell year-round. Eric signed up for an AOL account. One day at the end of August 1999, he entered a chat room about scuba diving. A recent University of Tennessee graduate named Amanda Graham happened to be in there, looking for information about rivers. A native of Williamsburg, Virginia, she had just started a job as a SeaWorld dolphin trainer and wanted to know the best places to dive.

Eric volunteered that he knew some good spots. As they chatted online Amanda noticed that he didn't force himself on her like other guys did. "Why don't *you* take me diving?" she finally said. He agreed, telling her, "I've seen it all, so I'll let you pick the place."

They planned to go that very Saturday. In the meantime, they talked on the phone and traded emails and photos, the sight of which unsilenced Eric like never before. "Wow! I got chills and butterflies in my stomach when I saw you," he told Amanda in one email. "I can't wait to see you in person..." One night after they had talked for hours on the phone, Eric emailed to say he'd been thinking about her so much he could hardly eat. He kept calling her answering machine and hanging up, just to hear her voice.

They chose the Santa Fe River, north of Gainesville, for their first date. "If you want lunch, bring what you want because I don't know what kind of weird stuff you eat," he emailed her. "If you like, later we can get dinner and a movie or something in Gainesville, if you're into that kind of thing." He said, "And by the way, with the affect [*sic*] you have on me, I might get excessively nervous and shy, so if you want a kiss or anything, go for it. I feel stupid saying that, but I know how much you like kissing and everything, and I don't want to screw up."

A hundred years earlier, loggers with axes and two-man saws cut cypress and longleaf pine from old-growth forests, lashed the logs together with metal "spike dogs," and floated them downstream to sawmills. Untold thousands of logs dropped along the way. The cool, low-oxygen water of the river depths cured the wood to fine-grained reds and browns. Deadheads, the logs were called. Salvaging them for money was called deadheading. Deadheaders looked for logs marked with the X's and cuts of lumbermen a century gone and raised them to the surface by cable and winch. The logs were ten times more precious than regular wood, the value rising by the year. Anyone could become a deadheader as long as he had enough money to buy a $5,500 state permit, and enough time, resources, and energy to find and pull the logs.

Chad Crawford, host of a TV series called *How to Do Florida*, once showed viewers how deadheading worked. He drove to the Choctawhatchee River in northwestern Florida to find a log he could make into a mantel, and met up with Rich Mitchell, a deeply tanned deadheader with a silver goatee who worked for Bruner Lumber Company. "I don't quite know what I got myself into," Crawford said the night before the dive.

"You got yourself into a whole lot of, ah, danger," Mitchell told him. In his estimation deadheading was "about an eight" on a ten-point danger scale. Some of these logs were five or six feet across, and weighed tons. "This is not something that we just let anybody come up here and do," he told Crawford. "You're fixing to get off into a river full of alligators, poisonous snakes—and beautiful lumber!"

The next morning Crawford met Mitchell and two other deadheaders at the river and was introduced to his dive partner, Lexie Cook, a burly guy everyone called "Bobo." "You've got to really know what you're doing, diving rivers, especially if they're swift-running rivers and deep," Bobo told Crawford. They piloted two fishing boats and a winch platform out onto the water. As Crawford and Bobo stepped into their wetsuits, the veterans looked over the newbie's getup. "What *is* all that?" Bobo said.

"I don't know," Crawford said, adjusting his gear. "I just bought it yesterday."

Mitchell reached for the orange flippers Crawford was holding, and said, "You're not gonna need those where you're going."

"Them there's for a swimming pool at home," Bobo explained.

Seconds before they jumped into the water, Crawford asked, "What happens if we get separated?"

Bobo said, "Good luck to ye."

Below the surface, sunlight instantly disappeared. Silty murk rose up to meet the divers as they descended, levels of diminishing visibility that Crawford said "redefined dark." The current bumped him along the bottom, where he neither saw nor felt any logs. When he surfaced, spooked, Mitchell met him with the boat and they moved to shallower water.

"You find ninety percent of 'em with your feet," Bobo explained. "I can tell you the difference in a pine, an oak, or a cypress with my feet."

Pretty soon they found a cypress log, about 15 feet long and 4 feet in diameter. After winching it to the surface, Crawford said, "I tell you what, there was something about being down there with that log—knowing that the last person that touched that log was probably a hundred years ago." He later told Mitchell, "That's what I love about what you guys do, man. You get to touch, feel, unearth history like this."

Eric had started deadheading to supplement his fossil income, partnering with a scrappy older diver and roofing contractor named Joe Kutis, who lived near Gainesville. Joe and his wife, Charlene, looked after Eric like a son. When Eric and Joe hunted fossils together, it wasn't always with pure intentions. One night, they spray-painted a couple of bicycles black, rode into a quarry, past No Trespassing signs, and dug till dawn.

Eric found that he could net $40,000 in ninety days pulling logs and spend the rest of his time on fossils. For years he'd owned an old second-hand pickup, but now made enough money to buy a new Ford F350, the vehicle he was driving when he showed up for his date with Amanda.

"He drives a *truck*?" Amanda's mom said when Amanda called home to say she had met the man she would marry.

"No, Mom, you don't understand," Amanda said. "This truck is like the price of a *Ferrari*."

Amanda was as blond and outgoing as Eric was dark and quiet. "Your constant smile and pleasant manner make everyone want to be part of your inner circle," one friend would someday tell her. Her mottos included *Go big or go home*. She wore her long hair straight, with big sunglasses often shoved on top of her head. Her collar was popped, her headband was Burberry. Her wrists were layered in what she called "good Virginia silver." Instead of keeping an inventory of earrings, she consistently wore one good pair, the narrow gold hoops her parents had given her on her eighteenth birthday. Instead of buying a dozen different handbags, she believed in investing in one good one, like the Louis Vuitton Neverfull tote. Amanda liked to say that good clothes and nice hair meant nothing if you carried a cheap purse, and that with the right bag, you could get away with sweatpants and flip-flops.

Amanda had one sibling, Jeff, who was finishing up at Duke, on his way to becoming a dentist in Richmond. Their father, Maurice, was a Williamsburg pediatrician. Their mother, Betty, was a copper-haired Tennessee native who wore headbands and ballet flats, and spoke in a smooth, throaty drawl. Amanda had inherited her father's dimples and upbeat outlook, and her mother's grace and elegant taste. Maurice and Betty had divorced not long after Amanda went to college, and Maurice had remarried, but Betty still lived in the house of Amanda's childhood, in Kingsmill, a gated resort on the James River.

By land there were two ways into Kingsmill, each marked by a guardhouse staffed around the clock. The two thousand or so permanent residents lived on streets with colonial names like Archers Mead and Winster Fax, and with access to tennis courts, swimming pools, playgrounds, a marina, fire pits, playgrounds, and a PGA golf course. The Grahams lived on a cul-de-sac, in a modest, two-story shake-shingle house outfitted in early American antiques and Oriental rugs. The heart of the house was the kitchen and breakfast area, with a redbrick fireplace and a large window overlooking the fenced backyard. At holidays Betty decorated mantel and hearth with dense arrangements of aromatic greenery and gilded candle-sticks, tucked through with glittering ornaments. She insisted upon cloth napkins and nice silver, and would sooner die than serve a takeout meal without first transferring the food to a pretty casserole dish. She often told Amanda, "Treat your friends like guests and your guests like friends."

The exacting standards occasionally wore on Amanda, but she admired her mother's hostessing skills and her talent for making ordinary moments feel special. For a football-themed party, Betty carpeted the house in temporary Astroturf. For a breakfast party, she sent a limo around to fetch all the children. For Amanda's "rock 'n' roll" sixteenth birthday, she allowed everyone to tag the walls with graffiti, then painted over it afterward. Amanda absorbed her mother's instinct for making every-thing feel pulled-together. Riding the school bus home, she'd spot certain drab houses and fantasize about giving them makeovers. When she envi-sioned her own dream house, she saw it white and stately, the driveway lined with palms.

At Tennessee, Amanda majored in psychology with an emphasis in animal behavior, thinking it would be cool to be the next Jack Hanna yet cheerfully admitting to friends that she enjoyed college more for the social perks than the scholarship. As a senior, she applied to SeaWorld in Orlando, making it clear that she wanted to be in the water with the ani-mals, not standing on the side of the pool. They assigned her to Discovery Cove, where she enjoyed interacting with tourists, preferring to address a crowd of hundreds over a cluster of three. One of her favorite things to do was grab a dolphin's fin and let the animal zoom her through the depths of the pool. Hair streaming, she imagined herself a mermaid.

As a graduation present, Amanda's parents had planned to give her a safari trip to Africa, but now that she worked at SeaWorld they'd given her a laptop. When she stumbled upon Eric in the chat room, she hadn't been looking for a boyfriend. She had seen women's ambitions ruined by the confinements of the wrong relationship and had vowed not to let that happen to her. Still, she believed that if you were in a place that you loved, doing work that you loved, you were bound to meet the right people. Eric was a fellow diver, an adventurer, a self-starter. "Mom, he has Dad's work ethic!" Amanda told her mother, and Betty had to think about that.

On their Santa Fe River date, Eric was not able to say how much he already liked Amanda, but she never hesitated to speak her mind. "You've been hiding *that* all day?" she said when he finally took off his shirt. After diving, they went out for Amanda's favorite food, cheap Mexican, and made plans to see each other again. "I miss you so much already," Eric emailed her two days later. "I love that you are always smiling and laughing, it makes me feel so good. I hate to be around people who aren't cheerful. I've never met anyone who is as happy as you all the time." He liked her face, her eyes, her sense of adventure, her "neat freak" tendencies, and her independence. "Not that I'm cheap, but I like that you are willing to pay your share when we go out," he told her. "You are smart and you seem to know a lot about a lot of different things. I am never bored listening to you talk; maybe that's why I don't say as much as you want."

Eric wanted to visit Amanda in Orlando, but first he had to do the Denver show. The second time she ever saw him, he was pulling out of town, towing a trailer. Weeks later he told her, "Before I met you I was scared I would never meet the right person and I would be alone forever. You are that person. I'm glad you see it as a partnership and not a pair of handcuffs. . . . I can't wait to see where our lives will go together."

CHAPTER 7

BIG GAME

AMANDA FOUND ERIC'S WORK INTERESTING AND MYSTERIOUS. He would come home covered in scrapes and bug bites, hauling the bones of creatures she couldn't begin to identify. The first time they went on a serious fossil dive together, she quickly found a mastodon tooth and thought, *Pfff, this is easy*, but she never found anything good again. She and Eric could be looking at exactly the same spot and she would miss the fossil hiding in plain view. Eric seemed unafraid of anything, except heights, and as far as Amanda could tell he could *do* anything except ice skate. She liked to say that he was happiest working in places where a killer might dump a body. "Eric is perfectly proportioned between Reality Guy and Cool Adventure Guy," she told friends, often recalling the time she watched him chainsaw a log while standing barefoot on that log in a river. "He's Superman," she'd say. "If you're ever stranded on a desert island, you'd better hope Eric is with you."

She liked that as a self-employed businessman Eric could set his own schedule—this would be important once they started having children. He was financially stable and spontaneous—if she mentioned Las Vegas, he'd say, "Let's go Friday." He loved the water as much as she did. They both liked to stay busy. Ten years hence, Amanda wouldn't be able to remember a single fight. Eric got sad but never angry, and he never raised his voice. At one point, she listed on paper all the reasons she loved him. He replied simply with, "I don't love you because you are beautiful, but thank you. I don't love you because you are always smiling, but thank you. And I don't love you because you love me so much, but thank you. I just love you because *you* are *you*."

Amanda had always thought of an engagement ring like a tattoo: you'd best be careful about letting someone else pick it out. So as marriage seemed inevitable, she described her ideal ring to Eric, then waited. In October 2000, they went on vacation to Atlantis, a Bahamas resort known for its scuba- and shark-diving adventures. Amanda was sure Eric would propose, but the vacation came and went, and nothing. "Oh, did you want to get engaged in the Bahamas?" Eric said once they were home. They went back to Atlantis and Eric proposed on the beach. Amanda left Paradise Island wearing six carats' worth of diamonds in a custom-made platinum setting.

They announced their engagement in the newspapers, listing Eric as a "commercial paleontologist," a job title Amanda thought was fine. *Paleontology* meant "the study of fossils," did it not? Eric knew all the Latin names and owned more paleontology books than Amanda realized existed. She had never known anyone who read scientific papers for fun.

As they tried to decide where to make their life together, they never considered anywhere but Florida. Gainesville, a couple of hours north of Land O' Lakes, felt familiar, right-sized, and affordable. They rented a house where they would live after the wedding, and in the meantime bought a patch of property and started thinking about building a place of their own.

In late September 2001, Eric and Amanda met their family and friends in Williamsburg. They had thought about postponing the wedding after 9/11, but decided to carry on. The night before the ceremony, they held a rehearsal dinner in historic Yorktown, on the York River, at the Watermen's Museum, which honors those who make a living by water. Eric had warned Amanda that public speaking made him anxious and that he might be too shy to give a toast, so Amanda stood before the crowd, holding a flute of champagne. "Now, I'm sure everyone knows Eric is pretty quiet, but that's one of my favorite parts about him," she told everyone. "You see, Eric is an amazing listener and close observer. He sees things that crazy, hyper people like me usually miss." She went on to say, "Not only does Eric not hold me back from my dreams, he makes them happen."

The next day, Amanda walked down the aisle on her father's arm at Williamsburg United Methodist Church, in a white strapless gown, her bouquet of orchids and baby pineapples studded with seashells that she and Eric had collected on diving trips. At the reception, at Kingsmill, a sand dollar crowned the bride's cake. The groom's cake was festooned with white-chocolate megalodon teeth.

Eric and Amanda honeymooned for weeks in French Polynesia, where *Survivor*, the reality TV show Amanda really wanted to be on, had just filmed its fourth season. When they got home to Gainesville, they moved into the rental cottage. Once Eric cleared their new land, it was so valuable they sold it and bought nine acres across the road.

There were now so many fossils on the market, it was possible to earn income buying and reselling other dealers' finds. When Frank Garcia decided to unload more of his collection, which he liked to call his "retirement fund," Eric bought it for $75,000 and resold it in parts. Two months after the wedding, he went digital and opened an eBay account under the handle *floridafossils*. He didn't enjoy the administrative tedium of photographing and posting shark tooth after shark tooth, or shipping the merchandise, or keeping the books, or dealing with customers, but eBay was an easy new way to grow a global clientele.

Tucson remained the linchpin, though. Tucson was where the international museum buyers showed up with entourages, looking for spectacular specimens. "Especially in what I call the good old days, the early 1990s, there would be people from the museum in Tokyo walking around the show with eight helpers," Andreas Kerner, another dealer, once said. "The first guy would point things out. The next guy would walk up and say, 'Hold this for us.' The next guy would be the one to write the checks. The next guy would pack it up." Private and overseas museums and collectors were buying big. One American *Diplodocus* sold to the United Arab Emirates to be displayed at a Dubai mall. To scientists' outrage, a complete *Stegosaurus*, found on private property in Wyoming, was sold to the Hayashibara Museum of Natural Sciences in Okayama, Japan, by a commercial company in Utah. (The company supposedly made the sale contingent on scientific access, but the caveat did little to assuage those paleontologists who were sickened by the fact that U.S. law allowed the sale and export of vertebrate fossils at all.) A major buyer was Sheikh Saud bin Mohammed al-Thani, a Qatar royal known for burning through well over $1 billion in fine-art and other purchases for the emirate's burgeoning museums. China alone was "building natural history museums like wildfire."

As Eric worked on Florida Fossils, Amanda started Everything Earth, an interior design company. Her girlfriends had been urging her to go into business because the spaces she decorated were such an inviting mash-up of safari, beach house, bohemia, and proper Williamsburg. Tucson, vast and eclectic, gave her endless ideas. Why hang a ho-hum print on your wall when you could display a framed sea fan? "*That's* art," she liked to say. At Tucson she bought skins, furniture, and imported corals, along with beads for making jewelry. It never occurred to her to ask how materials were sourced; if it was pretty and interesting, she simply bought it, envisioning Everything Earth as "*National Geographic* for your house." She acquired inventory with the intention of sending it right back out the door. She was the purger, Eric the keeper: he'd had to salvage Amanda's college diploma from the garbage and talk her out of selling her wedding dress. "Eric lives in organized chaos," Amanda once said. "I live in organized organization."

After a successful Everything Earth trunk show, Amanda opened a booth at the Bizarre Bazaar, a popular crafts market that takes place in Richmond in the spring and at Christmas. One of the first sights shoppers saw when they walked into the cavernous showroom at the Richmond Raceway Complex was Everything Earth: woven baskets, cow skins, mercury-glass jars, handmade jewelry, and miscellaneous antlers, seashells, fossils, and starfish. In 2004, her first year as a vendor, Amanda won Best of Show.

That year, Guernsey's, a New York auction company, held a natural history show at the Park Avenue Armory. The catalog featured a primer on paleontology. Amid the Lebanese shrimp imprints, Italian soft-shell crab, and the whale skeleton—one that P. T. Barnum bought during the Civil War—were dinosaur parts from Montana, Utah, and Oklahoma, but also from Argentina, China, and Mongolia, countries that banned the trade in fossils. In the acknowledgments, Guernsey's thanked seven commercial hunters by name. Eric made a note of it.

David Herskowitz was the kind of guy who, a hundred and fifty years ago, would have succeeded in a place like Deadwood, a hard, enterprising town

built in the forested folds of the Black Hills of South Dakota. At Tucson
and Denver, he tended to wear jeans, a frumpy T-shirt, and a faded baseball cap, his belly leading, his eyeglasses slipping to the end of his nose,
his New Yawk accent audible across the room. One could just as easily
imagine him on a street corner of the Old West in a slightly shabby top
hat, conducting business as the passing horsecarts misted dust onto his
three-piece suit.

Herskowitz, who was in his early fifties, lived with his phone to an ear
the way auto dealers watch the showroom door. The caller might be either
a buyer in want of a *Torvosaurus* or a hunter eager to sell. Herskowitz was
terrible with names to a comical degree, but he had established such an
exclusive clientele that people often thought of him first when it came
time to do business.

The natural history broker came from the city: Flushing, Queens.
Herskowitz's grandfather manufactured uniforms and blue jeans, his
father owned a liquor store, and his mother's family had a pharmaceutical
company in Manhattan. Herskowitz earned a college degree in hotel and
restaurant management but ended up in the advertising and newspaper
subscriptions business, where he did so well in the early 1980s that he
started investing in real estate. One of his tenants was a Korean woman
who didn't seem to have much money but always paid her rent on time.
One day Herskowitz asked a friend where the woman got her income and
learned the tenant sold diamonds in Korea.

Herskowitz knew nothing about diamonds, but his Colombian girlfriend's sister's office assistant was a Korean woman who had worked in the
diamond district of Manhattan. The assistant introduced Herskowitz to a
Forty-Seventh Street appraiser who schooled him a little in the trade. Herskowitz started buying diamonds and having his new friend appraise them,
then took the diamonds to Korea and sold them. When he found out the
Koreans also wanted emeralds, off he went to Colombia.

As the Soviet Union crumbled in the early 1990s, Herskowitz got interested in Russian topaz, alexandrite, rare green garnets, and amber. *Jurassic Park* was to debut in 1993, its blockbuster plot centering on dinosaur
DNA resurrected from amber, and Christian Dior was showing amber
jewelry. Sensing an opportunity, Herskowitz acquired a load of amber
from a girlfriend, who happened to be in Russia and who liked to shop at

flea markets. When he took the amber to Forty-Seventh Street to have it graded, the appraiser noticed an insect in one of the nuggets. Herskowitz pressed his eye to the loupe and stared, not believing what he was seeing: a black fly, as clear and intact as if it had died yesterday.

How could a *fly* get into a *gemstone*? Herskowitz couldn't figure it out. He wasn't even sure what amber *was*.

One day he told his aunt in Huntington, Long Island, about the insect in amber, which in paleontology is called an inclusion. The remarkable preservation allows scientists to describe new species and better understand evolutionary diversity, sometimes offering clues about how insects cared for their young. The aunt told Herskowitz there was a jewelry store nearby that sold that kind of stuff. Herskowitz immediately went and consigned some of the Russian pieces. A few days later, the jeweler called and told him to come get his money. A client had paid $375, and the jeweler wanted to keep a hundred. Was that okay? *Okay?* Herskowitz thought. He had paid six bucks! He sent his friend back to the Moscow flea market, telling her to buy all the insect-bearing chunks she could find. She started hauling them to the United States by the luggage load, which wasn't that difficult because amber travels light.

The jewelry store could unload only so much, so Herskowitz reached out to the London auction company Bonhams, which was planning a natural history auction in England. Bonhams (once known as Bonhams & Butterfields) accepted a couple hundred pieces of Herskowitz's amber—and sold them for over $14,000. Herskowitz knew almost nothing about fossils, but he knew profit margin, and in the marriage of amber and auctions he sensed global, untapped promise. At auction, the *market* decided what was fair. "People fight over stuff and pay what they want!" he once said. "So the fact that buyers would pay fourteen thousand dollars for something that cost me thirty or forty dollars? I loaded up the truck and moved to Bever-ly."

As Herskowitz began working London auctions he noticed that most of the major buyers were American. Wondering why natural history auctions hadn't made it to the States, he pitched the idea to New York City houses, whose leaders included Sotheby's and Christie's. London-based Phillips, which dated to 1796 and claimed clients including Marie-Antoinette and Napoléon Bonaparte, was the house that bit. "I must have been a good bullshitter because I'm nobody," Herskowitz later said. "I mean I just had an *idea*."

Hoping to authenticate the items before the auction, Herskowitz approached paleontologists at the American Museum of Natural History. No, they told him: they certainly would not assess scientific materials for commercial purposes.

But Herskowitz did learn a name.

A peculiar little shop existed a block away from the museum, and dealt in all things natural history. Maxilla & Mandible had opened in 1983 in a basement space on West Eighty-Second Street, calling itself the world's first and only osteological store—it sold bones. Human skeletons, reportedly imported from Europe, dangled along one wall like a macabre chorus line. The shop did so well so fast it expanded to Columbus Avenue, between Eighty-First and Eight-Second Streets, where it would remain for the next three decades.

The owner was Henry Galiano. As a child in Spanish Harlem he was so obsessed with nature that his father, who ran a beauty parlor, regularly took him to the museum, where Henry could name all the creatures. Later, he dropped out of art school in order to work at the American Museum, starting as a janitor and eventually becoming a curatorial assistant in vertebrate paleontology. Galiano was so skilled that one of his supervisors, a paleontologist named Richard Tedford, urged him to go back to school, then realized that "Henry was more of a free spirit." In the early 1980s, after selling off part of his personal collection of rat and pigeon skulls at the Canal Street flea market, Galiano decided to open his own shop.

Running Maxilla & Mandible, he discovered what he called a "niche within the public's innate interest in and attraction to bones." People enjoyed giving and receiving them as gifts. "If you've given someone the scarf and the gloves and the book and can't think of anything else, give a skull," Galiano once told *People* magazine. Another time he said, "Everybody lives in apartments. They buy plastic furniture, manufactured stuff, a stereo from Japan, a camera from Germany. After a while you lose sight of what you are. With bones, you're in touch with something that's real."

Galiano acquired his inventory through farmers, trappers, African game wardens, Fulton Street fishmongers, and Chicago meatpackers, altogether engaging with a supply network familiar with "slaughterhouses and game-processing plants." Whenever an "interesting" carcass came in,

Galiano got calls. He told the press he would never put a bounty on an endangered species but he might sell the body parts, if given the chance. Early on, he traded in African ivory, the loathsome poaching of which led to bans. "We scrounge everywhere," he told the *New York Times*. "I used to pick up road kills. The highways of the United States are great places to collect natural history. This is an opportunistic business."

The Maxilla & Mandible basement often reeked of formaldehyde and boiled skulls. A visiting journalist once found the subterranean network of cellars and corridors filled with "buffalo heads peering from dark corners, python bones stretched on a table in elegant sweeping arcs, horse skulls, wart hog tusks, giraffe legs, a complete black bear, and articulated rat skeletons." In an aquarium, thousands of beetles steadily nibbled at some dead thing's flesh. Galiano told the reporter, "There's no life down here."

And yet there was life, in death. Maxilla & Mandible's inventory found resurrection in classrooms, private collections, and museums. The store eventually expanded to other areas of the natural sciences, its staff advertised as "paleontologists, entomologists, osteologists, anthropologists, sculptors, and master craftsmen." Galiano, who maintained strong friendships with AMNH paleontologists, including Mark Norell, considered his business a vital bridge between the public and science. He attended the Tucson and Denver shows, and eventually acquired a quarry in the Morrison Formation of Wyoming, legendary for Jurassic dinosaurs.

In the summer of 2011, everything at Maxilla & Mandible was quietly marked half price. One Monday morning in late August, passersby found the door locked and the shop dark after twenty-seven years. The front window, which once showed skulls framed by warm white lights, was opaque with plain brown wrapping paper; a sign on the door read GONE DIGGING. Online, Galiano's customers mourned. Maxilla & Mandible "wasn't just a store; it was a unique escape from the city. Whenever you stepped inside you were whisked away to another time...," wrote one. Another lamented, "Now who will sell us dinosaur bones?"

The first time David Herskowitz tried to meet Galiano, he failed. Galiano didn't know Herskowitz, had never heard of Herskowitz. Look me up in Tucson, he said. There, Herskowitz found a slim Asian fellow with longish curly hair and eyeglasses, and an encyclopedic knowledge of natural history. When Herskowitz explained that he wanted to bring natural history auctions to the United States, Galiano was in. Galiano had the market expertise and the museum background, as well as the contacts. The fossil world is so insular, strangers stand out. Dealers weren't predisposed to accepting Herskowitz, but they knew Galiano.

What Herskowitz described as the first natural history auction in the United States took place on June 8, 1994, at the Phillips showroom in New York. The standing-room-only event fetched a disappointing $300,000. Yet Herskowitz said company officials told him they'd never seen the room fill up like it did for natural history; Phillips decided to try again.

Fossils' association with upscale auction houses stood to elevate them in the eyes of buyers. Old brands like Sotheby's conveyed highbrow respectability, a rarefied counterpoint to the earthier trade shows. To dealers, operating within a world where natural history equaled art felt validating in a way the industry had never seen. Whereas the price could always drop at a trade show, the price at auction could only go up, with no limit on the possible payout. But auctions also provided cover for the grayer areas of the trade. Illicit fossils and forgeries were streaming out of China, selling in the rough at Tucson, and later emerging at market as finished specimens. Chinese dealers advertised in the Tucson show guides. After the Chinese government got stricter, some dealers went on selling, claiming that their fossils had been exported before the new laws took effect. The average buyer had no way of knowing what was true, and dealers knew law enforcement had no way of proving anything. Certain dealers rationalized that there wouldn't be so many Chinese dinosaurs available if it weren't legal. Even paleontologists weren't sure of their information. "You talk to a paleontologist and they'll say, 'Oh yeah, it's illegal,' but very few people have looked at the legal paperwork from that country and know what the legal implications are—and furthermore they probably know very little about how law is enforced in that country," the Smithsonian's Kirk Johnson once said. "It was hard to understand what was going on—whether the laws were changing or simply being ignored."

Another confusing thing: paleontologists claimed to detest commercial dealers yet allowed them along on excavations as paying paleo-tourists. And some museums seemed to encourage fossil collecting. An invitation from Friends of the New Jersey State Museum had found its way to Eric in the spring of 2001, offering access to Liaoning, a northeastern province of

China so hot with fossils that scientists had to post guards at their dig sites. The New Jersey museum's letter was addressed to "Fellow Fossil Hunters" and described an upcoming "collecting expedition" to the "famous fossil fields at Sihetun, to the quarries which have produced the feathered dinosaur specimens." The tour guide was to be a University of Pennsylvania graduate student affiliated with the Institute of Vertebrate Paleontology in Beijing. Participants would be allowed to collect common fossils to bring home, "subject to approval by the Chinese scientists on site."

Eric didn't sign up for the museum trip, but so many people were talking about Chinese fossils, and so many Chinese fossils were circulating openly, he started buying them in the rough through Tucson contacts—mostly mammals like saber-tooth cats—and prepping them out and selling them. His eBay clients included science teachers, collectors, other dealers, and museums, all of whom left positive, public feedback: "ANOTHER AWESOME TRANSACTION BY THE CHINESE FOSSIL KING!!!!" For a decade Eric maintained a 100 percent positive rating on the site, slipping once, to 99 percent.

Occasionally, Eric heard of dealers who had been caught poaching or selling bad bones in other countries. He knew a guy who got locked up in Paris and another who got pinched in Uruguay. In Uruguay, the suspect's wife tried to soften the prospect of overnight incarceration by delivering a pizza and a warm coat to her husband in jail, but all the husband enjoyed that night was the sight of the guards eating his pizza and wearing his coat. Eric listened to the stories with amusement, saying very little at all.

———

Amanda wanted to try flipping a house: buy an outdated or run-down property, renovate it, and sell it at a profit. In the early fall of 2004, she and Eric found a bungalow on Southwest Second Avenue, near the university. The house, built in 1929, had a brick façade that had been painted white; two folksy porch columns the color of dried blood; and scraggly landscaping. So, perfect. A mortgage company loaned the Prokopis nearly $200,000 for the purchase and renovation, and after a shark-diving trip to the Farallon Islands, they got to work.

Contractors' estimates came in so high the Prokopis decided to gut the place themselves. "We had a ball at Lowe's, Home Depot, and Home

Expo, just having daily shopping sprees," Amanda wrote in her annual holiday newsletter. To install granite countertops, they Googled "how to install granite countertops." Whenever possible, they used materials salvaged from Eric's river dives, such as vintage bricks. Their taste almost always aligned. They had their first disagreement as a married couple after Eric installed a metal desk in their home office. "Uh-uh," Amanda told him. "This is not a tire store."

To save money they moved into the flip as they renovated. They bathed in the backyard, ate out for every meal. Amanda hosted two more successful Everything Earth shows, and at Christmas they drove up to Richmond, for Bizarre Bazaar, using some of the proceeds to "dive into adulthood" and buy Oriental rugs.

When Eric ran out of fossil prep space in the garage, he moved bones into the gutted kitchen. By now he was finding giant ground sloths—at Tucson, his room at the Ramada Inn-University featured a skeleton that loomed in a back corner, its skull grazing the ceiling. The work was going so well that in May 2005, he decided to take on a partner, a military veteran in his forties with eccentricities that Eric and Amanda found hilarious and endearing. This guy cleaned everything—even his dog's paws—with puffs of canned air. He washed and ironed his money in case the bills had ever been used in drug deals. He distrusted cell phones so he made passengers in his car leave theirs in the trunk. But the partner was loyal, and knowledgeable about fossils, and he always had "unusual" stuff. Eric bought a retail space with him in Micanopy, an historic town of six hundred just south of Gainesville. The two-story building of pinkish brick sat beneath Spanish moss in a picturesque stretch of antiques shops. Amanda would run Everything Earth in the front, and the guys would use the back rooms and upper floor for prep and storage.

The Micanopy deal ultimately fell through, but Eric went on working with the other dealer as he and Amanda pursued their projects. At the flip house they kept the white bricks, painted the shutters black, and added an awning. They replaced the hokey porch architecture with sleek Doric columns. A semi-circular driveway and fresh landscaping added curb appeal. The results gave the Prokopis confidence that they had a gift for flipping, though they wouldn't know for sure until they sold the bungalow at a profit.

For over a decade Eric had been working with shark teeth and Ice Age mammals, and with miscellaneous pieces like framed insects and giant bats. The Chinese saber cat skulls had been solid income: Eric usually bought them in the rough for up to $6,000 and sometimes sold them for $75,000 at auction. But marriage, the accumulation of property debt, and plans to start a family altogether forced him to think about pursuing projects with the potential for major payout. The big money was in big dinosaurs.

A handful of countries are known for large dinosaurs, but only one of those countries, the United States, allows commercial hunters to collect and sell whatever they find on private property. Formations like the Hell Creek, which traverses parts of South Dakota and Montana, were prime *T. rex* territory, but those hunting grounds were already spoken for and highly speculative. A hunter could spend all summer paying to scour a rancher's badlands and go home with nothing but more debt.

And the federal government appeared poised to tighten fossil-collecting laws. Recently, Congress had convened a joint subcommittee hearing on the proposed Paleontological Resources Preservation Act, or PRPA, legislation sponsored by Congressman Jim McGovern, a Democrat from Massachusetts. On June 19, 2003, McGovern had told the subcommittee that despite an "exploding" black market in fossils the United States still hadn't developed a "clear, consistent, and unified policy" on how to protect paleontological resources. The PRPA called for a standardized system of fossil-collecting permits across federal agencies, required that all significant fossils found on federal property be curated at museums or "suitable depositories" (such as a university collection), and recommended the enactment of tougher penalties for the theft or vandalism of significant fossils. The Society of Vertebrate Paleontology, a worldwide association of over two thousand scientists and a smattering of collectors and dealers, had endorsed the legislation, as had the American Association of Museums. Testifying on behalf of the SVP, the paleontologist Catherine Forster had said the "heightened public interest in dinosaurs and other extinct life forms" had given paleontologists an "unprecedented opportunity to share with the public the excitement of recent advances in this fascinating science that records the history of life on our planet" but also had imperiled fossils by making them black-market targets. Fossils of extinct groups are not renewable, Forster had reminded the lawmakers, saying, "More fossils will be discovered and collected, but always from a finite supply." Even the tiniest grains of sand may yield clues about anything from an animal's habitat to how long a species survived, which in turn could tell us something about the planet as it exists now or may exist in the future. "As paleontologists and geologists learn more ways to interpret ancient environments and ecological communities from fossil assemblages in their original context, this information becomes more and more valuable and important," Forster had explained. Because fossils help scientists understand how creatures change over time, "researchers must be able to compare new specimens with those previously unearthed," she had added, ending her testimony by saying that the increasing commercial value of fossils threatened to distort the scientific record.

Paleontologists and commercial dealers needed to find a way to work together, yet there was still no formal mechanism for bridging the worlds. Scientists and law enforcement got smidgens of insight into the market with each new poaching case. One such case had begun several months after the Congressional hearing, as a U.S. Department of Fish & Wildlife warden drove through the Oglala National Grassland, a swath of mixed-grass prairies and badlands in northwestern Nebraska. Volcanic ash had once blanketed the area, eventually mixing with sediments that became siltstone, mudstone, claystone, and sandstone, which preserved the remarkable remains of bear-dogs, alligators, tortoises, lizards, oreodonts, peccaries, deer, and false saber-tooth cats. Poachers often hit the formation, called the White River Group, and usually got away with it, because basically there was one federal officer to patrol over a million acres of grasslands.

This particular game warden had gotten lucky. He'd spotted several men out there in the middle of nowhere, acting strange. One of the men had fled, but the warden had stopped the other two and called in the Pine Ridge office of the Nebraska National Forest. A Forest Service paleontologist named Barbara Beasley had soon arrived to question the suspects, who were from Wisconsin. One, a naturalist in his early sixties, collected fossil bumblebees. His companions were a welder and former upholsterer who taught at Lakeshore Technical College, and his son, who was in his late twenties. Beasley split the suspects up for questioning. When she told one to empty his backpack he pulled out fossils wrapped in newspaper. The remains were brontothere, an enormous rhinoceros-like plant eater that lived around 43 million years ago in North America and Asia. The Sioux, who had long come across its bones on the prairie, called the animal "Thunder Beast."

When broken bones fossilize, crystals sometimes grow inside them. That's what had happened to these bones. The suspect had told Beasley the Wisconsin winters were long and that he had planned to make the crystals into jewelry. "He wouldn't admit to me that he had fossils, just kept calling the material laid out in front of him crystals," Beasley later said.

She understood the financial and thrill-seeking incentives that drove poaching, but the brontothere case inspired her to also understand the mechanism. The overwhelming majority of fossil sales were legal, she knew, but some of them were not, and even though the PRPA was under consideration, no single law enforcement agency had a comprehensive view of the situation. To get an idea of the potential scope of fossil sales, she emailed a well-known dealer in Colorado, Charlie Magovern, asking, "What is the percent difference between fossils sold at wholesale and fossils sold at retail?"

"It can be anywhere from twenty to four hundred percent depending on the initial cost," Magovern told her. "Very inexpensive items that were purchased for under a dollar could be marked up to three or four times the wholesale cost." And very expensive items, "say in the range of $100,000 or more, may be marked up as little as twenty percent, or even less," he added, "depending on the seller's cost of doing business." Auction houses assessed buyers and sellers twenty percent each over the hammer price "unless it is a very expensive item, then they assess fifteen percent of each," Magovern went on, describing a practice he called keystoning: "Generally the retail is, as with most business, double the price the reseller pays for the item."

As government officials considered the PRPA, the sales stream of Chinese specimens happened to slow. Then a source of dinosaurs appeared, with skeletons rivaling that of *Tyrannosaurus rex*.

Andreas Guhr, a German gemologist with a broad face and dark hair that curled at his shirt collar, lived in Hamburg, where he had studied at the University of Fine Arts and trained as a ceramist, painter, and graphic designer before becoming a dealer of natural history. A mineral wholesaler, Guhr was known for his enormous personal collection, which included the "biggest amethyst druse in the world." He had started a natural history museum in Hamburg and would coauthor a book called *Crystal Power*, wherein he attempted to trace the mythology and cultural significance of gems and minerals from Mesopotamia and ancient Egypt to Greece, Rome, and Europe. For his author photo he posed shirtless with a crystal the size of a fire hydrant.

Guhr would eventually own an interior design company called RedGallery ("Nature at Home"), selling zebra jasper, geodes, and furniture made with petrified logs from Arizona and Crooked River, Oregon. Like auction houses, private galleries now presented natural history with a posh lexicon and a splash of QVC: "This extremely valuable log... shows noble blueish tones, porcelain-like in appearance surrounded by a coffee-and-milk coloured bark..." Guhr promoted himself as someone who had "led expeditions to the most distant corners of the world." In the summer of 1992, he had returned for at least the second year in a row to Mongolia in search of dinosaur bones, an excursion recounted at length by the magazine *GEO*, once described as Germany's *National Geographic*. The cover story, "The Grave of the Dragons," appeared in July 1993, but photocopies of the article were still passed around at Tucson.

The article's author had followed the "expedition" south from Ulaanbaatar by plane, into the Gobi. "What a magical word!" he wrote. The Gobi was "still a huge space filled with secrets, in which new discoveries are always possible." The article called Guhr a "new kind of dragon hunter" who "operated a flourishing trade in the hope of being able to finance further scientific-paleontological excavations with the proceeds."

Guhr was traveling with paleontologists from the University of Hamburg, accompanied by Mongolian scientists, "foremost among them" the veteran paleontologist Khishigjav Tsogtbaatar of the national paleontological center of the Mongolian Academy of Sciences. At one point Guhr and the scientists excavated a couple of *Protoceratops* skeletons. "The find has no especially great scientific value, because the species has already been found very often—but its commercial worth is another matter," *GEO* reported. "Private collectors pay high sums for dinosaur skeletons." In the canyons of the Nemegt Basin, an important late Cretaceous site some 550 miles southwest of Ulaanbaatar, the earth was "positively peppered with fossils." In some areas "dinosaur heads loomed like allegorical figures in a façade out of the sandstone walls." There appeared to be thousands of dinosaur bones, the article went on, repeating the adage that there weren't enough paleontologists throughout the world to collect all the fossils. The lengthy feature altogether depicted an open collaboration between commercial hunters and Mongolian paleontologists, strongly suggesting that some arrangement allowed dealers to dig Gobi dinosaurs and sell them at market. So it wasn't much of a surprise, to Eric, to come across a Mongolian skull one September at the annual fossil show in Denver, at the booth of the dealer Tom Lindgren.

Lindgren specialized in the Green River fossil fish and plants of Utah and Wyoming, and had shown at Tucson since 1986. In the early 1990s, he became what he called "QVC's Indiana Jones," cohosting the show *The Fossil Exhibit*. "We would bring fossils that we could sell en masse," he once said. " 'Here's a shark tooth—we can sell a thousand of these. Here's a fossil fish, here's an insect in amber.' " The show premiered on the day *Jurassic Park* hit the theaters, selling out four thousand pieces of amber in forty minutes. Video footage showed Lindgren in the field and the prep lab, explaining his work. "Because as we know, it's the story that really sells most of these products," he later said.

But "the art of presentation" was best achieved at trade shows, and Lindgren enjoyed the events, especially Tucson. Billionaires wanted "their own private time" with him, he once said: "They like to be romanced." His expansive booth might feature a massive *Triceratops* behind stanchions and ropes, or a rare Green River snake, or, in this case, as Eric saw, a nicely mounted *Tarbosaurus bataar* skull with a mouthful of spiked teeth.

Lindgren also worked as the natural history broker for the Los Angeles branch of the auction house Bonhams. He had handled some of Eric's past items, and had found him to be polite and hardworking, with a lovely family. "You'd shake his hand and you didn't need a contract because his handshake was good," he said. Eric, in turn, knew Lindgren as one of the industry's most successful dealers. If Eric could get a skull like Lindgren's—and more of the dinosaurs that were coming out of Mongolia the way fossils had once poured out of China—he could achieve that golden ratio: acquire bones cheaply in the rough, then transform them into skeletons that sold big. A good prepper could turn a considerable profit by producing something close to art. Eric started asking around.

CHAPTER 8

MIDDLEMAN IN JAPAN

THE SAME YEAR *GEO* PUBLISHED THE ARTICLE ON ANDREAS Guhr's commercial exploits in Mongolia, Hollis Butts was a "happy tourist enjoying a stroll" around Medicine Bow, Wyoming, a tiny town founded in 1868 as a water stop along the transcontinental railroad.

Five miles east lies Como Bluff, a once renowned site in the "bone wars" between Othniel C. Marsh and Edward Drinker Cope, the East Coast paleontologists who nearly killed themselves (and each other) battling for dinosaur bones in the late 1800s. The rapid westward expansion after the Civil War had put more people on the ground in unexplored territories; more eyes meant more finds, and Marsh and Cope wanted it all. They spied and sabotaged, and they brutalized each other in the press. Their competitiveness left them more or less broke and alone at the end of their lives, but in the meantime they advanced the young field of paleontology, giving rise to American science, in large part thanks to the sweep of Jurassic rock near Medicine Bow. Documented dinosaur skeletons and trackways had been turning up in America at least since 1802, but it was Como Bluff that became the world's first major dinosaur discovery site—stegosaurs, camarasaurs, apatosaurs, allosaurs, and the massive *Diplodocus*. "It was *Jurassic Park*, finally," the paleontologist Robert Bakker once said. "These Jurassic critters are a world like none before, none after. The average size of a plant eater was...five, six, seven tons. Many of the plant eaters go a hundred feet, a hundred and twenty feet," the latter of which could walk onto a baseball diamond and straddle both first *and* second base.

In the late 1890s, as the querulous Marsh and Cope approached their deathbeds, the American Museum of Natural History sent paleontologists to Como Bluff to see what was left. Under the direction of the paleontologist Walter Granger, crews opened new sites, naming one the Bone Cabin Quarry, after a shepherd's cottage whose foundation had been built with local dinosaur bone. Nearby lived a middle-aged rancher named Thomas Boylan, who, in 1908, received homestead status just south of the low silhouette of Como Bluff. He opened a Texaco filling station there and made a stone cottage to live in, building it 90 feet long to echo the size of *Diplodocus*. In his spare time, Boylan collected dinosaur bones, hoping to build a complete specimen; he eventually gave up, telling a reporter that "erecting such a skeleton is a long and costly task for an individual to undertake, so I abandoned the idea and proceeded to use [the bones] the best way I could."

Boylan and his son Edward got out their hammers. Next door to the filling station they built a one-room cabin that measured 19 feet from the front door to the back wall, and 29 feet end to end. From a distance the façade appeared to have been made with river stone, but come close and you saw a replica of the shepherd's cabin: the "stone" was 5,796 dinosaur fossils—chunks of vertebrae, femurs, pelvis, all masoned together in a fantastical, functional mosaic. The "bone cabin" yielded easy marketing slogans: "The world's oldest building," "The building that used to walk." Boylan opened it as a roadside fossil museum in 1933, just as Sinclair, an oil company that operated out of the Wyoming town of the same name, began using dinosaurs as mascots. Boylan's "Dinosaurium" made Ripley's Believe It or Not! and survived for decades as a tourist attraction that he ran with his wife, Gracey. The construction of Interstate 80 eventually diverted traffic; by then Boylan was dead. Gracey sold the property and it remained in private hands, along with Como Bluff. Bone Cabin closed as a tourist attraction but visitors still made their way there, and to Medicine Bow, or what was left of it.

"With nothing much to do, I picked up a copy of the *Medicine Bow Post*," Butts, the "happy tourist," told his Facebook friends, describing his 1993 visit. The newspaper's articles included one about the West being the "true America," and another about the positive impact of the movie *Jurassic Park*. Describing an interview with Brent Breithaupt, a Bureau of Land

Management paleontologist and curator of the University of Wyoming Geological Museum, Butts said the newspaper noted—

> Breithaupt is active in a movement to create tougher legislation to prevent poaching of fossils on public land. Recently an almost complete Stegosaurus skeleton was found in Wyoming and sold to a Japanese collector. "It is unfortunate that it is gone," Breithaupt said. "Every fossil tells a story and since this one was almost complete, we could have learned a great deal, but it's gone. Science loses, the public loses, we all lose."

Butts wrote, "Well, I was the scumbag who had brokered that deal, so I knew the true story. I liked the way the article slyly suggested that the stegosaurus [*sic*] was an example of a poached fossil. But it actually came from near the Bone Cabin Quarry, off a private ranch. It was not sold to a Japanese collector but to a Japanese museum. It was not almost complete, but about 65% there... Science lost nothing."

Breithaupt would not have agreed. Wyoming's particularly magnificent and old fossils show the landscape's changes through time, he once explained: "The seas came in, the seas went out, the seas came in, the seas went back out again. The mountains rose up, they weathered down; the mountains rose up again, and we have animals and plants that reflect... these changes of environments. These are parts of America that are scientifically and educationally important. These are parts of America's heritage that are irreplaceable."

It was Hollis Butts's name that Eric learned when he asked how dealers were acquiring Mongolian dinosaurs. Vendors knew Butts as someone who showed up at Tucson in a floppy fishing hat and safari vest, carrying a backpack; he always had "wonderful things, beautiful things—they were gorgeous things to sell," one dealer recalled, "but everything was pretty much in Japan."

Butts lived about two hours northwest of Tokyo, in Saitama Prefecture, with his wife and beautiful daughters. His Facebook profile photo would later show an old black-and-white image of a lanky young man in a fishing hat, standing alongside a loin-clothed couple and children in what is now the Democratic Republic of the Congo. "I had a map, a sleeping bag in a small back pack [*sic*], a bag with a strap, a pair of trousers, mosquito coils, matches, a pair of socks, three shirts, a bar of soap, a small knife, 3 underwear, a hat, passport, and some money. That was all and it was enough," read the caption. "Young and fearless. I was never happier."

The available data on Butts provided only shards of a biography. He came from Garden Grove, California, in Orange County, south of L.A. The son of a World War II veteran, he graduated from Pacifica High School in 1968, where he participated in German Club. He studied physical anthropology at the University of California, Santa Barbara, and served in the military before settling in Japan, where he married, started a family, and worked as a restorer of "really old" Japanese furniture. He sold fossils on the side and had a booth at the annual Tokyo show. On Facebook, he professed to adore quizzes, tardigrades, economics, and garden creatures. The USS *Constitution* and the HMS *Victory* were two of his favorite ships. His home in Chichibu City, an ancient town of about 70,000, known for its silk, Buddhist temples, and vistas of Mount Buko, was, in fact, built "much like a ship," with "hand-carved joints, massive wooden beams." He enjoyed studying ancient Greece and was "especially interested in their warfare and triremes," the war boats with three sets of oars. Although he admired cockroaches he killed them. A self-described libertarian, Butts liked the Nobel-winning economist Milton Friedman, who has been called the "grand guru of the movement for unfettered capitalism." At one point, Butts posted about a Los Angeles woman caught between state and municipal laws that simultaneously required her to maintain her lawn and not to water her lawn because of drought. "Is life now so regulated by the government that it needs to supervise your lawn?" he wondered.

Butts's fossil sales turned up in the published acknowledgments of paleontology papers, and in the press. One story involved Dr. V. S. Ramachandran, the distinguished behavioral neurologist who directed the Center for the Brain and Cognition at the University of California, San Diego. As a child in his native India, Ramachandran was obsessed with magic tricks and fossils, having developed an intense interest in evolution and taxonomy; he enjoyed sketching seashells and mailing them to the American Museum of Natural History, asking, "Are these new species?"

After becoming a scientist and moving to the United States, where his peers knew him "for being able to solve some of the most mystifying riddles of neuroscience," he participated in fossil digs in South Dakota, discovering the Tucson show along the way. In 2004, Ramachandran browsed the event with his friend Clifford Miles, owner of a commercial fossil company in Utah called Western Paleontological Laboratories. Miles had founded the company in 1988, after working as a prepper for the Monte L. Bean Life Science Museum at Brigham Young University. He and his brother, Clark, had been cited as coauthors on papers in publications including the *Journal of Vertebrate Paleontology*. At Tucson, when Miles and Ramachandran came across an odd dinosaur skull, Miles told the neurologist, "You buy it, I'll name it after you."

Ramachandran paid $10,000 for the skull. Cliff and Clark Miles eventually published "Skull of *Minotaurasaurus ramachandrani*, a new Cretaceous ankylosaur from the Gobi Desert" in *Current Science*, a journal in India. The brothers described the animal as having a "bull-like appearance" and "flaring nostrils." They acknowledged Ramachandran as the buyer and thanked the skull's seller, Hollis Butts, "for making this fossil available to science." An article soon appeared on the website of the influential science journal *Nature*, headlined, "Paper sparks fury: Paleontologists criticize publication of specimen with questionable origin." Any Gobi fossil was by definition illicit, the article pointed out. Ramachandran responded that if someone could prove "laws were indeed broken," he would return the skull.

Eric arranged to meet Butts at the Denver show in September 2006. He found an older man, fair-haired and balding, with a ruddy complexion and a backpack containing photo after photo of Mongolian dinosaur bones in the rough. Eric browsed the images as if shopping for boots in the L.L.Bean catalog. The item that spoke to him was the disarticulated skull of *Tarbosaurus bataar*, similar to the one he had seen in Tom Lindgren's booth. A jawbone was missing, along with most of the brain case, but overall the skull was roughly 65 percent complete. Eric agreed to pay Butts $18,000 for it. In the spring of 2007, a commercial invoice arrived from Museum Imports Co., Ltd., in Chichibu City, and the skull soon followed.

U.S. Customs and Border Protection (CBP) asks several key questions about imports, and by law importers must answer truthfully: What's in this shipment? Where did it come from? How much is it worth? Fossil dealers were known to wax philosophic with their answers. *What is it?* Technically, a crumbly pile of trash rock. *Where did it come from?* Technically, from its last known port. *How much is it worth?* The price paid— everyone knows it's the work that goes into a restoration that gives a specimen its commercial value.

This shipment, whose country of origin was listed as Japan, was described on customs forms as "fossil stone pieces," with a declared value of $12,000. Once it cleared customs, Eric outsourced the prep work to a guy he knew through Tucson, because by now he was engrossed in another major project.

CHAPTER 9

HOLLYWOOD HEADHUNTERS

ALONG THE SIX-MILE ROAD BETWEEN GAINESVILLE AND Micanopy lies Paynes Prairie, a sweeping freshwater marsh and savanna named for the son of an eighteenth-century Seminole chief named Ahaya, "the Cowkeeper." Florida's first state preserve is one of those places that, when you get out of the car, just *sounds* hot. Visitors may see a long-headed toothpick (grasshopper), a scarlet skimmer (dragonfly), pirate perch (fish), and a canopy so thick, snakes stretch between treetops. Brightly striped banana spiders the size of a human hand weave golden webs as big as badminton nets, their spindly legs working like the fingers of an elegant old woman tapping ashes from her French cigarette. The only un-wild features of Paynes Prairie are the man-made boardwalks and the posted alligator warnings, one of which shows an illustration of a rabbit, a raccoon, and a toddler and reads, "At dusk if it moves, it's food."

Before Paynes became a state park, it abutted a plantation called Serenola. Eventually all that remained of the plantation was a two-story farmhouse, built in 1936. The man who inherited the estate decided to sell the land to developers in 2006 but hated to see the farmhouse destroyed, so he posted a sign that read FREE HOUSE and entertained offers. When Eric saw the sign, he talked Amanda into jumping the fence and having a look around. Serenola was white, with four front columns and an overall living space of 4,000 square feet. On one end was a side porch and on the other, a *porte cochere*. Renters had left the place so putrid with garbage and feces that Amanda had to step outside and vomit. To her, the house was too far gone, but Eric asked her to look beyond the superficials. Even despite the termites, the structure appeared sound and worth saving.

Amanda gave in. The Prokopis won the house by describing their first renovation. They promised to relocate Serenola to the nine acres they had recently bought, just down the road, and transform it into something special, where they could raise a family. Secretly, Amanda thought of the project as a second flip, but once Serenola was stripped to its studs to lighten it for the move, she looked at the high ceilings, the original windows, the heart pine floors, and told Eric, "Yeah, we have to live here."

One April morning, the house was loaded onto flatbed trucks and hauled one mile down US 441 to the Prokopi land. Eric and Amanda began redoing the house from the foundation up, keeping as much of the original woodwork and architectural detail as possible. The side porch would become a family den. They would add a laundry room and a garage. The appliances would be Viking. Using pinkish bricks salvaged from a demolished Burrito Brothers restaurant, they would build a long driveway, front steps and porch, and a graceful privacy fence with an electronic gate. When Eric heard that the rock icon Tom Petty's grandmother's house was marked for demolition, he and Joe, his old log-pulling buddy, went inside and took a particularly nice mantel rather than see it destroyed. Eric installed it over the living room fireplace at Serenola.

While the house was "free," the move and renovation were not. Eric and Amanda used a private lender because it was getting harder to borrow money from banks. The interest rate was 13 percent, far higher than the national average, but the Prokopis told themselves they'd refinance at a lower rate—which turned out to be impossible. Then they told themselves they'd get solvent again by selling properties and by adding dinosaurs to Eric's fossil inventory.

Eric had never mounted a dinosaur skull, but he figured it out the way he had figured out Pleistocene armadillos and ground sloths, by studying other specimens alongside photos and scientific sketches. He welded a display stand using materials bought at a local ironworks, then mounted the skull with its jaws slightly open, showing off teeth that once tore through the flesh of other Cretaceous dinosaurs.

He preferred dealing with Tom Lindgren as an auction broker because he considered him less finicky than David Herskowitz, but it was

Herskowitz who took the *T. bataar* skull, even though he found Eric difficult to read. At this point, Herskowitz worked for I.M. Chait, an auction house and gallery in Beverly Hills. The company was founded by Isadore Chait, an antiques dealer and jazz singer in his seventies with voluminous eyebrows, a white mustache, and a ponytail. People called him "Izzy." His signature look involved a fedora. Chait had been collecting Asian antiques and art since serving in Vietnam, as a cook in the marines. After the war he had studied anthropology and Buddhism at UCLA and supported himself by performing in nightclubs; he collected so much art that he had to sell some, later saying, "So I did gun shows, swap meets, the Rose Bowl, the Glendale antique fair—and people were buying." By 1970 Chait had opened his first gallery, on Melrose Avenue, selling porcelains, enamels, jades, carvings. "As a committed supporter of Asian art at a time when few others shared my passion, I had to create a market and demand where there wasn't one," he said. Chinese art became popular in the West after President Richard Nixon's diplomatic visit to China in 1972, Chait once told an interviewer: "Americans went crazy buying all this stuff from China and bringing it in. China was just getting over the trials of the Cultural Revolution and a lot of people were selling things."

The company eventually expanded into watches, jewelry, and "one of a kind fossils!" Chait's first auction dedicated primarily to natural history was scheduled for Sunday, March 25, 2007, in New York City, with Herskowitz handling the inventory. The 345 items included an "Egyptian mummy's hand; lion, hyena, and warthog skulls; a gold nugget weighing 62 troy ounces," the *New York Times* reported. The *T. bataar* skull appeared in profile on the catalog cover: "*Tyrannosaurus bataar*, Late Cretaceous (67 million years), Nemegt Formation, Central Asia." Chait declared the fossil "perfect for a New York City apartment."

Eric and Amanda headed to New York, taking a room at the Shelburne, a boutique hotel at Lexington Avenue and Thirty-Seventh Street, an easy walk to the Fifth Avenue auction venue. The night before the sale, they attended the preview party, where to Amanda's embarrassment Eric accidentally dribbled wine down his shirt. The next afternoon, the Prokopis watched in amazement as the offers for the skull shot past $100,000. The two most aggressive bidders were at the other end of phone lines, anonymously battling for the win. The sale quickly landed at $276,000, with the buyer's premium. Now $180,000 richer, the Prokopis went to dinner that night at China Grill, then returned to their hotel room to celebrate.

Only a few people knew the identity of the skull's buyer. The *Times* reported only that it went to a "private collector on the West Coast whom the gallery would not identify." Eric and Amanda later learned, as did millions of other people, that the warring bidders were the movie stars Nicolas Cage and Leonardo DiCaprio, both prolific collectors, and that Cage had won.

Right away, Eric fielded an order for another *bataar* skull, from a buyer whom the broker identified only as "my client." Eric didn't have a second skull, but he knew a dealer who did. This time, he did the prep work himself, working millimeter by millimeter until he had produced and mounted a piece much like the first. Satisfied with the results, he crated the fossil with its armature and shipped it to the broker in California. Then he and Amanda flew to Los Angeles, rented a minivan, fetched the skull, and headed to the address they had been given to assemble it.

Eric kept hearing that the buyer was DiCaprio. In L.A., he matched the delivery address to the one listed for the actor on a map of stars' homes. He and Amanda found the house in a showbiz enclave where the streets, known for "stunning views and extreme privacy," were named for birds. On Oriole Way, the Prokopis drove through a gate and were shown into the foyer of a house once owned by Madonna.

Eric assembled the skull in the entryway and then he and Amanda posed for a photo, and they were done. Before leaving, Eric couldn't help noticing a side room filled with natural history—the skull of a Chinese saber-tooth cat, a *Psittacosaurus* skeleton, a narwhal tusk, a framed collection of flying lizards. So much for DiCaprio's public image as an avid supporter of environmental and conservation causes, he thought. But then who was Eric to talk, as the guy taking the money.

News of the Cage–DiCaprio bidding war had renewed scientists' outcry against the sale of vertebrate fossils, but among collectors and auctioneers, the sale only generated more interest. Bonhams alone sold $3.5 million worth of natural history in three separate Los Angeles events that year, "up from none five years ago," the *Wall Street Journal* reported. The article, headlined "The Oldest Crop," heralded fossils as a new way of ranching, saying the going rate for a *Triceratops* skull was $250,000, "up from $25,000 a decade ago."

Eric had now sold two *T. bataar* skulls, unencumbered, to two of the world's biggest movie stars, for roughly half a million dollars. He and Amanda used the proceeds from the first skull sale to buy another flip house, across the street from the first, bringing their property purchase count to three. Whatever else Hollis Butts had in inventory, Eric wanted it.

Twice, he visited Butts in Japan. In Tokyo, he boarded a train to Chichibu, arriving to find that another American dealer had already claimed a large batch of Gobi dinosaur parts. Eric chose a block containing two *Tarbosaurus* jaws and various bones, intending to make another skull, ultimately giving Butts all the cash he had on him, five grand, and agreeing to wire another thousand. Then he rushed home for Amanda's thirtieth birthday. They flew to the Bahamas, treating a small group of friends to the vacation. One afternoon at the pool, Amanda ordered a virgin piña colada, thereby announcing her first pregnancy. She told everyone the baby was conceived in New York on the night Nic Cage helped launch Eric's future in Mongolian dinosaurs.

As Eric waited for the bones to arrive from Japan, he got word from Butts that their deal was off. Without explanation, Butts said he simply no longer wanted to sell unprepped fossils.

The Prokopis now had house renovations and mortgages to pay for, and a baby on the way—Eric had been counting on the Butts materials as income. He questioned the older dealer by email, argued with him, got angry with him, but Butts refused to sell him another bone. When Eric demanded to be reimbursed for the expense of traveling to Japan, Butts demanded to be reimbursed for the dinners he had treated Eric to in Chichibu. Eric got his money back, but Butts had made an enemy.

By December, Amanda was enormous with a boy, to be named Greyson. As her due date approached, she centered a giant "GP" in fresh garland on the front door at their house on Southwest Second Avenue. In the nursery, above the diaper-changing table, she hung a framed print of a Victorian curiosity cabinet filled with seashells, stingrays, and starfish.

She washed the car and got a pedicure, and on December 14, she went to the hospital and had the baby.

Greyson Prokopi had his mother's bright blond hair, his father's wide mouth. His first Christmas fell on his eleventh day of life. He came home to a house that smelled like stargazer lilies and fresh-cut pine. Packets of Nestlé hot chocolate filled a crystal bowl on the kitchen counter. The many toys beneath the perfectly trimmed tree included a large, plush dinosaur.

Two months later, in late January 2008, Eric and Amanda loaded up truck, trailer, and newborn, and drove to Tucson. One day a wealthy woman came into their showroom and sat for a while as her husband shopped for fossils.

"What do you do?" she asked Amanda.

"*This* is what we do," Amanda told her.

"For a living? You pay your bills with bones?"

"No," Amanda said cheerily. "*You* pay my bills with bones!"

Eric hadn't forgotten that Hollis Butts had cut him off. One day, still pissed and baffled about the behavior, he asked Tom Lindgren the identity of Butts's Mongolian supplier so he could buy from him directly. Lindgren, who had his own reasons for doing what he was about to do, gave Eric an email address and a name: Tuvshin.

With that, Eric stepped into something he in no way foresaw, and hand to God it started with Genghis Khan.

PART II

CHAPTER 10

THE WARRIOR AND THE EXPLORER

LONG BEFORE MONGOLIA HAD A NAME, IT HAD HUMAN inhabitants. Stone Age people left behind their tools. Bronze and Iron Age clans formed alliances and fought. A great wall went up, belting the broad land. The tribal kingdoms warred until, in the late thirteenth century, a leader united them in one of the most legendary military campaigns in history.

Genghis Khan—or Chinggis Khaan, as he is known at home—and his immediate successors conquered half the world, on horseback. They rode out of a vast swath of a landlocked territory to Germany, to the Adriatic, and almost to Vienna. In *The Mongols*, David Morgan wrote, "There was no reason to suppose that armies which had defeated the best that China and the Islamic world could throw against them would meet their match in Europe." And they didn't. For centuries, the Mongol Empire ruled Russia, ruled Iraq and China—ruled damn near everything, all the way to Hungary—in the largest contiguous empire ever created. As Morgan put it, the "empire was so huge that although its centre was in the Far East, it constituted for a century or more Europe's most formidable and dangerous eastern neighbour."

The empire eventually fractured, as empires do. China and Russia, meanwhile, grew stronger, and the power dynamic flipped. Outer Mongolia, as Mongolia was then known, came to be seen by its two enveloping neighbors not as a fearsome superpower but as a convenient buffer zone. (Inner Mongolia, meanwhile, was, and is, an autonomous region of northern China.) After Mongolia fell under the rule of China's Qing dynasty in 1691, Russia more or less waited for its chance.

Westerners started arriving in the 1800s as "adventurers, missionaries, or merchants," former ambassador Jonathan Addleton wrote in *Mongolia and the United States: A Diplomatic History*. The first American to receive a Mongolian passport to travel from China to Siberia did so in 1862. A young mining engineer named Herbert Hoover, twenty-eight years away from the U.S. presidency, would soon visit the capital, then known as Urga, while working in China. One American visitor declared the Mongolian steppe similar to "the rolling prairies of Kansas and Nebraska" but most found it "exotic" and "wild," despite the utilitarian food (mutton with a side of mutton).

In 1911, at the end of the Qing dynasty, Mongolia declared independence from China in what anthropologists have characterized as a crucial step toward survival. The long, extremely cold winters made it hard for humans (and some livestock) to live in Mongolia, as did the overwhelming lack of arable land and access to fresh water and grass. There, in one of "the world's most perilous environments," herder families' pastoral lives "centered on the drive to feed and water their animals," Morris Rossabi wrote in *The Mongols: A Very Short Introduction*. A Buddhist theocracy was established, with the Bogd Khan, or "Living Buddha," as the new head of government. Hoping to establish diplomatic ties beyond Russia and China, Mongolia contacted a host of distant countries, including the United States, to request "friendly cooperation." One Mongolian press secretary later said, "I don't think this country can be compared to any other. We are nomads, and our psychology is different from that of other people. Other countries can afford to play cards—China cards and Russia cards—and some of us would like to play those cards, too, but we don't have the trumps."

Washington declined to establish diplomatic relations. Mongolians' desire for independence only grew, along with an awareness of their own history. A scholar translated *The Secret History of the Mongols*—a "semi-mythical and semi-accurate work" believed to be the only surviving account of the life of Genghis Khan—from ancient to modern Mongolian. The text dates to sometime after 1227, the year Genghis died, and has been called "one of the great literary monuments of the world." *The Secret History* presented a legend that inspired Mongolians to decide for themselves whether to think of the founding father as a democratic hero or a genocidal terror, or both: Genghis Khan was born Temüjin, around

1162, near what is now Ulaanbaatar, the ultimate product of a "bluish wolf" that mated with a "fallow doe." He exited the womb clutching a blood clot, signifying his destiny to rule. Promised in marriage at age eight or nine, he wed at sixteen. After his father, Yesügei, a nomad chieftain, was poisoned by Tatars, his mother, Höelün, taught him about tribal warfare and politics, impressing upon him the importance of alliances. As a leader, he favored meritocracy and rewarded the loyal. He advocated religious tolerance and protection of the environment; created Mongolia's first written laws; and encouraged development of the Silk Road, fostering trade between Northeast Asia, Muslim Southeast Asia, and Christian Europe. "Splendid Iranian histories, beautiful Chinese textiles and porcelains, and exquisite Roman gold vessels were some of the products of such cultural interrelationships," Rossabi wrote. As a military strategist, Genghis enjoyed analyzing the psychology of his enemies. He didn't boil his captives alive in giant cauldrons, as one leader was said to have done, but his tactics were horrific enough that Europeans tended to think of Mongolian warriors as "fantastic monsters" or "a punishment sent by God." By 1215, Genghis had invaded, captured, and ravaged the settlement that became Peking, now Beijing. Within twelve years he was dead and buried in an unmarked grave in a secret location that archaeologists attempt to find to this day.

During the Russian Revolution of 1917, as the Communist Party leader Vladimir Lenin rose to power, the "Reds," aka Bolsheviks, seized control north of the Mongolian border. Czar Nicholas II was executed with his family the following summer, as Russia descended into civil war. Chinese forces still occupied Mongolia, but the Bolsheviks soon took Urga with the help of a "Red Mongolian" named Damdiny Sükhbaatar, founder of the Mongolian People's Army. Urga's name was changed to Ulaanbaatar, "Red Hero."

Into this remote, volatile world stepped a young Wisconsin native and New Yorker with what some called a "flamboyantly crazy scheme" to explore the Mongolian Gobi for his employer, the American Museum of Natural History. His name, soon to be known throughout the world, was Roy Chapman Andrews.

In 1837, a group of Yankee pioneers headed west from New Hampshire on behalf of the New England Emigrating Company to settle the Northwest

Territory. Just above the Illinois border, a couple of hours north of the new city of Chicago, they stopped at a location that reminded them of home. On a bluff of the Rock River, the Yankees founded what became the town of Beloit, Wisconsin. Their factories came to make bicycles, plows, paper, waterwheels, and much more, but, nostalgic for the New England culture they'd left behind, the settlers also created organizations devoted to science, religion, and "all the adjuncts that contribute to happiness, thrift, and the elevation of society"—public parks, churches, a Philharmonic Society, an opera, and Beloit College, a progressive institution with ties to Harvard.

In January 1884, Cora Andrews and her husband, Charles, a druggist, had a baby boy, naming him Roy. The Andrews family lived in a two-story house near pastures and creek-cut forests, and west of town they kept a cabin where they often spent their weekends. Cora, who enjoyed books about travel and history, read *Robinson Crusoe* aloud to her son over and over—

> He told me it was men of desperate fortunes on one hand, or of aspiring, superior fortunes on the other, who went abroad upon adventures, to rise by enterprise, and make themselves famous in undertakings of a nature out of the common road...

The best-loved stories of Roy's childhood involved wild animals and scientific exploration. As he tramped around with a camera and a notebook, he recorded whatever he saw. From the *Handbook of Birds of Eastern North America*, his "bible," he learned about migration patterns. With a book on taxidermy, he set upon the local fauna. The Andrews attic soon became a small *Wunderkammer* of "minerals, fossils, stuffed animals, insects, bird skins, Indian artifacts, and dried plants," Charles Gallenkamp wrote in *Dragon Hunter*, a biography of Andrews. Becoming a naturalist felt less like a decision than an identity preordained. Andrews later wrote, "I was born to be an explorer."

At Beloit College, where Andrews majored in zoology, the school's Logan Museum of Anthropology often brought in guest lecturers. During his senior year, an assistant curator of geology from the American Museum of Natural History came to talk about the eruption of Mount Pelée, the kind of event that could not have excited Andrews more. The AMNH, founded in 1869, was by now nearly forty years old. Major names in industry and finance—Rockefellers, Vanderbilts, Astors—supported the institution, often gathering at black-tie fund-raisers. The Dinosaur Hall had opened only recently, on a February afternoon in 1905, featuring an enormous fossil *Brontosaurus* that the AMNH paleontologist Walter Granger had found at Como Bluff, Wyoming. The museum had spent years preparing and mounting the skeleton, using cast *Apatosaurus* parts, a stand made of "repurposed pipes and plumbing fixtures," and the bones of four different specimens. Nothing so large had ever been mounted at the museum, or anywhere. Luminaries such as Nikola Tesla and J. Pierpont Morgan gawked at it at a four o'clock Dinosaur Tea. The tail alone of the first sauropod ever mounted measured 31 feet long. "It didn't feed on flesh, but my, I wouldn't want to meet it!" a reporter heard one woman say.

Half a million people visited the museum each year. Details about research expeditions filled the pages of its *American Museum Journal* (later renamed *Natural History*), to which the Andrews family subscribed. By the time the vulcanologist appeared at Beloit College, Andrews had identified the institution as the place where he most wanted to work. He stalked the visiting curator and made him go look at some deer heads and birds he'd taxidermied for Moran's Saloon, then asked him for a job. At the curator's suggestion, Andrews wrote to the museum's director, Hermon Bumpus. In return he received a polite rejection—there were no openings, but Bumpus said to stop by the museum if ever he visited New York City.

Andrews was barely out of his cap and gown before boarding an eastbound train. On July 5, 1906, he arrived in Manhattan aboard the Twenty-Third Street ferry, with thirty dollars in his pocket. "The magic city" was "more beautiful than anything of which I had dreamed," he later wrote. "I knew it was my city."

At eleven the next morning he called on Bumpus. The museum occupied twenty-three acres between Seventy-Seventh and Eighty-First Streets at what is now Central Park West. The neo-Romanesque towers on the east and west corners of the southern façade gave an impression of "baronial splendor." Twenty-five scientists worked at the AMNH, overseeing an extensive collection that included "birds, mammals, reptiles, fish, insects, botanical specimens, minerals, fossils, and anthropological material," Gallenkamp wrote. Nine years earlier, the museum had bought the "bone wars" maven Edward Drinker Cope's collection of ten thousand fossil mammals, and Henry Fairfield Osborn, a vertebrate paleontologist who had studied under Cope at Princeton, had recently named *Tyrannosaurus rex* based on bones that the prolific fossil hunter Barnum Brown had found in the Hell Creek Formation of Montana several summers earlier. AMNH expeditions were "penetrating some of the earth's remotest areas in search of scientific data and collections," Gallenkamp wrote.

Visiting the museum surely made Andrews want the job all the more, but again, he received disappointing news. Pressing the director, he said it wasn't a position he wanted, but rather a home, if only mopping floors. And so it was that Andrews was hired as a custodian in the taxidermy department at an institution "in which men worked who to me were as gods."

Before long, Andrews was off the mop and pursuing an advanced degree at Columbia University. He studied with Osborn, who had become the museum's director. Osborn had founded Columbia's zoology and paleontology programs, and had arranged an academic partnership between the museum and the school, tying scientific research to the art of conservation and curation. He also had a pedigree. His father had founded the Illinois Central Railroad. His uncle was J. P. Morgan. His connection to Childs Frick, a steel-fortune heir and museum trustee, generated a small staff of hired fossil hunters who helped the vertebrate paleontology department build out its collection. It was Osborn who had sent Barnum Brown to Montana, leading to the discovery of *T. rex*, and to Canada, where Brown competed affably with the prolific Kansas hunter Charles Hazelius Sternberg for vertebrates. Osborn also commissioned increasingly sophisticated, lifelike museum exhibits that drew large crowds, widening the public's interest in the natural sciences.

"But one had to take him in context," wrote the AMNH curator Edwin H. Colbert, an Osborn protégé. "He had grown up in the lush days of the Robber Barons, and he viewed society as a highly stratified arrangement, in which he occupied a top stratum." That stratum, as it concerned Osborn, included some disturbing ideas that future admirers would find difficult to reconcile. For instance, he would praise a 1915 book, *The Passing of the Great Race*, by Madison Grant, a "pseudo-scientific work of white supremacism that warns of the decline of the 'Nordic' peoples," Jedediah

Purdy wrote for *The New Yorker*. The book "influenced the Immigration Act of 1924, which restricted immigration from Eastern and Southern Europe and Africa and banned migrants from the Middle East and Asia," Purdy noted. "Adolf Hitler wrote Grant an admiring letter, calling the book 'my Bible,' which has given it permanent status on the ultra-right." Osborn believed in evolution but rejected the idea of humankind's descent from apes, suspecting that Asia, not Africa, would prove to be "the evolutionary 'staging ground' in which both the dinosaurian and mammalian life of the planet had evolved and dispersed," Douglas Preston wrote in *Dinosaurs in the Attic*, a history of the AMNH.

Andrews, meanwhile, focused on whales. There he was, in the summer of 1912, age twenty-seven, peering off a page of the *New York Times*—fit and clean-shaven, with a receding hairline, a cleft chin, and the creamy complexion particular to photographs of the age. The article placed him among a rising generation of adventurers, mostly Ivy League men who appeared determined to "solve the geographical, anthropological, zoological, and botanical mysteries that have lain veiled for ages," the *Times* reported. Over a hundred major expeditions were launching that very summer. While his peers were bound for places like the Amazon jungle and the Congo, Andrews was headed for "the unknown section of North Korea, never visited by white man." The expeditions altogether represented what the newspaper called "the most vivid example of the mighty effort of science to make the enigma of life somewhat clearer."

But Andrews also craved adventure. After Gallenkamp published his biography, in 2001, a reviewer observed that "Andrews's career was a straight line from Beloit, Wis., to the cover of *Time* magazine," noting:

> George Lucas denies the rumor that he modeled Indiana Jones after the explorer-zoologist Roy Chapman Andrews, but readers of the enormously entertaining *Dragon Hunter* will certainly be inclined to believe it. On his first journey to East Asia in 1909, when he was 25 years old, Andrews spent two weeks stranded on a deserted island; fended off sharks after his boat was capsized by a finback whale; survived typhoons, heatstroke, poisoned bamboo stakes, headhunters, and 20-foot pythons... He delivered two babies, pulled several teeth, and amputated a man's mangled hand.

He also sampled opium; befriended Mother Jesus, Yokohama's most famous madam; enjoyed the pleasures of Shimonoseki, "the hardest-drinking port in the East"; and along the way collected 50 mammals, 425 birds, and a new species of ant. And that's just in the first 35 pages…

To little surprise, Andrews fell in love with a fellow naturalist and adventurer. On October 7, 1914, he married Yvette Borup, a photographer and the sister of George Borup, who in 1909 had helped Robert Peary claim the North Pole. The couple made their home just outside New York City, in Bronxville, but planned to spend most of their time in the field—Korea, Borneo, "Darkest China"—beyond the knowledge and reach of most of the rest of the world. By 1918, Andrews was venturing into Mongolia, looking for bighorn sheep, the "supreme trophy of a sportsman's life." There, he delighted at the similarities between Mongolian roebuck and Virginia deer. He learned not to travel at night or camp near villages lest he be attacked by brigands. He declared the traditional Mongolian home, the *ger*, a genius piece of architecture, and the Gobi a natural wonder—towering sand dunes in one place, "almost as smooth as a tennis court" in another, and, often, "the most desolate waste of sand and gravel."

Yet the Gobi crawled with life. Andrews saw a herd of antelope so vast that from a distance it appeared to be a field of yellow grass. Crested lapwings "flashed across the prairie like sudden storms of autumn leaves."

And the city! "The world has other sacred cities, but none like this," Andrews wrote after visiting Urga—

It is a relic of medieval times overlaid with a veneer of twentieth-century civilization; a city of violent contrasts and glaring anachronisms. Motor cars pass camel caravans fresh from the vast, lone spaces of the Gobi Desert; holy lamas, in robes of flaming red or brilliant yellow, walk side by side with black-gowned priests; and swarthy Mongol women, in the fantastic headdress of their race, stare wonderingly at the latest fashions of their Russian sisters.

The Chinese quarter felt like a frontier outpost, reminding him of Wisconsin. "Every house and shop was protected by high stockades of unpeeled timbers, and there was hardly a trace of Oriental architecture save where a temple roof gleamed above the palisades," he wrote. The main city square was an "indescribable mixture of Russia, Mongolia, and China. Palisaded compounds gay with fluttering prayer flags, ornate houses, felt-covered yurts, and Chinese shops mingle in a dizzying chaos of conflicting civilizations."

Andrews knew he had to return to Mongolia, and to the Gobi in particular. Marco Polo had traversed the desert en route to the Chinese court of Kublai Khan in the 1270s, but afterward "came centuries of silence, as though the desert had disappeared," noted Troy Sternberg, a researcher in geography at Oxford University. Andrews wrote, "There is no similar area of the inhabited surface of the earth about which so little is known." Back home in New York, he plotted a scientific expedition more ambitious than any on record, pitching it as a way to prove his boss's belief about the origins of humankind: he intended to search the Gobi for bones.

As far as anyone knew, knowledge of eastern Asia's fossils rested "almost entirely upon the report on a small collection of teeth and fragmentary bones purchased in the medicine shops of Tientsin [China] and described by a German named Schlosser." When Western colleagues teased that the AMNH explorers would find everything "obscured by sand," Andrews argued that while Mongolia had been "crossed and recrossed by some excellent explorers, mostly Russian," none of the country had been "studied by the exact methods of modern science."

———————

The proposition of a Mongolian expedition was dangerous, given the region's political instability. Outsiders were suspect to the Chinese, the Mongolians, *and* the Russians. Mongolia wasn't just politically volatile, it was remote—mounting an expedition would be expensive. The country presented "unusual obstacles to scientific research," such as forbidding terrain, unpredictable weather, and armed bandits. Summer temperatures easily reached 110 degrees in the shade (if shade could be found), and winter's sharp winds dropped the temperature to 50 degrees below zero, with the power to freeze livestock where they stood. One could not travel to Mongolia from Siberia or China by "roaring train." The only roads were dirt trails that nomads and merchants had followed by habit and instinct for thousands of years. People crossed the Gobi by horse or Bactrian camel, a double-humped "relic of the Pleistocene" that fascinated Andrews as a moody spitter whose "great flat feet" were "natural road-makers."

A friend, Charles Coltman, had recently traveled from China to Urga by car, giving Andrews a new idea. No scientist had ever attempted a major expedition by automobile. A car caravan could carry the research team, while camels would haul the fuel, food, and gear to prearranged points. Over lunch with Osborn in New York, Andrews proposed the idea for what he called the Central Asiatic Expeditions, explaining that the project would run for the next five to ten years:

> We should try to reconstruct the whole past history of the Central Asian plateau—its geological structure, fossil life, its past climate, and vegetation, and general physical conditions, particularly in relation to the evolution of man. We should make collections of its living mammals, birds, fish, and reptiles. We should map the unexplored parts and little known regions of the Gobi Desert.

Andrews envisioned an interdisciplinary team and support staff—paleontologists, geologists, topographers, paleobotanists, Mongolian guides, Chinese taxidermists. "As we sat in the mess tent at night discussing the day's work, it was most interesting to see how puzzling situations in geology would be clarified by the paleontologist; how the topographer brought out important features which gave the key to physiographic difficulties; and how the paleontologist would be assisted by the paleobotanist or geologist in solving stratigraphic problems," he later wrote. James "J. B." Shackelford, a cinematographer, would document the expedition with still photos and film, and Andrews, who had already published a popular book called *Whale Hunting with Gun and Camera*, would write about the experience for a general audience. By basing the operation in Peking, the party could enter the Gobi from the south by way of the Chinese city of Kalgan, the gateway to the Great Wall and the Mongolian plateau.

Altogether the project would cost at least $250,000. The museum could afford to spend only five thousand a year, but Andrews believed he could raise the rest of the money. Already he was known as someone who felt as comfortable in a room full of the rich and powerful as he did in the backcountry, and who could coax big checks from millionaires by turning the "dry achievements of science into something with popular appeal," as Helena Huntington Smith put it in *The New Yorker*. Andrews, she noted, had an unusual advantage: "Scientific capacities, it seems, are not too commonly combined with social ones."

Andrews quickly lined up backers such as the railroad magnate and statesman W. Averell Harriman and the soap-and-toothpaste manufacturer Sidney Colgate. The Dodge brothers provided customized open-body cars with heavier springs, larger fuel tanks, stronger tires, and pull-hooks bolted onto the front and back chassis. Dodge's advertisements for "the covered wagon of the Gobi Desert" featured Andrews as a celebrity explorer, which suited him fine. "Most people derive a thrill from public applause, but they often feel called upon to hide their delight behind a vast pomposity. Roy Chapman Andrews has better sense than that; he enjoys it without pretense," *The New Yorker* noted, adding, "Probably few men are having a better time."

The museum announced the Central Asiatic Expeditions as a scientific endeavor focusing on the origins of ancient man. *Asia* magazine and the American Asiatic Association pledged their support, with a unified goal of making a "contribution of large value to the development of the scientific knowledge of the world, an instrument of further establishing friendly relations between China and the United States and an important factor for increasing the interest of the people of the United States in the people and affairs of the Orient."

China controlled border access to Mongolia, but Andrews thought of a way to win over the government. In exchange for being allowed in the Gobi, the Americans would teach the Chinese how to explore, and give their scientists cast copies of whatever fossils they found, to start a natural history museum. The project promised to forge "friendly relations" and make discoveries "destined to increase the prestige of the United States in the world of science."

The expedition made its headquarters 400 miles south of the Mongolian border in a compound within the high walls of the Forbidden City palace complex. "Soon it became a small city in itself, devoted to the multiple interests of the Expedition," Andrews wrote. "There were the living quarters of my own family, garages for eight cars, stables, a house for the storage of equipment, an office, laboratories, and a complete motion-picture studio." The arrangement required complex negotiations, including bribes—what Andrews called "squeeze." He wrote, "There is almost unending bargaining: Middlemen with their 'squeeze,' the police with their squeeze, all the squeezes of the contractor, the squeezes of those in control of the water, the electric light, and the telephone, and dozens of others, until one feels as though one had been squeezed to death." Andrews redid the 161 rooms of his rental into 40 larger ones, and kept at least 20 servants, including a head butler. "It is a delightful Aladdin's Lamp sort of existence," he wrote. "You say what you want and things happen. It is best not to inquire *how* they are to be done."

In March 1922, seventy-five camels filed through the Great Wall, each carrying nearly 400 pounds of supplies. Seven cars followed in late April, the men dressed in khakis and riding boots and wide-brimmed felt hats. Summiting the pass that led to the Mongolian plateau felt like reaching "the roof of the world."

The team's maps were barely maps, inked with little more than dotted lines and other markings that indicated caravan routes, oases, and mountains where often there turned out to be none. Descending the rocky plateau, they entered the grasslands. The grass thinned out and coarsened. Camel sage studded the plains. Badlands appeared. "Red hills and buttes showed prominently against the skyline," Andrews wrote. "It was ideal country in which to search for fossils..."

On day four he was relaxing at his tent when two cars arrived in camp. Out jumped Walter Granger, the expedition's chief paleontologist. "I knew that something unusual had happened because no one said a word," Andrews wrote. "Granger's eyes were shining and he was puffing violently on his pipe. Silently he dug into his pockets and produced a handful of bone fragments." Granger told Andrews, "Well, Roy, we've done it. The stuff is here."

CHAPTER 11

THE FLAMING CLIFFS

THE "STUFF" WASN'T HUMAN. BUT IT WAS PREHISTORIC REMAINS. One afternoon at the end of the first field season, the expedition stopped near a couple of *ger*s to ask directions. Shackleford, the photographer, walked a few hundred yards out to look at a strange formation. He soon stood at the edge of what Andrews later described as "one of the most picturesque spots that I have ever seen," overlooking a "vast pink basin, studded with giant buttes like strange beasts, carved from sandstone." Mongolians called the area Bayanzag, but the Americans decided to name it the Flaming Cliffs, because "when seen in early morning or late afternoon sunlight it seemed to be a mass of glowing fire," Andrews wrote. The sight wasn't unlike that of the western badlands recorded in 1800s America, for "there appeared to be medieval castles with spires and turrets, brick-red in the evening light, colossal gateways, walls, and ramparts."

Shackleford descended the slope, planning a brief search. Then, "almost as though led by an invisible hand he walked straight to a small pinnacle of rock on the top of which rested a white fossil bone," Andrews wrote. The sandstone had weathered away, leaving balanced there a skull—"obviously reptilian" but unlike anything any of them had ever seen, with horns, a parrot-like nose and mouth, and a frill like an Elizabethan fan collar. Granger would name the dinosaur *Protoceratops andrewsi*.

The party made camp and spent the rest of the day searching the ravines and gorges, finding bone after bone glowing in the red sandstone. Granger came across what appeared to be part of a fossil eggshell, which everyone assumed to belong to a bird, though the stratigraphy told them the locality dated to the Cretaceous. The site begged intensive study, but all they had time to do before hurrying back to Peking, to beat the winter, was mark the eggshell's location and plan to investigate the following year.

In the off season Andrews wrote about the expedition for a wide audience, but just as often he was the story subject, pictured kitted out with his revolver or Mannlicher rifle and cartridge belt. In public lectures, he captivated listeners with tales of bandits, killer dogs, flash floods, and Gobi sandstorms ferocious enough to strip a camp to its tent spikes and rip the clothes off a man's back. There would be an amusing incident involving accidentally shooting himself in the leg, and a terrifying one about the night venomous snakes slithered into camp as everyone slept: lighting their lamps, the men found vipers coiled around cot posts and nesting in boots for warmth. One expedition member later suggested that Andrews's love of adventure sometimes led him to embellish: "Water that was up to our ankles was always up to Roy's neck." His field journals gave "the clear impression that what primarily motivated him was not the advancement of science or the discovery of fossils but a lifelong yearning for adventure and remote, dangerous country," *Smithsonian* later wrote. "In his heart he was an explorer first and a scientist second." *The New Yorker* noted, "Mr. Andrews is at bottom that ancient type, the hero of the chase, the Nimrod. All his life he has been a hunter and collector, and what he is doing now is simply a glorified version of what he did as a boy, when he roamed the woods around Beloit, Wisconsin."

If Andrews was the storyteller, Granger was the story. The son of a Civil War veteran and lifelong roamer of the Green Mountains of Vermont, he was twelve years older than Andrews and had joined the museum at seventeen as an assistant taxidermist. He had never gone to university and would not hold a degree until 1932, when, at age sixty, he was awarded an honorary one by Middlebury College in Vermont. Andrews got the glory, but Granger got the bones. "I hope you know that it was Walter Granger and not Roy Andrews who was primarily responsible for almost all the American Museum's paleontological discoveries in Asia," George Gaylord Simpson wrote to the Polish paleontologist Zofia Kielan-Jaworowska in 1970, soon after she published *Hunting for Dinosaurs*, about her own groundbreaking Gobi expeditions. "I knew them both very well, and have always regretted that Granger's modesty and Andrews's egotism led to a misunderstanding of their accomplishments." As Vincent Morgan and Spencer Lucas put it in a biography project on Granger written for the New Mexico Museum of Natural History and Science, "Granger represented science, Andrews represented saga, and it was science that was important." Granger was the one "universally held in esteem as the man who reconstructed the evolutionary sequence of the North American horse, the man who helped initiate the search for Peking man, the man who discovered the fossil rich Bone Cabin quarry in Wyoming," Morgan and Lucas wrote. While Andrews handled the Gobi logistics, Granger oversaw the paleontology, often coaxing bones from near dust. "At Granger's core lay the master craftsman," Morgan and Lucas wrote. In his field notebooks Granger cited each fossil's physical position *in situ*, its stratigraphic location, its relationship to other fossils; and the circumstances of the discovery, making the Central Asiatic Expeditions one of the best-documented projects in the annals of science.

One day, Granger cracked open a concretion and at first thought he was looking at the skull of a tiny dinosaur. Later, in the lab, the skull was found to be that of a mammal. Since the late 1800s, a major evolutionary question had involved the origin of placental mammals (mice, humans, manatees) as opposed to marsupials (koalas, opossums, kangaroos). Placentals were far more successful in number and distribution—their species lived around the world, while marsupials lived mostly in Australia and South and Central America. These Gobi remains looked placental. And they were itty-bitty. All of which was odd, because the fossils came from Cretaceous layers—which suggested that mammals evolved much earlier than scientists had thought. The museum told Granger, "Do your utmost to get some other skulls."

He got hundreds. Andrews described a single day in which Granger found nearly two hundred jaws and skulls of carnivores, rodents, and insectivores.

And there were dinosaurs—enough to rival Como Bluff and then some. In addition to *Protoceratops*, the crew discovered *Velociraptor*, once described as a "lapdog-sized predator covered in feathers," and *Psittacosaurus mongoliensis*, an early *Triceratops* ancestor that measured about 6 feet long and probably ate plants and small animals. In the first field season alone, the expedition collected nearly two thousand specimens, many of them new to science.

The expedition returned the next summer in 1923, heading straight for the Flaming Cliffs. On the afternoon of July 13, an assistant, George Olsen, came in from hunting and said he had found fossil eggs. Walter Granger had found a shell shard a year earlier, but nothing had come of it, and now everyone scoffed that Olsen's eggs were probably concretions. But they followed him to where he'd been searching, and sure enough, three cylindrical objects lay next to a sandstone ledge. Each measured about 8 inches long and several inches across, resembling enormous cracked potatoes or fat, stale baguettes. Others just like them were poking out of the earth.

Until that point, no one knew how dinosaurs procreated. Did they give birth to live babies like humans? Lay eggs like turtles? No clues existed, other than some intriguing shell fragments a Catholic priest had found years earlier in the French Pyrenees. "With mounting excitement they began to brush the sand away from the ledge, exposing more of the fossils," Preston wrote in *Dinosaurs in the Attic*. "Conclusive proof shortly emerged." Remarkably, they found "the fragmentary skeleton of a tiny, toothless, unknown dinosaur" on top of the eggs.

But what dinosaur? Over a hundred skulls and skeletons of *Protoceratops*, a plant eater, had been (or would be) found at the Flaming Cliffs, leading the paleontologists to deduce that the eggs were *Protoceratops*, too. Granger excavated the entire block and shipped it to New York, where preparators revealed thirteen eggs laid in two layers of concentric circles, the narrowest end of each egg pointing toward the center—a nest. Yet the dinosaur found on top of the eggs was a meat eater, leading Henry Osborn to conclude that it died while in the act of attacking the nest. He named the dinosaur *Oviraptor*, "egg thief"—a case of mistaken identity that wouldn't be corrected for the better part of a century.

Now that Andrews had seen what the Gobi offered, he knew he needed ten years in the field. This would require more fund-raising. It was decided that the expedition would recess in 1924 and return to the States, where he would raise the quarter of a million dollars necessary to continue the work. He would deliver the first dinosaur eggs known to science personally to the museum.

"What are the darned things worth?" asked the baffled Shanghai agent of Lloyd's of London when Andrews went to insure them.

"Scientifically, they're priceless," Andrews told him. "Commercially, they would be worth only what someone would pay for them."

Andrews decided to insure the eggs for $60,000. The next question was how to secure them on the Dollar Line vessel that would carry him home. He ultimately packed the eggs in a "good, tight" suitcase, waterproofed it, and wrapped it in a homemade life jacket.

By the time the ship docked at Victoria, British Columbia, the world knew about the eggs. Reporters clamored for exclusives. "I will give you fifteen hundred dollars for the exclusive use of the dinosaur egg photograph for a week," a *Seattle Post-Intelligencer* representative told Andrews. Another offered three grand. A San Francisco paper upped it to five. "I was aghast," Andrews wrote. "From the foreign correspondents in Peking we knew that the dinosaur eggs had 'caught hold' all over the world but expected nothing like that."

In New York, Andrews received a hero's welcome. Adolph Ochs, publisher of the *New York Times*, invited him to lunch. John D. Rockefeller Jr. attended his welcome-home reception and pledged one million dollars for the American Museum's endowment. A sketch of Andrews's hangdog face ran on the cover of *Time*. Newspaper executives bid for his writing, with William Randolph Hearst alone offering a quarter of a million dollars for a Roy Chapman Andrews exclusive. Standard Oil pledged twenty thousand gallons of gas, five hundred gallons of oil, and candles for the next round of field research.

By Thanksgiving weekend, the Mongolian dinosaur eggs were on display at the American Museum of Natural History, drawing hordes of visitors. Thousands of people jammed into Andrews's public lectures, where he "spread the gospel" about the museum's fieldwork while barely mentioning science. Hiring the same agent as Will Rogers, he gave more than a hundred talks in one four-month period, ending the tour looking and feeling like "a sucked orange." As he and Yvette walked down Fifth

Avenue one night, Yvette pointed out the neon R.C.A. sign, whose letters stood for Radio Corporation of America, and said, "That's you."

Publishing book after book, Andrews galvanized the public imagination with stories that attempted to put readers, including children, directly into his experiences. Where the eager public saw heroics, critics came to see exploitation. Expeditions rarely bothered to credit local participants by name. As Andrews tried to raise money he pointed out that New Yorkers had funded the first round of research and that he hoped the rest of the country would kick in for the next phase, an "all-American expedition made up entirely of Americans and carrying the American flag, American ideals, and American inventions into a part of the world of the utmost importance to scientific progress." In truth, the expedition had included nine Mongolian assistants, a representative of the Mongolian government, and nine Chinese assistants—hardly an "all-American" venture. "Imperial objectives were an integral part of the expeditions," the science historian Ronald Rainger wrote in the 2004 book *An Agenda for Antiquity.* "For them, Asia was fertile ground for economic development and exploitation. Projects such as the Central Asiatic expeditions not only followed up the openings made by political and economic expansion but embodied the same attitudes and objectives." Critics saw traces of theater in the iconic photos of Andrews in the Gobi in his dust-coated boots, wearing a "ranger hat, complete with a feather." These were the ways Americans, "especially white Anglo-Saxon Protestant Americans," made their presence known "throughout the world," Rainger noted, observing that it was no accident that Andrews titled one of his most important books *The New Conquest of Central Asia.* The title "embodied the sense of priority, superiority, and the right to take control of knowledge that characterized these expeditions."

At the time, though, Andrews was as famous as the celebrities who courted him as a friend.

———

The Gobi fieldwork of 1922 and 1923 had produced what Andrews's boss, Henry Osborn, saw as a "paleontological Garden of Eden." Andrews still hoped to find evidence of early mankind in Mongolia, but for now relished the attention he received from having opened the world's greatest known fossil field. The museum had nine tons of Gobi fossils to work on, including seventy *Protoceratops* skulls, plus thousands of specimens of modern creatures.

Yet all anyone could talk about were the eggs. The actor John Barrymore, a prolific collector of oddities, begged Andrews for one. Osborn said the eggs did "more than anything else in the whole history of paleontology to make the 'man on the street' dinosaur-conscious."

The eggs were so popular, Andrews decided to sell one.

The "grand publicity stunt" of an auction would help the museum raise fast money and allow the public to feel intimately involved in exploration. "They believe this is only a rich man's show," Andrews told Osborn. If people knew that even small contributions helped, they would give. "Every news story could explain that we've got to have money or quit work."

"Roy, it's a great idea—a ten strike," Osborn reportedly told him. "Let's do it."

Andrews had forty reporters in his office by that afternoon. The American Museum of Natural History as a policy did not sell its discoveries, but Andrews explained that marketing a Mongolian dinosaur egg would ensure future expeditions. "We have got a perfectly good 'corner' on dinosaur eggs and I cannot possibly conceive of the 'corner' ever being broken," he told the press. "While a good many dinosaur bones and skeletons have been found in the Western part of the United States and in other parts of the world, the museum's explorers are the only ones to uncover the petrified eggs of the ancestral reptile. We have felt there is no good reason why we should not sell one of these eggs." After all, the museum had twenty-five of them. "The majority are in excellent condition," he said. "There is no desire on our part to make any money for the museum, but only to help defray the expenses of the Asiatic expedition, which will start out again next summer." A *New York Times* headline on January 8, 1924, read, "Dinosaur Egg 100,000,000 Years Old for Sale; Museum Asks Bids to Aid Explorers' Fund."

No real price could be assigned to the eggs because "none had ever before been bought," the *Times* noted. Yet Andrews set the auction reserve at $5,000 (about $71,000 today)—a price that he said would surely be "greatly exceeded by the highest bidder." To bid, prospective buyers would send written offers. As the offers arrived, Andrews would notify other bidders until the bidding stopped.

A day after the news went out, the "Illustrated London news cabled an offer of two thousand dollars," Andrews later wrote. "The National Geographic Society upped it to three thousand. A museum in Australia bid thirty-five hundred. Yale University offered four thousand. The publicity was enormous and, true to their promise, the newspaper men included a plea for money in every story. Checks began to come in by every mail." Lord Rothschild, of the British Museum, made an inquiry, along with magazine owners from Washington to Australia. The winner, at five thousand dollars, was Austen B. Colgate, of Colgate & Co., whose grandfather founded the company in 1806. Colgate donated the egg to the university that bore his family name, in the upstate New York town of Hamilton. The school displayed the egg for a while, but moved it to the university vault after someone tried to steal it.

A novelty manufacturer suggested that Andrews make autographed casts of the famous Gobi egg to be sold as paper weights and desk sets. "The original Easter egg!" he told Andrews, suggesting a first run of a million copies. The egg replicas could sell for twenty-five cents each and Andrews would of course take a royalty. This idea went too far, in Andrews's opinion. He had already declined an offer from an oil and mining company to attach a petroleum prospector to his expeditions. If he started mass producing Gobi eggs, everyone would assume the money "went into my pocket," he explained. "Moreover, the expedition would be stamped as a money making venture in the eyes of the world. Science camouflaging business."

Hearing of the auction, the Mongolian government complained, believing a false rumor that Andrews had sold the egg to the British Museum for a million dollars. That year, in its new constitution of 1924, the Mongolian government declared "all lands and resources within their subsoil, forests, water, and the natural resources within them" national property. Government leaders reminded Andrews that he had signed a contract promising to eventually return the fossils that he had borrowed for study. "My eggs touched off an explosion against us," he wrote, saying the sale now threatened the museum's Gobi permits. "Nothing else so disastrous ever happened to the expedition. Up to this time the Chinese and Mongols had taken us at face value. Now they thought we were making money out of our explorations," he wrote, adding, "Why should the Mongols and the Chinese let us have such priceless treasures for nothing?"

Yet he also understood the assumption: "They couldn't know that the five thousand dollars was a fictitious value engendered by publicity, or that any purely scientific or art object has a market value of just what it can be sold for to someone who has a special reason for desiring to possess it."

Mongolia now had a constitution and self-identified as independent, but, remembering two hundred years of Chinese rule, the country ducked beneath Moscow's wing, becoming the second Communist nation in the world. The State Department had finally installed a U.S. consulate, locating it in the Chinese city of Kalgan, near the Great Wall. Western companies like Standard Oil and the British American Tobacco Company were there, and Mongolia appeared to hope the United States would become the first Western power to recognize the nation. But after Prime Minister Dogsomyn Bodoo proposed "friendly relations, especially for trade," and followed up on the idea with a piece in *The Nation* titled "Mongolia Speaks to the World," a political rival accused him of treason and he was executed. So was the minister of finance, a Harley motorcycle enthusiast with a fondness for capitalism. The "lesson was obvious enough," Ambassador Addleton later wrote. "Working with Americans or, for that matter, any other foreigners, might easily prove fatal." The Americans pulled the consulate, wary of the regional instability and unconvinced of Mongolia's value to the West.

Andrews had sensed that human life was "worth less than that of a sheep" in Mongolia. The AMNH crew had come across the ravaged remains of temples and villages, finding weathered corpses still wearing lama robes and military uniforms. One day, a Mongolian general who was close with Russia described to Andrews an attack on a Chinese encampment: "We rode at full speed through the camp, killing everyone we saw. Then we rode back again. The Chinese ran like sheep and we butchered them by the hundreds." Andrews wrote, "Except for the modern weapons, the tale might have been a thousand years old."

By the third expedition, in 1925, the "dashing horsemen" and "strange costumes" of Urga had been replaced by "Russians and swaggering Buriats," Mongolian tribesmen who lived in Siberia. The city square "filled with squads of awkward Mongols being drilled as soldiers." The Soviet secret police tailed the Americans, suspicious of plots to "annex" Mongolia for the United States. Or maybe Andrews was scouting for oil. "In short, one was treated as a spy and a generally undesirable character," he wrote. Such an enormous enterprise as the Central Asiatic Expeditions surely had not come merely for science. After all, Andrews had the perfect cover for a spy: as a museum explorer he roamed the countryside for long periods with "high-powered binoculars, a variety of weapons, camping equipment, supplies, and a small party of assistants," plus a wife.

Yet the Mongolians' suspicious were somewhat correct. In the summer of 1918, Andrews had sworn an oath to serve the U.S. Office of Naval Intelligence as a civilian informant, code-named "Reynolds." The job paid eight dollars per day and required no training: "Reynolds" was simply to send information or observations by letter to the naval attaché in Peking when he thought it important, with the "salient lines penned in invisible ink." The espionage deal was apparently fleeting. Andrews lost the intelligence contract after Yvette wrote a letter home that the U.S. government deemed "highly indiscreet," in which she demanded that Andrews's family be "informed as to the nature of his mission."

Andrews went on reporting his observations publicly in magazine and newspaper dispatches that provided Westerners rare accounts of what was happening in eastern Asia. By 1926 he was writing about the fighting around Peking, where "the murder of white residents and mysterious disappearances of others happened frequently." Ships sailing from China were "packed to the rails with missionaries, merchants, and others who had resided for years in a country which had suddenly gone mad as though smitten with an attack of the rabies." By 1928 the Chinese were intercepting the expedition's finds. By 1930 Andrews could get no closer to Outer Mongolia than Inner Mongolia. The Mongolian minister of war had been executed. The former minister of the interior had been dragged out of his home and shot. Chinese soldiers had executed Andrews's old friend Charles Coltman, who had inspired him to conduct his expedition by car. Andrews wrote, "Murder and sudden death stalked ahead upon the streets. It was an exceedingly good place to leave."

The American Museum of Natural History pulled out of Eastern Asia in August 1932, just before the purges began in Mongolia, wherein the leader Khorloin Choibalsan ultimately ordered an estimated thirty thousand intellectuals and Buddhist leaders killed, some thought on the orders of Lenin's successor, Joseph Stalin. Andrews had already auctioned off almost everything in his Peking compound and departed, saying he was through with China.

"Is it surprising that I was filled with regret as I looked for the last time at the Flaming Cliffs, gorgeous in the morning sunshine of that brilliant August day?" he later wrote, remembering his last moments at Bayanzag. "I suppose I shall never see them again!" And he never did. In fact, the American Museum of Natural History would not be allowed in Mongolia for the next sixty years.

CHAPTER 12

MARKET CONDITIONS

IN THE SPRING OF 1985, MIKHAIL GORBACHEV, LEADER OF THE Soviet Union, began talking about glasnost and perestroika, signaling the end of the Cold War. On January 27, 1987, bilateral relations between the United States and Mongolia were finally announced, in a signing ceremony in the Treaty Room of the State Department, beneath a portrait of Thomas Jefferson. Joseph Lake, a foreign service officer from Texas, was named the first permanent ambassador to Mongolia and soon opened an embassy in Ulaanbaatar. Twenty years hence, on the anniversary of the agreement, Mongolia's postal service would issue two commemorative stamps, one featuring Abraham Lincoln, the other Genghis Khan.

Another pivotal moment came on December 7, 1988, as President Ronald Reagan prepared to exit the White House and President-elect George H. W. Bush prepared to enter it. In a landmark address to the UN General Assembly in New York, one that would contribute to his winning the Nobel Peace Prize, Gorbachev declared, "It is obvious...that the use or threat of force no longer can or must be an instrument of foreign policy. This applies above all to nuclear arms, but that is not the only thing that matters." He went on: "All of us, and primarily the stronger of us, must exercise self-restraint and totally rule out any outward-oriented use of force. That is the first and the most important component of a non-violent world..." Gorbachev spoke of the importance of "rule of law" and the "highest standards" for individual rights. He pledged to work with the United States most urgently to reduce arms and eliminate chemical weapons, saying, "I would like to believe that our hopes will be matched by our joint effort to put an end to an era of wars, confrontation, and regional conflicts, to aggressions against nature, to the terror of hunger and poverty as well as to political terrorism."

Mongolia had been a Soviet satellite for the better part of a century, and depended upon socialism for basics like shelter, food, and education. Some eleven thousand graduates of Mongolian high schools and vocational schools had attended college in the Soviet Union and Eastern Europe. Thousands had learned English there. Exposed to opposition movements and democratic ideals, they had learned that it was wrong for citizens to be jailed for, say, disagreeing with their government. They were returning home to Mongolia with new ideas about how to define freedom, the Soviet Union having "unwittingly sown the seeds of its own destruction on fertile ground," Christopher Kaplonski, a University of Cambridge anthropologist, wrote in *Truth, History and Politics in Mongolia*.

A secret pro-democracy group called Shine Ue ("New Epoch") organized. Members made leaflets on mechanical typewriters and left them on people's doors in the night. Ulaanbaatar was now a city with nearly four hundred broadcasting relay stations, and over 90 percent of the population could read and write—one of the world's poorest countries had one of the world's highest literacy rates. In other words, Mongolia had both an infrastructure and a ready audience for the proliferation of information.

In June 1989, as pro-democracy demonstrations swelled in Beijing, the Chinese government declared martial law. The military opened fire in Tiananmen Square, killing hundreds, if not thousands, of people. Over seven hundred miles north, Mongolia's democratic revolutionaries took note but went on with their plans, forming the Mongolian Democratic Union, the nation's first independent political organization.

After seventy years of Soviet control and centuries of Chinese rule before that, it was hard to know how to become a democracy, but holding multiparty elections seemed like a good place to start. Only one party had ever existed in Mongolia though it had gone by two names: the Mongolian People's Party (MPP) in the early 1920s and the Mongolian People's Revolutionary Party (MPRP) since 1924. The Democratic Party would be the first opposition party in history.

Pro-democracy demonstrations began in Ulaanbaatar on December 10, 1989, on the steps of the Youth Cultural Center. A young political science professor named Sanjaasuren Zorig led several hundred people in demands for wholesale reforms—open elections, a free press, respect for human rights, and a market economy. Each new demonstration attracted more participants than the last, despite winter temperatures of 22 degrees below zero. Demonstrators were partly stoked by a recent *Playboy* interview in which the expatriate Soviet chess champion Garry Kasparov joked that the Soviet Union could cash in on Mongolia by selling it to China.

By mid-January, over five thousand people were demonstrating, some of them standing outside the Lenin Museum for hours in winter battle dress of felt boots, fur hats, and *deels* lined in sheepskin. Surprisingly, the politburo allowed leaders to broadcast their demands on radio and TV, and to take as their symbol Genghis Khan—the "perfect warrior"—the mention of whose very name the Soviets had long banned. By March, the rallies had grown larger and moved to Sukhbaatar Square. When Democratic Union members launched a hunger strike, supporters settled in with them, declaring their intentions to "sit day and night on the stone cold square" until the politburo resigned. The rallies, though illegal, never drew a police presence, never got out of control or required the presence of tanks. Sitting atop a friend's shoulders, Zorig spoke into a megaphone, keeping the crowd calm.

To everyone's shock, the politburo stepped down. They were gone by the middle of March. As the first multiparty elections were scheduled for summer, news of the changes filtered to the West. "An isolated and little-known country between Soviet Siberia and China's Inner Mongolia region, the once hard-core Communist Mongolian People's Republic, has quietly but decisively spawned an experiment in Soviet-style reform...," the *Los Angeles Times* reported.

Democratic transition usually takes time, and often spills blood, but in Mongolia it happened quickly, peacefully. That summer, a staggering 98 percent of the nation's nearly one million citizens turned out to vote in the election for State Great Hural (parliament), some riding all day on horseback to reach the polls, dressed in their finest *deels*. The MPRP, the old Communist party, won most of the seats and stayed in power, but opposition parties were being formed by the dozens. The news media proliferated. Mongolians started talking about resurrecting the beautiful old vertical script, which had been replaced with a horizontal Cyrillic alphabet under the Soviets. Genghis was back, his whiskery image appearing on vodka labels and cigarette packages, and in a mass-marketed portrait that hung in homes the way some Americans display Jesus and JFK. As the country clawed toward a market economy, the visage of Genghis became a "21st Century marketing phenomenon," deployed in the names of hotels, banks, and, in one case, an Irish pub. Lenin statues came down; Genghis statues went up.

The borders were open.

In the early days of the democratic momentum, Michael Novacek, a vertebrate paleontologist and the American Museum of Natural History's dean of science, received a visitor in his office in New York. Novacek was a California native in his late thirties, with a UCLA zoology degree, a Berkeley PhD, and, during field season, the beard of a lumberjack. He got into paleontology through a childhood love of dinosaurs, via the books of Roy Chapman Andrews. His adult interests had broadened to the "greatness and complexity of history" that fossils represented, and he now specialized in fossil mammals, "blueprints of our own heritage."

His visitor was Sodnam (many Mongolians go by one name), president of the Mongolian Academy of Sciences, the country's highest academic organization. Sodnam wasted no time in saying he hoped Mongolian and AMNH scientists could collaborate for the first time since the days of the Central Asiatic Expeditions. Within weeks of Mongolia's fast break toward democracy, a delegation reappeared at the museum, to reiterate Sodnam's words and invite AMNH scientists to return to the Gobi. By June 1990, a scouting expedition, led by Novacek, the newly hired Mark Norell, and their colleague in the vertebrate paleontology department, Malcolm McKenna, was en route to the desert. They had permission to spend a field season assessing whether the desert still held enough paleontological promise to warrant a full expedition and the recruitment of sponsors.

The team headed straight to the Flaming Cliffs, some 400 miles southwest of Ulaanbaatar. The last American scientists known to have set foot there were the crews led by Andrews and Granger, whose names had long faded from public consciousness and whose Gobi finds remained in New York. The scientists, accompanied by members of the Mongolian

Academy of Sciences, expected the Flaming Cliffs to be tapped out, having been worked by the Americans in the 1920s, the Russians in the 1940s, Polish-Mongolian teams in the 1960s and '70s, and at least one Sino-Swedish expedition. Their goal was to advance the science and make new discoveries, not simply rework old territory. If the new AMNH team could prove the Gobi was still worth exploring, they could secure funding for a series of joint expeditions, picking up where Andrews and his colleagues had left off.

The Americans' key Mongolian colleagues in the joint venture were Demberelyin "Dash" Dashzeveg, Altangerel Perle, and Rinchen Barsbold, paleontologists whose work was published in Russian and wasn't widely available in the West. Dash—"lean and hungry" like a "Siberian wolf," as Novacek described him—knew the Gobi "perhaps better than any person alive." Barsbold, an expert on Gobi theropods, the group of dinosaurs that walked on two legs and ate meat, directed the Geological Institute, which oversaw the national paleontology center and laboratory under the Mongolian Academy of Sciences. Perle worked on dinosaurs, too; he had a laugh and a flubbed English phrase for everything, and could be temperamental to the point of terrifying blind rage. At night as the crew sat around the campfire, he "sang passionate Mongolian love songs and related the highlights of his country's long history," Novacek wrote in *Dinosaurs of the Flaming Cliffs*. On this inaugural trip, Perle proved a facile guide, bargaining for "sheep, fifty pounds of potatoes, rice, and a small chunk of fresh ginger from China" and altogether smoothing the way with provincial officials for the scientists to do their work.

Norell was fascinated with the prospect of birdlike dinosaurs, Novacek with Cretaceous mammals, some of them "small enough to curl up on a teaspoon." The Gobi appeared to be extraordinarily valuable to science because it revealed "biological empires in transition." The Americans soon found the desert promising enough to sign a three-year renewable exploration contract. Everyone agreed that any found fossils would be the property of Mongolia but could be taken to New York for study. A Mongolian scientist would be assigned to each expedition, and Mongolians would be hired as drivers, preppers, and field crew. Ecstatic, the Americans returned home and began planning. They bought a customized fleet of four-wheel-drive Mitsubishi Monteros—beige for McKenna, red for Norell, green for Novacek—and outfitted the rigs with heavy-duty winches and racks. They shipped themselves plaster, lumber, shovels, ropes, cable, camping gear, toilet paper, walkie-talkies, cameras, laptops, and GPS devices. They stocked up on freeze-dried and canned food, intending to supplement it with items bought on the ground in Ulaanbaatar, but when they arrived in Mongolia for the 1991 field season, they found frightening food and gas shortages now that the country had lost its Soviet subsidies. "It seemed an absurd predicament for a country with only two and a half million people and over twenty-five million domesticated animals," Novacek wrote. "But the infrastructure of the young democracy had collapsed." The scientists rationed supplies as they headed east from the city, following the "Big Gobi circuit," a huge loop studded with known dinosaur sites.

In the 1940s, the Russians, encouraged by the Americans' success in the Gobi in the 1920s, began their own major expeditions. A paleontologist and bestselling science fiction writer named Ivan Efremov led the crew in 1946. After visiting the Flaming Cliffs, he turned southwest toward the Chinese border. Two hundred hard miles later, his party entered a "broiling isolated depression," Novacek wrote. The Russians had "audaciously penetrated much farther into the white-hot core of the Gobi than Andrews was in a position to attempt, places where the winds blew harder and the sun was more scorching. This was the 'outback,' even to nomads." There, in the Nemegt Valley, they found what Novacek later called the "grand canyon" of fossil beds and a "wonderland of fossil vertebrates on a scale far beyond anything Andrews encountered."

After finding seven massive hadrosaurs in one place, the Russians named one locality of the Nemegt Formation "Dragons' Tomb." The skeletons turned out to be the remains of a plant-eating duckbilled dinosaur that grew up to 40 feet long and 25 feet tall and which they named *Saurolophus angustirostris*. They also found the remains of what are now known to have been therizinosaurs and ankylosaurs.

Surely an apex predator fed on these dinosaurs, the Russians realized. They soon discovered a likely suspect in the form of a skull and the cervical vertebrae of a large carnivore, then hauled the fossils north in their heavy-duty military trucks, to what was then the USSR Academy of Sciences in Moscow. The carnivore's remains were cataloged as "type specimen 555-1." By the end of the decade, the scientists had collected at least a dozen more specimens like it, leaving some with the Mongolian Academy of Sciences.

A Russian paleontologist named Evgeny Maleev eventually examined 555-1 in order to determine a species, and in 1955, he published his description, "Giant Carnivorous Dinosaurs of Mongolia." The fossils represented an animal from the Tyrannosauridae family, one that appeared remarkably similar to North America's *Tyrannosaurus rex* and that existed contemporaneously with *T. rex*, if in an altogether different place. The Mongolian dinosaur stood a bit shorter than *rex*, with the same disproportionately short forearms and the same lethal hind feet equipped with sharp, curved claws, but was longer by over 6 feet. Both skulls measured about 4 feet long, though the Gobi dinosaur's head was narrower, with a slenderer snout and more teeth. The teeth continually replenished themselves, kind of like a shark's, and, unusually, the animal could lock its jaws, a likely biological adaptation that helped the animal tear through swaths of flesh. "Everything about it suggests power and agility, an animal capable of lunging its massive body at a hapless hadrosaur and disemboweling it in an instant," Novacek wrote. The Gobi dinosaur appeared more primitive than *rex*, which raised an interesting question: had the animal crossed the land bridge that once existed between Asia and North America and evolved into *Tyrannosaurus rex*?

Maleev named the creature *Tyrannosaurus bataar*, "tyrant king hero." The word is actually spelled *baatar*, but Maleev's misspelling lived on, even if the name did not. Scientists eventually realized that sufficient differences existed between *Tyrannosaurus rex* and *Tyrannosaurus bataar* to warrant separate identities. So the first part of the name was changed to *Tarbosaurus*—"alarming reptile."

The Americans had returned to the Gobi questioning what, if anything, remained. They needed fresh finds, and in the third field season, they found them. In the summer of 1993, the joint AMNH-Mongolian expedition came upon what Mike Novacek described as a "forgotten corner of the Nemegt Valley." Mongolians called the reddish hills Ukhaa Tolgod; the Americans thought of it as Xanadu. "In an area the size of a football field we had found a treasure trove that matched the cumulative riches of all the other famous Gobi localities combined," Novacek wrote. Ukhaa Tolgod contained enough fossil dinosaurs, mammals, and plants to occupy scientists for generations.

At one point, Novacek saw Norell running up, breathing hard, saying, "I found something..." They hurried to the flats and saw dinosaur skeletons "scattered across the surface," plus "an extraordinary abundance" of eggs and egg fragments. In one broken egg, they found the "delicate bones of a tiny dinosaur that looked like an intricate Chinese carving"—the first embryonic carnivorous dinosaur on record. Another team member found a dinosaur skeleton with eggs beneath it, similar to the configuration the Andrews expedition had discovered in the 1920s at the Flaming Cliffs. The fossils appeared to vindicate *Oviraptor* as an egg thief, for the Nemegt assemblage showed that instead of raiding a nest, the dinosaur had been tending it. The Mongolians renewed the AMNH contract, and everyone looked forward to many years of returning for still more "clues to one of the most extraordinary cycles of life, death, and burial ever recorded."

The Gobi now crawled with outside scientists. The Italians were there. The Japanese and French were there. Teams routinely hired local crews of forty or fifty Mongolians, trained them in excavation techniques, and showed them how to hunt. The Germans included the commercial dealer Andreas Guhr, who arrived in 1992. The commercial allowance described in *GEO* alone contradicted the latest Mongolian Constitution, which in 1992 had been updated but still protected all "historical, cultural, scientific and intellectual heritages of the Mongolian people." Somehow, deals were being made, and it wasn't hard to guess why.

———

Imagine yourself a citizen of a vibrant, very old civilization. Artifacts have been passed down in your family as important links to your personal and cultural history. Now imagine a shift in your country's political leadership. There is but one governing political force, one supreme leader to admire, one set of rules to obey. The Party determines what possessions you may own and how much money you make. The Party demands that you maintain a "non-bourgeois" path in life; you are no longer permitted to wear your traditional ceremonial clothing or practice Buddhism. You may not study or celebrate your country's history, such as the Mongol Empire and the life of Genghis Khan. To protect your possessions—a painted chest, a wooden bowl, a hand-made saddle, a bone-handled dagger, a blue-glass snuff bottle capped with precious red coral—you decide to send them away lest they risk being destroyed. Maybe they wind up in museums in Russia or Denmark or France. The museums of Mongolia won't take the items because they are little more than propaganda machines. Throughout the country, there are about fifty such museums, the first of which opened in 1924, three years after Mongolia escaped Chinese rule with Russia's help. The State Central Museum opened in 1956, consisting of two parts: natural history and history/ethnicity, the latter of course limited to what Communist leaders want the people to see.

Then comes the democratic revolution of 1990. Museums suddenly may reimagine themselves. The two largest institutions are the National History Museum and the Natural History Museum, which now has its own location, in a stately white building that once housed a school. The collections rooms are small, with no shelving or archival boxes. The materials are stored in a jumble rather than organized by makeup or theme, some described in terms as vague as "made of soft material." Temperature, relative humidity, and dust are not controlled. Restoration and reconstruction are almost unheard of. Important artifacts are crumbling, fading, becoming insect food. The Natural History Museum's collection alone houses more than eight thousand items, many of them deriving from the AMNH expeditions of the 1920s, curated by the Institute of Sutras and Scripts, the original name for the Mongolian Academy of Sciences. Exhibits initially consisted of photos of fossils—the fossils themselves were kept in storage—but now the real thing is on display, though it's unclear what the specimens are.

Each museum soon has a purchasing committee. You may now sell your family heirlooms, if you have any. The minister of culture creates new rules on conducting archaeological and paleontological research and excavation: paleontologists and state-affiliated institutions may receive government permission to collect, and the Natural History Museum may now participate in fieldwork conducted by the paleontological institute for the first time.

The new laws also allow antiques shops to exist. Suddenly there is a market for collectibles. There is a market for everything. If you are like most Mongolians, you look around to see what may be sold, because chances are, you and your family need money to survive. If you live in Ulaanbaatar, you probably live in housing the Soviets built. If you are a herder, you have relied upon Soviet-provided fodder structures, where livestock shelter during severe weather, and on collectives, which coordinate feed. These benefits vanish after the revolution, and at the worst possible time, as an unusually warm series of summers and droughts lure novice herders into the countryside only to be hit by the worst *dzud*, a particularly lethal winter storm, in decades. After tens of thousands of animals die, herders return en masse to Ulaanbaatar, already a city of too few jobs. Many settle in the poor *ger* districts that fringe UB, where people live without running water and burn whatever they can find for heat, including tires, which only worsens the already apocalyptic pollution. Ulaanbaatar's rickety old coal plants, which supply much of the country's electricity, struggle to produce enough fuel to last the winter. The Soviets have withdrawn the technicians who knew how to fix them. At one point, the U.S. embassy advises the handful of Americans living in Ulaanbaatar to "keep their passports, money, and other essentials ready" in case they need to evacuate, to survive winter in what is often called the coldest capital on earth.

The first wave of foreign aid is incoming, including medicine and other healthcare supplies that were amassed for the Gulf War, but the penicillin, anesthetics, and iodine will soon run out. "Imagine in the 1930s in the United States, during the depths of the depression, trying to create North Dakota as a separate country. That's what the Mongolians were facing," one ambassador writes. When the fledgling economy hits an awful new low, experts compare Mongolia's suffering to the Great Depression but "almost twice as severe." Every Mongolian may buy "lesser assets of the state" like "cars, small shops, and livestock" with government-issued vouchers, some of which can be exchanged for shares in the corporations that are being carved out of large state enterprises; those shares can then be traded on Mongolia's new stock exchange. But what most Mongolians want is food.

The American Museum of Natural History teams realized they would need to approach field preparation in Mongolia like a "trip to Antarctica or a distant planet." Only 3 percent of the nation's roads were paved. Almost no one owned cars. The rail service was limited. The planes were not great. There was no communications network to speak of. Two hotels existed in Ulaanbaatar, one where Mongolians stayed and another, the Ulaanbaatar Hotel, where everyone else stayed. The Ulaanbaatar Hotel was so "epic in scale," one visiting American said it looked like "something out of an old copy of *Soviet Life*." One reporter arrived to find that the hotel's guests included "a BBC television crew, several print journalists, foreign diplomats, staff members of United Nations agencies, circus talent scouts, oil prospectors, an investment banker from London, a British agronomist, an American Peace Corps official, a South Korean highway engineer, a Dutch veterinarian, two English trekkers, an American promoter of outdoor-advertising signs, and hunters of both the rifle and the bow-and-arrow variety." Despite the catastrophic economy, Mongolia's budding tourism industry showed promise. Ten thousand tourists had visited in 1990, five hundred of them big-game hunters who spent something like $3 million, many stalking the "Marco Polo sheep" once coveted by Roy Chapman Andrews.

Mongolia already had the means for its own survival, but wasn't using it. The nation sat on a vast, untapped fortune in gold, copper, coal, oil, uranium, and other in-demand minerals and rare-earth elements, with little to no means of extracting, refining, and exporting it. The country's overall natural resources were estimated—conservatively—at $750 billion. Texas companies were already there, conducting seismic surveys and looking to develop two former Soviet oil fields in the desert, Zuunbayan and Tsagaan-Els. The Zuunbayan site had operated from 1950 to 1969 before the Soviets shifted exploration to western Siberia and the Caspian Sea, leaving scads of Gobi oil just sitting there. Mongolia relied 100 percent on Russia.

The capable handling of Mongolia's natural resources promised to stabilize a lot of lives. More than a third of the nation's two and a half million people still lived below the poverty line, earning roughly the equivalent of $12.15 a month. They endured food shortages and other deprivations by leaning on their social networks or by finding their bread, milk, vodka, and information on the black market. Some families were so poor that if a parent died, the surviving spouse was known to turn stepchildren out into the streets or abuse them until they fled. In Ulaanbaatar, more than three thousand children were homeless. Many lived underground in holes and tunnels that held the city's heating system, sleeping on flattened cardboard.

Meanwhile, at the other end of the transitioning economy, people were getting "fantastically wealthy," with living standards approaching "those in the West." They ate in nice restaurants, drank French wine. During spending sprees in South Korea, they bought cars and computers. The new shops of Ulaanbaatar sold microwaves and VCRs. The buying frenzy happened even as the Asian economy collapsed. The state bank would soon fail because it had loaned too much money. Security seemed a long way off. Desperate and opportunistic, people made money however they could manage it, even in the most unexpected and taboo of ways.

It's been said that in ancient Mongolia, if someone was dying, you gave them a little food and a little milk tea, and left them alone to get on with it. You never fussed over the corpse for fear of being infected by spirits. "Under no circumstances will [Mongolians] touch or disturb a skull or a skeleton," Roy Chapman Andrews wrote. "As soon as a person dies the body is dragged off to a considerable distance and left to be devoured by the dogs, wolves, and birds." In early Urga, corpses were often abandoned at a dump; a Swedish missionary living in Mongolia in 1893 wrote that "no encampment is safe from the possibility that one of the family dogs may come in dragging a human leg or arm."

There were bones in the Gobi, meanwhile, that correlated to no known animal. Their size suggested monsters the enormity of which no one could comprehend. According to lore, the toes of herders' leather boots were made to curl upward to avoid jabbing the earth and antagonizing whatever lived down there. (The style also allowed riders better grasp of the stirrup.) But now Westerners were rushing into the Gobi in search of the very objects Mongolians shunned. The strangers handled the big bones of the desert carefully, like treasure, and packaged them up and took them away. Companies started capitalizing on these finds by marketing dinosaur tourism. Nomadic Expeditions, one of Mongolia's earliest tourism companies, offered a "Dinosaurs of the Gobi" package. An Australian company ran a "Gobi Dinosaur Dig" tour. Another offered the "Dinosaur Native Land" tour, which advertised the opportunity for tourists to "dig and find your own dinosaur bones." Another offered the "Roy Chapman Andrews Mongolia Gobi Desert Overland Expedition," wherein tourists could pay $1,433 per person to follow Andrews's camel tracks and ponder the question, "So what was it like for Mongolia's first American Explorer?" Mongolia Quest would offer "Walking with Mongolian Dinosaurs." The itineraries included all the top dinosaur discovery sites, including the Flaming Cliffs. Companies sometimes paid paleontologists to join the trips as experts, the scientists having found that a tourism sideline helped as grant competition intensified or funding got cut. Besides, tourists provided free field labor. "As long as people are interested we can train them to look for fossils in a few days," said the Canadian paleontologist Philip Currie, who once led Nomadic's dinosaur tours. "Some of the nicest finds are made by complete amateurs." Packages were often advertised in partnership with the national paleontology institute. One tour guide bragged online about having organized expeditions with the AMNH's Mike Novacek and Mark Norell.

By the mid-1990s, the Mongolian-AMNH expedition consisted of thirty people in twelve vehicles, including a *National Geographic* film crew. Many of the hired hands on these and other expeditions learned how to excavate and jacket—a quick study could grasp the crude mechanics within a day. The crews went and told relatives and friends about the work, and word spread of this intriguing activity in the Gobi. Paleontologists sometimes glanced up from their excavations to see a distant figure on horseback or motorcycle, watching them through binoculars.

Mongolia's Natural History Museum occupied half a city block at the corner of Sukhbaatar Street and Zaluuchuud Avenue, a large intersection behind Government House, home of parliament. The building was white, three stories tall, with six columns embedded in its flaking façade.

It saddened paleontologists to see the leaky roof, and to know that the fluctuating temperatures and crumbling interior threatened the very fossils the building was supposed to protect. These included, treasure of all treasures, the "fighting pair," a stunning scene of mortal combat between a *Protoceratops* and a *Velociraptor*, found by a 1971 Polish-Mongolian expedition led by Zofia Kielan-Jaworowska. Even as they recognized the museum's impossible financial situation, outside paleontologists lamented the lack of proper facilities to store one of the world's most glorious discoveries.

Catercorner behind the museum sat a redbrick structure, low and long like a chicken coop. Originally built as a garage, the building had been the Mongolian Academy of Sciences's paleontology center and laboratory for the last thirty years, despite its leaky roof and sputtering electricity. The museum and lab were together a sort of compound whose yard was used as a staging area for joint expeditions. The American Museum of Natural History teams stored their automobiles and gear there, in color-faded shipping containers lined up like sunbaked beach cabanas.

The Mongolian paleontologists wanted badly to build a new headquarters and laboratory. They envisioned updated facilities befitting the importance of Mongolian fossils, where foreign scientists would feel comfortable. The government couldn't afford refurbishments—the government could barely afford *science*. Before the democratic revolution, Mongolia had claimed a hundred scientific research institutes, three thousand researchers, and an "annual influx of scientists from other parts of the East Bloc," *Science* reported. The Soviets had built seismological stations to monitor nuclear tests in China and funded a hilltop full of telescopes "to observe U.S. spy satellites through Mongolia's clear skies." After the revolution, the number of institutes dropped to twenty. At least a third of Mongolia's scientists had altogether abandoned research. The Ministry of Enlightenment spent around $3 million a year on science. As inflation spiked by over 300 percent, the money bought precious little. Scientists lamented that they were paid only with "whatever is left over after all the other programs are funded."

A private natural history museum in Japan, the Hayashibara Museum of Natural Sciences, eventually built the paleontologists a new headquarters, a two-story brick building that abutted the old laboratory. In the narrow atrium a *T. bataar* skeleton stood blocked off as if walking through a flower bed. The updated lab had an observation window so visitors could watch the preparators work. Tsogtbaatar, the center's chief, worked upstairs in a corner office, where foreign scientists gathered to ask for access to the Gobi and plot expeditions. An enormous topographical map of Mongolia hung on Tsogtbaatar's wall, along with portraits of Mongolian dinosaurs and Genghis Khan.

A skinny young man in large eyeglasses could often be seen at the paleontological center and in the Natural History Museum's dimly lit galleries. He had dark, floppy hair and tended to dress in jeans and preppy madras-plaid shirts and argyle sweaters. One day, he made his way to the prep lab, where he met Chultem Otgonjargal, "Otgo," a micropaleontologist and the lab's chief preparator. Stocky, with straight hair and protruding teeth, Otgo had been a paleontologist since the 1980s. He routinely participated in joint Gobi expeditions and had prepared numerous Gobi dinosaur specimens, most recently a *Saurolophus* skull, two *Protoceratops* skulls, and a complete *Gallimimus*.

The visitor introduced himself as a businessman and a collector who, now that the laws had changed, hoped to buy fossil-bearing land and open his own museum. His name was Tuvshinjargal Maam. One of his relatives had worked at the Natural History Museum, and Tuvshin hoped to volunteer as a field hand, at one point saying that if the "government paid the poor Gobi people to look for fossils, they could fill a museum with good fossils every four or five years."

Tuvshin told Otgo he owned a travel company called Chinggis Khaan Ltd., and asked if he wanted a side job as a private guide showing his clients the Gobi dinosaur sites made famous by Roy Chapman Andrews and others. Otgo's lab job paid roughly three hundred dollars a month. He was married, with a son who planned to go to college in Australia. A lot of people were taking side jobs as tour guides, and Otgo saw no problem with it. In his opinion, it was the government's fault if scientists went freelance—employees should be paid enough to take care of their families and do their work. The paleontology center had existed for half a century, but its staff still couldn't afford major solo expeditions in their own country. Mongolian research happened largely in collaboration with foreigners.

"We need more scientists, more study, more money," Otgo later said. "And government is not able."

To him, it seemed wrong to forbid tourists to pick up "lesser" fossils like random or damaged materials that paleontologists had passed over in their quest for something different or better. "In other countries everything is under control—regulations. Everything is described in the law. But in Mongolia it's not," Otgo said. "We are free people. If we want to collect anything, we collect it." He remembered one eastern Gobi site loaded with petrified wood. "All the locals brought this petrified wood, every piece, for sale," he said. "Now this forest disappeared. Now you can't find any petrified wood in that place. It's all brought to China." He explained, "The locals don't look at it from historical or scientific importance. They just think they have found some pieces, and they look at it as the source of money. They go, 'Ah! I got a fortune now! I heard it gets great price! Can you ask your brother in the city how much it could be, or does he know any dealers who deals with this?' Like maybe they say to someone, 'I want to sell this bones,' and the person maybe say to another person, 'You know, I have seen bones,' and the other person says, 'Okay, wait, I will ask my brother or my relative.' One by one."

He went on, "But really who is involved in this trade is rich people. Because dinosaurs are too big to go out through the borders, maybe some officials even can be involved, to get through customs, from both sides. Common people, they can't do it."

To Otgo, Tuvshin appeared "prosperous." He soon discovered that the businessman owned and lived in a mint-green building on Peace Avenue, in an apartment filled with minerals and geodes and an impressive collection of framed coins. By using his vacation days to work for Tuvshin, Otgo reckoned he could make a hundred dollars or more a day, plus all the vodka he could drink. He decided to make a demand: tourists must not take anything important. But if they wanted to pick up fossils that scientists didn't want or need, so be it. "I'm not a policeman who can say, 'You mustn't do that,'" he later said. "I don't have that right."

Figuring that Tuvshin would simply hire someone else if he turned him down, Otgo accepted the offer. But he kept the arrangement to himself, deciding that what he did on his own time was his business.

CHAPTER 13

"GO GOBI"

Once Eric knew the name of the Mongolian fossil supplier, he wasted no time getting in touch. Tuvshinjargal sent friendly emails in response. He seemed to write just enough English that they could communicate, and from the way it sounded, he had half the Gobi at his disposal. Eric couldn't tell whether "the Mongol" was working on behalf of someone else or if he was the top boss. He didn't much care, as long as there were dinosaurs.

Photo after photo of skeleton after skeleton arrived in Eric's AOL inbox. Some were nearly whole and others were in parts—femurs in crates, vertebrae in black garbage bags, other bits still *in situ*, surrounded by brushes, chisels, glue bottles, and men's sandaled feet. Other bones lay on laminate or concrete flooring, or on a blanket or tarp. Eric recognized *Gallimimus*, a small-headed, long-necked, ostrich-like meat eater, this one curled up in a field jacket, its legs overrunning the plaster. There was *Oviraptor* and *Protoceratops*. *Saurolophus* bones had been arranged in a lifelike floor puzzle, in more or less the correct anatomical order. One photo showed a set of therizinosaur hands pieced together on bare earth, a cigarette pack thrown in for scale; their size was staggering—*Therizinosaurus* had forelimbs and hands as disproportionately gigantic as *T. bataar*'s and *T. rex*'s were small. The middle claw measured 30 inches, nearly the length of a yardstick or a Major League Baseball bat.

Eric was interested in a largely complete *Tarbosaurus bataar* skeleton—an upgrade, as he saw it, from the two skulls he had already done, and the closest he would probably ever come to *T. rex*. The United States was the only country where a commercial hunter could get a *rex*, but it was hard to do. First you had to own or lease property where a *rex* might be found; then you had to find one that was largely complete; then the bones had to be in good enough shape to restore. It took a lot of time and often a lot of money to find a *T. rex*, plus luck. (Roughly fifty *rex* skeletons have been unearthed, a decent showing in the fossil record.) Then it took time, money, and skill to prep and mount the specimen well enough to attract a serious buyer.

Tuvshin made it easy, though. He was like a mail-order catalog for Gobi dinosaurs. He had started moving bones to the West via the Tucson and Tokyo shows. Hollis Butts, the dealer in Japan, remembered him carrying four to six "large travel roller bags or suitcases full of unprepared or partly prepared fossils, weighing a ton." Tuvshin would fly anywhere, Butts said, "like a traveling salesman."

Around 2006, he switched to shipping containers. Certain dealers were eager to work with him, while others stayed away. As Tuvshin tried to legitimize his museum idea, "he was trying to get international people involved, and he was going to trade material," another dealer later said. "It was always the idea to trade material from one museum to the next, so he could legally sell Mongolian materials." Tuvshin carried a business card bearing the name of a private-museums association yet seemed "very secretive" about what he was doing. "I'd ask him a question about it and he'd feign that he didn't understand English."

Eric knew it would cost a small fortune to become a client of Tuvshin's, but the successful sale of the two *T. bataar* skulls had given him confidence. He told himself that if he didn't buy Tuvshin's dinosaurs, someone else would. He decided to go to Mongolia, even though it was October, that frenzied time between the Denver and Tucson shows, and Ulaanbaatar was a very long way to go. Mongolia was 7,000 miles away and thirteen hours ahead in time—Mongolia was literally the future.

The easiest way to enter Mongolia has always been from the south. The plane from Beijing to Ulaanbaatar crosses 800 miles of Gobi. Nighttime is a surreal way to go in by air because the desert is even blacker than the ocean on a moonless night. Then a pinprick of a headlight may appear below, or the grid of a town may wink pinkly and disappear, as if transmitting a fleeting clue to dimension and scale. On summer days, if the weather is clear, one sees a swath of the whole terrestrial show—puckered brown terrain, a sheen of green, patches of redness where there's iron in the earth. The crisscrossing dirt roads spider out to the horizon, converging here and there like the impact points of stones flung against a windshield. Because 30 percent of Mongolia's current population of three million live in or around Ulaanbaatar, a person could travel for many miles in most of the rest of the country without ever seeing another human being. As Mike Novacek once put it, the Gobi is "one of the great empty spaces on earth."

Chinggis Khaan International Airport reminded Eric of a regional airport, like the one in Gainesville or St. Augustine. Tuvshin had promised to be waiting at customs. Eric looked for the nerdy, skinny man he had seen in the emailed photos, and when they spotted each other, Tuvshin motioned him to the front of the line, where a customs agent quickly stamped him through.

At baggage claim, one of Tuvshin's employees, a driver whose name Eric understood to be Ulzii, was waiting. They loaded the bags into Tuvshin's Toyota Land Cruiser, then made the half-hour drive northeast, to the city. It excited Eric to be in a place he had never seen before, and though he tried to get an impression of Ulaanbaatar, all that registered was the wild traffic and ugly Soviet architecture. He had booked a cheap room at Seoul Hotel, in the city center, which turned out to be appended to a Korean restaurant and a strip club or massage parlor. Exhausted, he went straight to bed.

The next morning, Tuvshin picked him up and drove him down Peace Avenue, the city's main corridor. In places, Ulaanbaatar appeared to still be a faint mash-up of the ancient nomadic culture and Soviet uniformity, overlaid with newfound capitalism—an old temple next to a new apartment building, not far from a Louis Vuitton store or a Kenny Rogers Roasters. Nearly twenty years of a market economy had lined the street with restaurants, banks, cell phone stores, clothing shops, cashmere outlets, and nail salons. Before 1990, most Mongolians could not own cars, and they'd been making up for it ever since. Wrestling, archery, and horse racing are the nation's three "manly" sports, but one of the bravest things a person can do in Mongolia is travel by motor vehicle or walk anywhere near traffic. The streets teemed with maniacal Kias and Hyundais; they tore past pedestrians, close enough to flip the hem of a coat.

Much of Ulaanbaatar was paved, but plenty wasn't. It was a really dusty town. Human street sweepers stood amid the madcap traffic, swiping at curbs with huge witchy broomsticks. The dirt only drifted and settled elsewhere. Land Cruisers were equipped with an "air snorkel" affixed alongside the windshield, an appendage that resembled a large vacuum cleaner accessory and protected the engine from dust and high water. Coal factory smokestacks spewed in the near distance, a visible reminder that in one of the world's most polluted cities, the air quality was a felt presence on the tongue and in the lungs.

Tuvshin arrived at a strip shopping center of four-story buildings fronted by a narrow parking lot and a bus stop. He parked around back, in a gravel lot, and led Eric up a few exterior stairs, through a rear door of his mint-green building. Inside, they climbed four flights of stairs and went into Tuvshin's apartment, where he lived with his wife and children. Eric counted four bedrooms and two bathrooms, plus the kitchen and living spaces. There were aquariums and museum-quality collections of framed coins and natural history. Tuvshin's travel business was on the same floor, at the front end of the building. His wife, whom Eric was invited to call Bobo, worked there. An older son was away at college in Japan; a teenage daughter and younger son lived at home, and spoke and wrote fluent English, whereas Tuvshin's English was minimal. Eric estimated that Tuvshin and Bobo, who was slim, with straight, shoulder-length hair, were in their late forties. (In fact, Tuvshin was forty.)

Tuvshin took Eric back outside and around to a basement door. Half the basement was a garage; the other half was a storage area filled with enough Gobi dinosaur skeletons to start a small standalone museum. When Eric started asking prices, Tuvshin, instead of answering him, put him back into the Land Cruiser and drove him out of the city.

They were going north, maybe, Eric thought. Or east. It didn't occur to him to feel nervous about traveling alone with a stranger to an unknown destination in an unfamiliar country where he spoke not one word of the language. As they passed through rolling grasslands, Tuvshin chain-smoked, and they chatted about Tuvshin's father, who was Russian, and his sister, who lived in Germany. Eric gathered that Tuvshin traveled often and had a lot of girlfriends.

They came to a small house in the countryside, in a kind of neighborhood—a summer house. The one-car garage contained more dinosaurs, plus display cases filled with rows of teeth and claws. Finally, Tuvshin was ready to name a price. The whole collection would cost $100,000.

But the bones weren't for sale! Tuvshin had promised them to another buyer. Eric asked him who but Tuvshin wouldn't say. Mongolia was a long way to travel to be rebuffed—the Korean Air ticket alone had cost over $2,000—but Eric wasn't angry. He had made his connection; now he would wait.

Back in the city, Tuvshin took Eric to the Natural History Museum. Online someone said the museum resembled an elementary school biology lab—"Honestly, if you're not a paleontology junky, don't go"—but Eric studied the mounts anyway. If Tuvshin agreed to sell him what he wanted, a *T. bataar*, the specimen would be the largest he had ever assembled.

The mount is where engineering meets metalworking and scientific accuracy meets art. The AMNH preparator Amy Davidson, for instance, was a sculptor before becoming a preparator, and had worked season after season in the Gobi. There were no formal degrees or training programs in fossil preparation, but it helped to be patient and detail-oriented, and to have experience in professions that required nimble fingers, like dentistry, jewelry making, welding. Taking certain courses didn't hurt—art conservation, anatomy, paleontology, vertebrate evolution, and geology were recommended—but there was "no substitute for aptitude and a genuine interest in preparation," the AMNH pointed out. Prep work wasn't a one-and-done job, either. Museums had been refurbishing exhibits to reflect new understandings of prehistoric biomechanics—*T. rex* is now shown not as an upright, tail-dragging sluggard but rather as a dynamic hunter whose body stood almost horizontally, the long tail outstretched to counterbalance the enormous head.

Eric would need to weld a large steel frame in a realistic pose. The bones would need to be removable, as he had seen done in Western museums. The armature, which might include plates, bolts, screws, posts, and wire, had to be strong enough to hold fossils that weighed hundreds of pounds yet graceful enough to disappear. Early dinosaur preparators often mounted skeletons by running steel rods up the hollow bones or by drilling holes in the fossils, yet embedded in the work was a "bona fide hidden metals craft," as one Carnegie Museum project manager once put it. Phil Fraley, who built or rebuilt some of the world's top dinosaur exhibits, once said, "Really, what an armature is doing is replacing all the tendons and ligaments—the soft tissue that used to hold the animal together." The best preparators created mounts so realistic that even the naked scaffolding evoked an impression of life. Siobhan Starrs, a Smithsonian exhibit developer, found early dinosaur framework so beautiful that she had always wanted to make an exhibit consisting solely of armature.

The museum dinosaurs of Ulaanbaatar had been crudely restored and posed. There were anatomical gaps in the skeletons; the rods and wires were as obvious as old-school braces on crooked teeth. Tuvshin seemed to find all of it glorious. After stopping by the prep lab to introduce Eric to his friend Otgo, he suggested that they "go Gobi."

In hard numbers, the Gobi Desert measures 1,000 miles long and, at its broadest point, 600 miles wide, with a total area of half a million square miles. A few paved highways now ran through the desert, though the harsh weather extremes made them hard to maintain. The surface of the moon is smoother than a Mongolian road, through no fault of the builders: the cycles of freeze/thaw taffy the asphalt and resettle it like peanut brittle. Given all that could go wrong with an impromptu trip to the desert, Eric didn't see how it was possible to pull it off. Tuvshin wanted to go eight hours out, a gamble considering how often motorists blew tires or got stranded by sudden changes in the weather. By now it was Saturday morning; Eric's flight departed on Monday afternoon at four. But he wanted to see the Gobi and couldn't say no.

Ulzii drove Eric and Tuvshin. They took Tuvshin's Land Cruiser south, out of the city, crossing a stream where the first crusts of winter ice were forming. Whenever they came to an *ovoo*, a cairn altar to the spirits of mountains or rivers or trees or stars, Tuvshin stopped and added tugriks to the prayer ribbons and stones. The grass disappeared. They passed through a light snow. In a village, they stopped for lunch, the restaurant a cold-storage room containing a half-butchered horse. Eric had always liked foods that other people hated—airplane food, cafeteria food—and now he found that he liked mutton, especially in *buuz*, dumplings.

After lunch, their young waitress joined them on the trip, with no explanation. The desert ran hard and bare. Twin ruts guided them for long miles over wiry grass and marmot holes, pushing toward the long line of the horizon. The ruts sometimes crossed and converged with other ruts, in some unspoken agreement about routes. The roads were always changing because the weather was always rearranging them. Conditions allowed for high speeds in one place, barely a crawl in another as the vehicle skirted a distortion that from the air must have had the embolistic appearance of a snake swallowing a goat. Nighttime magnified the remoteness, the darkness cut only by the limited reach of the headlights.

Late that night they came to a *ger*, which Eric was excited to see. The word was pronounced *gair*; another word for it was *yurt*, but that was the Russian term, and Mongolians did not use it. All *ger*s were alike in that each was a portable, round, windowless room framed with a collapsible latticework supported by flexible wooden poles and covered in canvas or felt. On the landscape a lone *ger* might resemble a mutant mushroom cap or vanilla cupcake. The roof has a center hole with a stovepipe up through it. The lower flaps of the felt or canvas can be raised and lowered like a window shade to control heat and ventilation. All *ger*s face south, toward the best sun. Each has a short wooden door, usually painted a cheerful color like orange. The stove burns dried camel dung, *argal,* for cooking and for heat. Families often use old tires and ropes and lumber scraps to hold down the roof against the Gobi wind. A *ger* can be disassembled in minutes, packed onto camels or lorries, and moved easily as the herders shift their animals to better pastures. A lot of *ger*s now had satellite dishes and generator-powered radios, TVs, and electricity. The left side of the *ger* is the "man's" side and holds items like tackle; the right side is the "woman's" side and holds items like cooking utensils. A narrow bed lines either wall, and it is okay for a guest to sit on the bed, even if someone is sleeping in it. The most important area in the *ger* is the space directly opposite the door, behind the stove, where a painted chest holds the family's treasures.

Eric had read enough to know that there were all kinds of *ger* rules— leave your hat on; never step on the threshold; always accept what you're offered. Ducking through the doorway, he saw that the man of this *ger* was tall and skinny and younger than him, with a mustache. To his surprise, he was wearing a University of Florida sweatshirt. He and his wife had two young children. Another guest was sitting on the floor by the warmth of the stove; Eric heard his name as Batta, but it could have been Baht or Bat. Eric, Tuvshin, and Ulzii joined him on the floor and were offered bowls of *suutei tsai*, salted milk tea, which Eric drank to be polite. Then came the national favorite, *airag*—fermented mare's milk, sour and intoxicating. Mongolian families kept a large barrel of it inside their *ger* and took turns pumping the liquid with a large wooden paddle, to keep the cultures alive.

Around midnight, Tuvshin stood abruptly, pointed to a spot on the floor, and told Eric to sleep there. Then he and the waitress spent the night in the Land Cruiser with the engine running.

The *ger* was so cold, Eric could see his breath. All he could hear was the occasional hiss of the stove and the wind. At one point he got up to relieve himself. The custom, as he understood it, was to walk a reasonable distance away from other people and just go. He couldn't find words to describe the Gobi darkness, even with the backlight of his camera to guide the way. The stars hovered as they do in deserts, in one silvery panoramic sweep. Once his eyes adjusted, he could make out some sheep, but otherwise the Gobi on a moonless night was like entering negative space, a not-unpleasant dimension of nothingness.

In the morning, the guests were served tea and mutton. When Eric went outside with the men, he could see that they were in a landscape of sand dunes and low grasses. In addition to sheep and horses, there were goats and camels. Someone asked Eric if he wanted to ride a camel and while it embarrassed him to act like a tourist, he said yes, so they put him between humps and one of the children led him around by a rope.

Tuvshin announced that the family would now slaughter a sheep in their American guest's honor. Eric would have rather this not happen, but again he deferred to custom. Mongolians used almost every part of the animal, the way Native Americans used every part of the buffalo: they ate what was edible; made clothes, felt, and *ger* coverings of the wool and skins; and burned the dung for heat and cooking. A couple of the men chased down one of the sheep, wrestled it to the ground, then carried it back to the *ger* by its legs. As they pinned it on its back, Eric braced for a gory death. One guy held the sheep while another pulled out a knife and cut an incision in the abdomen. The sheep said nothing. The man stuck his hand inside the animal and, it looked to Eric, pinched an artery near the heart. Instantly, the sheep was dead. The men left the carcass with the wife to be butchered and went driving.

Eric now realized his host was one of Tuvshin's diggers. They were in the Djadochta Formation, which encompasses the Flaming Cliffs, he later figured out by looking at maps. The digger showed everyone a *Protoceratops* skull that he had failed to excavate because it was so badly weathered, then showed them an equally crumbly hadrosaur pelvis. By the time they returned to the *ger*, the hunter's wife had boiled the flesh of the sheep along with some of the organs, and everyone ate together out of a communal bowl, their fingertips and lips shining with grease.

That afternoon, the visitors got back into Tuvshin's Land Cruiser and headed for Ulaanbaatar, stopping only to drop off the waitress in her village. Tuvshin drank vodka the whole way home and rambled about a brother who had died in a car crash. It was well past midnight when they reached the city. Eric wanted to talk business before leaving, but the next morning he waited and waited at his hotel and no one ever came. Just as he decided to find a taxi, Ulzii showed up and drove him to Tuvshin's apartment, where Tuvshin was still passed out from the night before. Bobo woke him, but he was too drunk to talk, so Ulzii drove Eric to the airport. He made it onto the plane just before the gate closed.

As photos of more dinosaur bones arrived, Eric organized them into files and folders on the family iMac in Gainesville. "Galli skeletons" and "big ankylo skull" and "tarbo arms," he slugged them. In some of the photos, Tuvshin posed with the bones like an angler with a trophy fish. He now agreed to sell to Eric, but refused to ship directly to the United States, preferring to route everything through Europe or Japan.

Eric didn't see the problem with direct shipments or with the preferred terminology. He adopted certain known semantics and rationalizations. Why not call fossil dinosaurs "reptiles" on customs forms? Was that not what they were? Why not cite the value as, say, $10,000? It was the prep work that added value. Whenever he pressed Tuvshin about the export permits he'd mentioned, Tuvshin said he was working on it. "It was difficult, but Tuvshin had connections," he said later. "I had no reason to think he couldn't get them."

Eric briefly considered making up with Hollis Butts, to have a shipping partner in Japan. Then a friend suggested partnering with Chris Moore, a well-liked veteran hunter who lived in southern England and was well known at Tucson and Denver. Eric approached him and Moore agreed to go in.

By now Amanda was six months pregnant with the second Prokopi child, a girl, due in May 2009. The Prokopis joined Facebook. In between posts about children, Earth Day, the March of Dimes Walk, *Real Housewives of Atlanta*, and fried Christmas goose, they marketed Florida Fossils and Everything Earth. They had sold the first flip house for a nice profit and now lived at Serenola, where the renovations were coming to an end. Recently, they had put in a swimming pool shaded by expensive Canary Island date palms. Eric hadn't swum competitively in well over a decade, but still loved the water. He hoped his own children would become swimmers, just as he hoped they would love fossils. Rivers Prokopi was born on the fourth of May, with the white-blond hair of her mom and the brown eyes of both her parents. She left the hospital wearing a tiny bracelet of good Virginia silver.

Eric and Amanda installed a gravel playground and a CedarWorks jungle gym. Amanda joined the Junior League. They hosted swim parties and barbecues. Amanda dressed Rivers in hair bows, and bought her a rainbow of tutus. Eric, bare chested and tanned, rode a four-wheeler through the yard with Greyson in his lap, teaching him how to steer. "I mean, how cute is that?" Amanda said as she filmed them on "the digger," a machine used to remove scrub. "Father and son, digging up trees! Grey has the coolest daddy ever!" The Prokopi cameras were always on.

They captured Grey's fears of the dark, sharks, sleeping alone, humans dressed as sports mascots, and University of Florida cheerleaders. They captured Rivers's fear of nothing. Their days were pumpkin patches, bouncy houses, petting zoos, piñatas, and IMAX theaters. Eric woke the children in their cribs—"Good morning!"—and Amanda soundtracked the first years of their lives. *What's a train say? Are you blowing bubbles? Happy, happy girl. Is today your first day of school? Whoa! Cool! Can you say "shark"? See the butterflies? Are you having fun at the museum? Take a bite of your dinosaur cookie! Is this your first trip on a ferry? Whoa, Grey's driving the boat! What a cool kid! Are you finding fossils? Did you put your sunglasses on? Touch the turtle! What does a pirate say? Look at the ducks! Grey, is this your backyard? What a lucky guy you are!*

In Grey's bedroom were an inflatable *T. rex*, framed dinosaur illustrations, and the dinosaur lamp from Eric's childhood bedroom in Land O' Lakes. On the armoire sat a framed I.M. Chait auction catalog cover showing the *T. bataar* skull that had sold in New York the day Grey was conceived. Rivers's bedroom was soft cowhide skins, a twinkly chandelier, a daybed fluffed with monogrammed pillows, and an enormous print of a pink octopus.

Few commercial fossil hunters lived in this sort of comfort. Yet when you looked closer, the Prokopis were still living the feast-to-famine life. They had no significant savings, no health insurance. They maintained a medical savings account and paid out of pocket for everything, including childbirth. Amanda's father served as the family doctor, her brother as the family dentist. But the lifestyle had been established and now had to be maintained, along with the mortgages, the renovation expenses, and the costs of doing business. In 2008 alone, Eric had spent nearly $71,000 on fossil prep, almost $10,000 on shipping, and over $5,000 on trade show fees.

A batch of dinosaurs would put everything straight. Tuvshin had emailed shots of a small ankylosaur skeleton; the vertebral column and articulated ribs of a huge unspecified dinosaur *in situ*; an enormous *T. bataar* foot; a broken dinosaur egg with a clearly visible embryo; a table laden with *Gallimimus* skulls and parts; a hadrosaur tail as perfect as the white spine of a catfish picked clean; and at least two nearly complete *bataar* skeletons. Feeling confident about his new revenue potential, Eric made plans to return to Mongolia.

———————

Later, once the name Eric Prokopi was synonymous with international fossil smuggling, people would imagine that he had slipped into Mongolia, dug dinosaurs under cover of darkness, and sneaked them across the border. Mongolian officials would speculate that the contraband went out in loads of salt, the least-checked export at customs. "Criminal mastermind" conjured certain clandestine images, which made Amanda snort. She said, "It was more like *Harold and Kumar Go to Mongolia*."

For this trip, in the summer of 2009, less than a year after his first, Eric invited his friend Tony Perez along. Tony owned his own welding business near Venice, Florida, where Eric had pocketed that first shark tooth as a child. They had met years ago when their mutual friend Andreas Kerner brought Tony to the Prokopi house in Land O' Lakes to look at fossils. Tony had a family and sold shark teeth on the side, and he and Eric had become close friends, which mostly involved hanging out at Tucson and pranking each other whenever possible. Once, after Eric teased Tony about being so hairy, Tony manscaped and mailed the fuzz to Eric in an unmarked envelope; Eric kept the hair, and the next time he saw Tony, he sneaked it into the air conditioner vents of his car.

They booked Mongolia for eighteen days in late June and early July, and got to work outfitting themselves with sleeping bags, tents, GPS-equipped walkie-talkies, ready-to-eat meals, guide books, and a satellite phone. Tony bought a new safari hat and a fanny pack. Eric bought motocross goggles, in case of sandstorms. The night before the flight, he drove to Land O' Lakes and stayed with his parents, and Doris and Bill drove him to the airport the next morning.

When Eric and Tony arrived in Ulaanbaatar, Tuvshin was once again waiting at customs, and they were expedited through. Eric double-strapped his backpack and pushed a luggage-laden cart out into the summer night, the red and blue lights of Chinggis Khaan International glowing behind him. The next morning, Ulzii picked them up at their hotel and took them to the mint-green building on Peace Avenue, where the basement was now crowded not only with dinosaur skeletons but also with geodes and shovels and cases of bottled water. Everyone took turns kneeling with bones for photos.

It isn't easy to rent a vehicle in Mongolia without also hiring a driver, but Eric managed to secure a Land Cruiser. They packed that rig and Tuvshin's, and everyone loaded up: Eric, Tony, Tuvshin, Ulzii, Tuvshin's daughter, and Batta, whom Eric had met in the Gobi eight months earlier. The seventh person was Otgo, the chief fossil prepper at the national paleontology lab. Eric had agreed to pay him fifty dollars a day as a guide, plus all the vodka he could drink.

Eric had a map of the "Big Gobi circuit"—a loop through the southern desert—and for two weeks most of what they did was drive and take photographs. A gas station. A horse cart. Old men on a purple bus bench. A guy wearing a tweed jacket, a surgical mask, and a white baseball cap branded with a golden spray of cannabis, riding a moped with a hooded eagle on back, the eagle tethered to a tasseled box. The distant glimmer of the Chinggis Khaan Equestrian Complex, a massive silvery statue of the ancient warrior mounted on horseback—Mongolia's version of Mount Rushmore or Rio Jesus. Flat endless land. Sheep. Blue sky, low clouds. Eric peeing. Tony peeing. Otgo peeing. Tuvshin peeing. A deserted-looking village. Badlands. Sagebrush. A cluster of camel dung that resembled a pile of Idaho potatoes. Every night in the desert Eric erected his new blue one-man tent 25 yards away from Tony's green one so that he wouldn't have to hear him snore. Just when they thought there was no one around for miles, they'd see the distant star-point of a passing motorcycle. The sun rose at five; the Mongolians slept till ten, coverless and pillowless on the hard ground. Tiny creatures crept through the camp at night, leaving tracks in the sand.

As Otgo and the others stayed in camp, Eric and Tony walked for hours finding nothing much more than endless desert. At an old hadrosaur site they came across broken lengths of femur and the claws of a freshly dead eagle, its carcass still tufted in feathers. Eric and Tony took turns crouching and grimacing over an odd stone the size of a grapefruit, pretending to have shat it. They got stuck in the sand and had to dig out. They blew a tire. They chased a wild camel on foot, for the hell of it. In a village they hosed themselves down at a public well and bought Otgo more vodka; Eric switched from Chinggis brand to something cheaper when he saw how much the old boy could drink.

Tuvshin's floral shorts read HAWAII, his baseball cap BOSTON. Eric's T-shirt read EVOLUTION. They came to a village with deserted Soviet apartment buildings, the windows gaping. They saw a high pile of animal bones. They passed a mummified hedgehog, a Chinese oil well, a man leading a horse across the parched, pale earth. Eric found a tiny lizard and held it in his large hand. They climbed a trail to see a frozen waterfall. They scaled a "singing" sand dune. When they came across roadside tables full of agate and scrap fossils, nomads appeared, as if out of nowhere, to sell Eric a little pile for five dollars.

In the Nemegt Formation, where the Russians found *T. bataar*, they made camp and walked around for two days. Tuvshin returned to a couple of *Protoceratops* eggs he'd seen earlier and dug them out of the cliff face. Eric found an ankylosaur scute. They collected whatever they could easily carry. At Bugiin Tsav, a well-known dinosaur locality that was on all the tourist itineraries, they found what was left of a *Gallimimus* skeleton.

At a military post outside the city, Tuvshin walked up to the base and asked if they could drive a tank. A tank wasn't available, but they were allowed to shoot AK-47s and an RPG as long as they paid for the ammo. In the city, Tuvshin took them to where he kept a shipping container full of the enormous skulls and horns of argali sheep, and to a cookout, where they drank beer and had a barbecue. Later, Tuvshin pulled his Land Cruiser into a parking lot to meet with a guy selling an enormous tarbosaur tooth, which Eric bought.

Before flying home, Eric placed a large order on behalf of himself and Moore, his new business partner in England. Back in Florida, he wired Tuvshin $100,000 as a down payment. It wasn't long before a shipping container departed Mongolia aboard a coal train headed south. The container was packed with traditional Mongolian *gers*, plus chairs, tables, and chests. The train crossed the Chinese border and headed southeast, past Beijing, stopping at Tianjin, a port city of 14 million on the Bo Hai gulf. The container was loaded onto the *CMA CGM Vela*, a brand-new container ship nearly four football fields long and 49 yards wide, sailing under the flag of Germany. Tucked among the *gers* were the crated skeletons of

Gobi dinosaurs. Eric had entrusted an enormous amount of money to a stranger on the other side of the world, with no legal recourse if the deal went bad. His only way of monitoring the "investment" was to watch the *Vela*'s progress online, on a website that showed real-time maritime traffic in digital blips. On his computer at Serenola he identified the neon ship-shaped speck that represented the *Vela* as it made its journey of over 12,000 nautical miles. Checking back daily, sometimes hourly, he followed the vessel as it moved from Yellow Sea to East China Sea to Hangzhou Bay to Philippine Sea and on, relieved to see it finally make port at Liverpool.

CHAPTER 14

THE GHOST OF MARY ANNING

THE SEA THRASHED AT THE SOUTHWESTERN SHORE OF ENGLAND for eons with no one to notice, but around the year 774, a village appeared between two of the oldest cliffs. The first mention of Lyme emerged when King Cynewulf granted a manor to Sherborne Abbey, which owned salt mining rights near the river Lym. By the time William the Conqueror surveyed England in 1085, Lyme consisted roughly of nine villagers, thirteen salt workers, four ploughlands, a meadow, and ten acres of woods. The lords of Lyme came and went—Wulfgeat, Aelfeva, Bellett—and in 1284 King Edward I granted the town its royal charter, making it Lyme Regis.

The port at Lyme Bay featured a serpentine breakwater called the Cobb, which protected the harbor like half an embrace. Made of boulders and oak trunks, the breakwater shielded vessels from the storms that whipped through the English Channel from the North Sea. Despite Lyme's reputation as an inhospitable harbor—"verrie daungerous in tyme of wynter and tempestes"—the town traded with France, 20 miles across the channel, "wool out, wine in."

Cottages stood along the foreshore just above the sea. Narrow streets wound steeply inland, lined with two- and three-story buildings with shops at street level and apartments above. "While most of the town slept at night, another age-old maritime industry thrived," wrote the novelist John Fowles, who lived in Lyme for forty years. In the late 1700s and early 1800s, smugglers worked the beach. In their fast, armed cutters, they used the French towns of Cherbourg and Guernsey as their pickup points, then sank their silk, brandy, and tobacco offshore until "the coast was literally clear."

Tempests had been known to take out half the village. Sections of land and construction occasionally lurched and disappeared: a church meadow, a whole harbor. The battering knocked strange objects loose from the coastal cliffs that loomed for 96 miles, their layers a towering vision. Locals called the objects "verteberries" and "ladies' fingers" and "thunderbolts," for the way they were shaped, or just plain "curios." The rocks were so distinctive, smugglers disembarking at Lyme in the dead of night could confirm their location by touch.

The coiled rocks called "snakestones" looked especially beautiful as decorations for jewelry boxes and picture frames, or displayed as a conversation piece. They were popular with the wealthy travelers who had started vacationing at the coast on the advice of a Brighton doctor who praised the health benefits of ocean bathing. Lyme, in the county of Dorset, became a spa destination as Romantic-era travelers sought out "the beauties of 'wild' nature." The streets were narrow and dirty, and there wasn't much of a path to the sea, but the village had a reputation for having kept two of its citizens alive past age ninety.

A cabinetmaker named Richard Anning moved to Lyme with his bride, Mary, called "Molly," in September 1793. They rented a timber-framed house on Bridge Street, at Gun Cliff, where the town cannon ever pointed toward France. At home at the buddle of the river Lym, the Annings lived so close to the sea, the waves beat at their windows and often washed through their rooms. They shared tiny cobblestone Cockmoile Square with a shoemaker and the jail, around the corner from the Three Cups Inn.

The Church of England banned dissenters from certain jobs and privileges, which put Richard out of contention for both university and the military. He was a Congregationalist living in a country where war had worsened poverty. For extra money, he sold the curios that turned up at the beach. Only two other "fossilists" were known to work the area, a fellow everyone called "Captain Curi," and a former Piccadilly coal merchant named John Crookshank who jabbed at the cliffs with a long pole.

The year after moving to Lyme, the Annings had their first child, also Mary. Then they had a second, Joseph. In the winter of 1798, amid record-low temperatures and coal and firewood shortages, the little girl got too close to the stove, and her clothes caught fire. She became the first of seven Anning children who would not survive childhood.

Molly was pregnant at the time of her first daughter's death. She gave birth on May 21, 1799, and once again named her girl Mary. The surviving Mary Anning was a sickly baby, said to be listless and dull. Fifteen months into her life, on August 19, 1800, she was taken to an equestrian show in town by a nurse and neighbor, Elizabeth Haskings. Around a quarter to five in the afternoon, it started raining. Haskings was standing beneath a large elm with her friends Fanny Fowler and Martha Drower, holding Mary in her arms, when lightning struck the tree. "As the spectators' senses returned they were aware of a brief stillness interrupted only by the incessant rain," one historian wrote. "A person pointed to the base of the steaming, burnt tree and began to run to it."

Haskings and her friends were clearly dead. Baby Mary appeared unconscious. Someone grabbed her up and rushed her home to Richard and Molly, who put her in a hot bath and revived her. Mary not only lived, but was said to have been transformed by the lightning into an energetic, engaged child.

Lyme Regis had become a larger and more important port even than Liverpool. When travelers came for the bathing machines and the bracing channel waters, they often returned home with souvenirs bought from Richard Anning. Brits of a certain class liked to display their nature collections in specially made cabinets, and Anning supplied both—cabinet and curiosity. When the young writer Jane Austen visited Lyme with her family, she consulted the cabinetmaker about fixing the lid of a broken chest, walking away when she was quoted a price that she considered "beyond the value of all the furniture in the room altogether."

One day in 1807, Richard walked a mile or so north to Charmouth, to sell fossils outside the Pilot Boat Inn, a stagecoach stop between Exeter and London. Returning home, he fell down a cliff, severely injuring his back. Soon after that, he made a bad business decision and accrued debt. Soon after that, he came down with tuberculosis. He died a few years later, at age forty-four, leaving Molly pregnant again and deeply in debt at a time when the people of Lyme were so desperate for food and money, some were selling their hair. As Molly applied for parish relief, her oldest children, Joseph and Mary, took over their father's fossil trade.

Quality fossils for scientific research were hard to get, and a smart hunter could earn a decent living by selling to the gentlemen scientists and university professors who were working out the new discipline of geology. The fertile cliffs and outcrops of Dorset would become known as the Jurassic Coast, for nowhere on earth is such a lengthy, raw swath of prehistory so vividly visible. The coastline altogether reflects not just one period of the Mesozoic but rather all three: the Triassic, Jurassic, and Cretaceous—one shoreline, roughly 186 million years of earth history. Because the coastline constantly shifts, signs and maps warn of rock-falls and mudflows. A careless beachcomber could die in any number of ways—trapped by the tide, dragged out to sea, dashed against the rocks.

Mary Anning hunted anyway.

Shortly after Mary turned eleven, her brother Joseph came across an odd skull and dug it out of a cliff. The skull was 4 feet long, with room for two hundred teeth, and orbital sockets that suggested eyes the size of prize onions. Joseph was by now apprenticing himself to an upholsterer and had less time for fossils, but Mary went on searching for the rest of the skeleton. Working in boots and a bonnet or cloak, and with the entanglements of voluminous ankle-length skirts, she carried a stick for balance and a basket for collecting smaller items. Within a year she had located a skeleton near where the skull had turned up. It was said that she had the bones hauled to Cockmoile Square piece by piece, where she cleaned them, revealing a creature nearly 17 feet long, with a mouth like a crocodile, a snout like a swordfish, and flippers like a dolphin.

A local landowner and enthusiastic natural history collector named Henry Hoste Henley bought Mary's strange "stone crocodile" for about 23 pounds, enough to feed the Anning family for six months. Henley, in turn, sold the specimen to William Bullock's Museum of Natural Curiosities in London, presumably for much more than Henley had paid Mary. Specimens similar to this one had already surfaced, but Mary's was the most spectacular because it was the most complete. Sir Everard Home, in "Some Account of the Fossil Remains of an Animal," a paper for

Transactions of the Royal Society, began a series of articles describing the creature in 1814, ultimately crediting Henley, the buyer, as the discoverer, without mentioning Mary Anning at all. The British Museum in turn bought the skeleton from Bullock's at auction for 45 pounds, more than double what Mary had been paid for it not even five years earlier. The creature was named *Ichthyosaurus*, "fish lizard," and its discovery electrified the scientific community, adding to the ongoing debate about the age and origin of Earth. Men like William Buckland, the first geology professor at Oxford, and Henry Thomas De La Beche, who founded the groundbreaking Geological Survey of Great Britain, became Mary's connection to science, and she a conduit to their success.

Mary specialized in invertebrates because the Jurassic Coast is so full of them, but she prized vertebrates because they fetched more at market. By the time she turned eighteen, she had a reputation as a talented fossil hunter who cared deeply about the scientific discipline, which, in 1822, would finally get a name: "palaeontologie." Yet she was as poor as ever. When one of her best clients, Thomas Birch, a Life Guards officer, visited Lyme Regis in 1818 he found the Annings so broke that they were planning to sell their furniture. Outraged, Birch decided to auction his fossil collection to benefit the Annings; after all, the family had "found almost all the fine things which have been submitted to scientific investigation."

The auction began on May 15, 1820, drawing attendees from Paris, Vienna, and Germany, among them Georges Cuvier. The French naturalist would change the world with his revelation that some species went extinct, a conclusion he would not have reached without fossils. His presence at Bullock's made it clear that he considered the young hunter of Lyme a vital contributor to science. Fetching 400 pounds, the auction gave the Annings financial security for the first time in their lives.

No other fossil hunter anywhere in the world was making as many discoveries as Mary Anning. She went on to find three species of *Plesiosaurus*; cephalopod ink chambers, which led to understanding of animal defense mechanisms; the first pterosaurs found outside Germany; the shark-ray *Squaloraja polyspondyla*, a vital transitional fossil that helped Cuvier prove extinction; and coprolites, or fossil feces, which are way more important than they sound. An early edition of the *Bristol Mirror* praised the "persevering female" who for years had searched for "valuable relics of a former world" that were "at the continual risk of being...destroyed by the returning tide."

The science fascinated Mary as much as the fossil sales episodically sustained her. She cut creatures apart in an attempt to understand how they worked. She copied out scientific papers by hand, adding detailed technical illustrations. She wrote to the British Museum, asking for a full list of its collection. She tried to learn French, in order to better communicate with Cuvier. By 1826, when she was twenty-seven, Mary received so many fossil-obsessed visitors that she moved from Cockmoile Square to a larger home on Broad Street, the village's hilly main road, with a view of the sea but not so directly a taste of it. She lived with her mother and brother in the back rooms and kept the rest as Anning's Fossil Depot, where an ichthyosaur skeleton filled the front window.

A trio of well-off sisters named Philpot had moved to Lyme from London and Mary had become friends with them, often visiting their cottage on Silver Street. Despite their differences in age and class, Mary and one of the sisters, Elizabeth, hunted together. The Philpots compiled a meticulously labeled fossil collection that drew geologists from around the kingdom. Mary used their visits to build on the early lessons she had received from the Congregationalists, who believed in educating the poor. "The extraordinary thing in this young woman is that she has made herself so thoroughly acquainted with the science that the moment she finds any bones she knows to what tribe they belong," a Londoner, Lady Harriet Silvester, wrote in her diary after meeting Mary. "She fixes the bones on a frame with cement and then makes drawings and has them engraved... It is certainly a wonderful instance of divine favour—that this poor, ignorant girl should be so blessed, for by reading and application she has arrived to that degree of knowledge as to be in the habit of writing and talking with professors and other clever men on the subject, and they all acknowledge that she understands more of the science than anyone else in this kingdom."

––––––––

Every other Saturday a ship called *Unity* departed Lyme for London. Mary often sent her fossils north, but in the summer of 1829, it was she herself who boarded. The geologist Roderick Murchison and his wife, Charlotte, had invited her to the city, her first visit. Mary had recently sold a complete plesiosaur to the British Museum, and she wanted to see it. She had never set foot inside the museum that showcased her unattributed labors, or inside any museum at all, for that matter. She had never *been* anywhere at all. She also hoped to tour the museum and the Geological Society, whose first president was her old friend William Buckland, who grew up hunting fossils at Lyme.

The *Unity* cruised up the coast to the Thames River, landing at London Bridge, where Mary disembarked into a city of nearly two million people, the world's largest metropolis. She made her way toward Regent's Park, near where the Murchisons lived. No record of the visit exists, not even of her stop at the Geological Society, but the geologist William Lonsdale was known to have given Mary a private tour of the place that had gladly accepted her discoveries but not her. A friend of Mary's, Anna Maria Pinney, once noted in her journal that "men of learning have sucked her brains, and made a great deal of publishing works, of which she furnished the contents, while she derived none of the advantages."

Anyone important in the emerging sciences of geology and paleontology eventually found their way to Anning's Fossil Depot. The groundbreaking geologist Charles Lyell contacted Mary to ask how the cliffs were holding up. Adam Sedgwick, who taught at Cambridge and whose students included Charles Darwin, became a client and friend. The renowned zoologist Louis Agassiz became the first to name species after her, memorializing Mary—finally—with *Acrodus anningiae* and *Belenostomus anningiae*. "Her history shows what humble people may do, if they have just purpose and courage enough, towards promoting the cause of science," Charles Dickens later wrote in *All the Year Round*. He added, "The carpenter's daughter has won a name for herself, and deserved to win it."

Still, more often than not, it was the gentlemen scientists who got the thanks, along with the dukes and the lords. Of all the important fossil creatures found at Lyme, not a single one carried Mary Anning's name.

When Mary was thirty and her brother thirty-three, Joseph married Amelia Reader and moved to St. Michael's Street, near the Annings' first cottage on Cockmoile Square. Mary and her mother remained on Broad Street. Mary went on making discoveries and selling them to the British Museum. She was so good at finding fossils and so bad at being paid well for them that she got a London agent, George Brettingham Sowerby I, a King Street conchologist and an artist of natural history. But as the economy failed, she once again ran out of money. Her friend Henry De La Beche devised another a fund-raising idea: an amateur artist, he painted a scenic watercolor of prehistoric Dorset based largely on Mary's discoveries, calling the picture *Duria Antiquior*, "A More Ancient Dorset." Creatures swam and flew and walked on land. *Ichthyosaurus vulgaris* bloodily snapped the neck of *Plesiosaurus dolichodeirus*, which, in its death throes, expelled coprolites. *Ichthyosaurus tenuirostris* gobbled up *Dapedium politum*. *Pterodactylus macronyx* wheeled through a pasty sky. Overall the piece depicted a crowded little aquatic scene of Jurassic predation, most of which was known to science—known at all—because of Mary.

Duria Antiquior was the first attempt at a scientific scene depicting deep time. De La Beche commissioned hand-colored lithographic prints and sold them to friends and scientists, giving Mary and her mother the proceeds. Teachers later enlarged the piece and used it in classrooms, allowing Mary's influence to endure as an inspiration for early scientific illustration.

In October 1833, a cliff came sloughing down, barely missing Mary but killing her beloved terrier, Tray. Five years later, more bad luck: she invested her life's savings, about 200 pounds, with a "gentleman" who died suddenly, leaving her broke. Mary, now forty, found herself "reduced to straitened circumstances" once again, the *Dorset County Chronicle* reported, "while her health was impaired from the hardships which she had exposed herself, and the distress of mind consequent on her loss."

This time it was William Buckland who stepped in to help, persuading the British Association for the Advancement of Science, and the British government, to award Mary what was known as a civil list pension, for her "many contributions to the science of geology." She would receive an annual income of 25 pounds, providing for her and her mother. Molly Anning lived to see her daughter receive the honor, but died shortly thereafter, leaving Mary alone for the first time in her life.

But Mary pressed on. When King Frederick Augustus II of Saxony visited Lyme and bought a six-foot-long *Ichthyosaurus* at Anning's Fossil Depot, Mary signed her name in his physician's book, informing him, "I am well known throughout the whole of Europe."

On March 9, 1847, just before her forty-eighth birthday, Mary died of breast cancer after suffering with the disease for a year. Her brother had died some time earlier, and she was buried near him in the St. Michael's Parish churchyard on a high hill overlooking the sea. Henry De La Beche, soon to be knighted, sent a stained-glass window from London to the vicar of Lyme, to be installed at the church in Mary's honor. At a Royal Society gathering in London, De La Beche read aloud a eulogy—the first to honor a non-member or a woman—hailing Mary Anning as someone "who though not placed among even the easier classes of society but one who had to earn her daily bread by her labour, yet contributed her talents and untiring researches in no small degree to our knowledge."

On Cockmoile Square, a redbrick museum eventually went up on the spot where the Annings first lived. What's now known as the Lyme Regis Museum stood three stories tall, overlooking Lyme Bay. For years the museum housed a cramped memorial to Mary Anning at the top of a winding staircase. The room consisted of an aisle flanked by glass cabinets filled with fossils. Inside a case at the back of the room hung photographs of the vocational heirs of Mary Anning, the latest generation of Jurassic Coast fossil hunters. Among them was Eric Prokopi's new business partner, Chris Moore.

Black Ven, the name of the towering cliff at Charmouth, sounds like a Marvel Comics character and consists of the thready, dark marls that Mary Anning called "beef." English law allowed hunters to collect whatever they wanted as long as they notified the authorities of potentially important finds, in case the crown wanted to buy them. Government officials in jackets inscribed with FOSSIL WARDEN patrolled the beach, mostly to answer tourists' questions and keep them safe, for twice in recorded history massive sections of the cliff have collapsed into the sea. Black Ven—and the rest of the Jurassic Coast, now a UNESCO

World Heritage Site—is one of the least stable stretches of geology on the planet.

Chris Moore found Charmouth in 1968, as a boy on holiday with his family from the northern town of Manchester. Wherever Moore's family vacationed, he hunted fossils—an uncle, a teacher, had inspired his interest—but the Jurassic Coast was the place to be. After moving to Charmouth in the late 1970s, he earned a living as a boat builder and hunted fossils in his off hours, prepping them in his garden shed. In 1991, Moore sold part of his collection and established Forge Fossils, a commercial hunting and prep business headquartered in an old potter's space on Charmouth's main village road, The Street. The company stationery featured the image of an ammonite, a fossil nearly as ubiquitous to the Jurassic Coast as shark teeth were to Venice, Florida. Sir David Attenborough, Great Britain's most beloved naturalist, once called Forge Fossils a "delightful shop."

Moore was a tallish, weathered Englishman with dark hair and a graying beard; other dealers at Tucson and Denver knew him as an affable vendor who traded in excellent fossils. His best finds included a 3-D ichthyosaur and a huge slab of ammonites now in the permanent collection of Tokyo's National Museum of Nature and Science. He discovered two new species of ichthyosaur, one of which was named for him, *Leptonectes moorei*. But Moore's favorite finds were in storage, including another ichthyosaur species, plus, according to one news report, a "complete fossil shark, an eight-food slab of crinoids, and an enormous ichthyosaur head." While his portrait graced the Mary Anning wall in Lyme, Moore's real goal was to open a museum of Jurassic Coast fossils. Remarkably, one does not exist.

Around September 2009, not long after Eric's second trip to Mongolia, Moore and his son, Alex, were featured in "Paleontology in Charmouth," a *Shoreline* newsletter package marking the bicentennial of Charles Darwin's birth. Lately, Moore had been working on a "stunning" group of symmetrical crinoids in limestone, telling a reporter, "I always wanted to put fossils in a different environment, to display them as natural sculptures, which, to my mind, they are." Soon, he would have a workshop full of Gobi Desert dinosaurs.

Eric and Moore barely knew each other, but they had friends in common. Vendors often partnered up to do deals, so it wasn't unusual for them to connect. Eric was happy to work with Moore because he had been in the business forever and had a good reputation, and because their gambit seemed easy enough.

When Eric learned that the Mongolian delivery had cleared customs in the United Kingdom, he flew to London and took a train south to Dorset. Moore picked him up at Axminster station and drove him to his home, Askew House, a shabbily grand Victorian at a bend in the road, behind a high box hedge and gate, just up the hill from Forge Fossils. To accommodate the new batch of Gobi fossils, Moore had rented a workshop outside of Charmouth, at Befferlands Farm, a cluster of one-story buildings with terra-cotta roof tiles, on a leafy, narrow lane named Berne. The largest of the structures was used for boat building. In the next building were the offices of a tour guide and author named Nigel Clarke, who led Lyme Regis fossil walks and self-published pocket-size tide reports and paperback guides such as *Mary Anning 1799–1847: A Life on the Rocks*. Moore worked in the same building, in a room filled with unprepped ammonites and, now, Mongolian fossils and *gers*.

Eric and Moore decided to split the bone inventory by value. They assigned each specimen a dollar amount: twenty-five grand for a *Protoceratops* skeleton and so forth. Moore made two columns on notebook paper and they flipped a coin to see who would choose first. Moore won. They started choosing:

Eric		Chris*	
BIG TARB	25	PROT	25
TRAB SKULL	15	ANK	24
PLACY	8	DUCK	8
ANK	22	HAD	6
TURT	4		
TRAB	5		

"TURT" was a slab of turtles. "HAD" was a largely complete hadrosaur. "DUCK" wasn't necessarily a duckbill. "It was this complete little dinosaur that had this duckbill on it," Eric later said. "I think it ended up being something that we didn't know what it was." They went round after round—OVIR, EGG NEST, LEGS. As he made the list, Moore often transposed letters—TRAB for TARB. Afterward, they swapped some items around.

One skeleton they decided to share: that of a large and nearly complete *Tarbosaurus bataar.*

By late February 2010, Eric either owned or was expecting more valuable fossils than he had ever possessed or even seen at one time outside a natural history museum or trade show. The garage at Serenola overflowed with inventory, Pleistocene to Cretaceous, Florida to Mongolia, with more en route from Tuvshin by way of the Jurassic Coast of England. It wasn't unusual for Amanda to come home and find an *Oviraptor* leg mounted on the kitchen counter or a giant ground sloth towering inside the front door.

The Serenola renovations were not paid for, but they were done. The front room was outfitted with a red leather sofa, skins from Tucson, and the "grown-up rugs" Eric and Amanda had bought in Richmond after the sale of the fossil they now thought of as the Nicolas Cage skull. Built-in shelves with tall glass doors flanked Tom Petty's grandmother's salvaged mantel. The dining room table was made of vintage champagne barrels. A tall antique case of glass and oak that once displayed department store wedding dresses now stood near the front door, filled with fossils and casts. Amanda had painted the rooms in muted greens, yellows, and grays, with names like "Mineral Deposit" and "Mountain Smoke." The *Gainesville Sun* published a large feature on Serenola's transformation, the writer mentioning Eric's fossil-collecting travels to Mongolia and other countries. "A home is more than just filling a place with furniture," Amanda liked to say. "Your home should tell a story."

The renovation of Serenola had cost more than Eric and Amanda had expected. When the fossil shipment arrived from Charmouth in late March, the Prokopis were well over $400,000 in debt, with dinosaurs as their prospective way out. The customs forms listed the import's "manufacturer" as Forge Fossils, the contents as "fossils," the value as $15,000, and the country of origin as England. An itemized commercial invoice read—

2 large rough (unprepared) fossil reptile heads
6 boxes of broken fossil bones

3 rough (unprepared) fossil reptiles
1 fossil reptile skull

Eric still owed Tuvshin part of the $250,000 overall payment for the dinosaurs, but he went on ordering.

———————

Hollis Butts, the dealer in Japan, resurfaced, suddenly suggesting that he and Eric partner up. Now that Eric was buying directly from Tuvshin, Butts suspected Tuvshin of "playing both sides" to generate a bidding war. Eric humored Butts by email, telling him, "I was happy buying from you. You just wouldn't sell to me anymore, that's why I started going direct."

Butts, who liked beginning his emails with "Greetings," replied, "Greetings." He wished to explain what had happened. One winter, he had sent Tuvshin $30,000, then $20,000, then another twenty, because Tuvshin said he needed money "badly." Eric's decision to travel to Mongolia in the summer of 2009 had presented a problem for Tuvshin: if Tuvshin sent Butts a container, he wouldn't have anything left over for Eric. Butts figured that Tuvshin had given Eric *his* container and compensated him with a couple of skulls. "In that way, he kept my money and a lot of yours," Butts told Eric.

The whole enterprise felt shaky, did it not? Had Eric never wondered why Tuvshin was always out of money, even when he'd been paid thousands? "He does not like the fossil business. It is very risky," Butts told him. "He is involved in various other businesses and finances it with fossil money." In the early days Tuvshin would simply send shipments; now he was "pushing for the customer to visit and buy," Butts went on. "This is so that if there is trouble, he can say that he knows the buyer but is not really involved in the business. 'The foreigner came and bought fossils from people.'" Butts added, "It is a safety net that worries me."

Butts proposed that Eric tell Tuvshin he no longer needed fossils because his clients had backed out. "This should give him a shock and keep prices low." He proposed a scheme that called for everything to go through Japan, "giving nobody else a chance." Butts said, "For Tuvshin, the more buyers the merrier but that works against us both and I think we should try to get everything."

One problem with Tuvshin, though: he was unpredictable. Eric was supposed to have seen him recently at Tucson, but Tuvshin never showed, never answered his phone; Tuvshin later emailed Eric a photo of blistered bare feet, saying he'd gotten too drunk on the flight from Mongolia and had wandered around barefoot in L.A. before he could figure out what to do. Often, he was unreachable for days or weeks. Butts said, "My fear is always that he has been arrested."

Tuvshin had gone dark this time probably because Butts hadn't sent money that Tuvshin expected. "I often feel like I am his bank," Butts told Eric. He'd considered "sending off a few emails that would bring his whole business to a sudden end," he added. "But that wouldn't help me." Butts had four guesses about Tuvshin's most recent disappearance: "Summer is a busy time when specimens are turning up so he is out gathering things OR he actually is in trouble OR he has an order from another buyer and is going to screw us both OR he is having trouble arranging shipping." Butts's other concern was Eric's new partnership with "the Brits." "Any chance they are now buying directly from Tuvshin, now that you have shown them how?" he asked.

Eric tried to keep the conversation neutral. He changed the subject to a *Microraptor* he had ordered through dealers he knew from Tucson. Customs had intercepted the fossil, and while Eric had briefly fought the confiscation, he'd ultimately surrendered. "US customs has been holding a lot of fossil shipments lately, and are demanding species lists and proof of origin," he told Butts. "I don't know what is going on but it is troubling."

Occasionally some of these cases made the news. On Christmas Eve 2007, a customs officer in Chicago had X-rayed express-mail shipments of "shoes" and "gifts" from China and found a fossil saber-tooth cat and the skulls of Chinese dinosaurs. In Virginia, a traveler tried to bring twenty-four dinosaur eggs through Dulles International Airport on a flight from Japan without proper papers. In Los Angeles, federal agents confiscated what had been advertised as a Chinese *Oviraptor* egg nest with visible embryos after it sold for over $420,000 at a Bonhams & Butterfields auction. The nest was a forgery: the seller had purchased the eggs separately, implanted them in a slab of imported sandstone, and advertised them as having been lain by a single creature. After the dealer declined to contest the seizure, the eggs were repatriated. Challenged by law enforcement, the dealer had walked away. Mostly, though, illicit fossils got through. At one point an ICE official told the media, "Fossils are a niche within a niche when it comes to customs enforcement."

Butts told Eric, "One thing is clear: You need to avoid Chinese and Argentine fossils. I wouldn't touch either."

Distrusting Butts, Eric had lied about Tuvshin—they had been in contact all along. "Do you have more photos?" he emailed Tuvshin one day in late September 2010. "Do you need more money?"

"We might find two of the things you want," Tuvshin replied. The diggers were working in the Gobi as he spoke. "We have many Ovi, Proto, and Gali," Tuvshin told Eric. "We need money so much."

The Prokopis now owned a long, goose-necked cargo trailer, a Ford F550 truck, and a flat-bottomed boat, even though Eric rarely hunted in rivers anymore. Amanda drove a Lexus SUV with a leather interior. They took out another line of credit via a wraparound mortgage while also owing some $60,000 on Bank of America credit cards, nearly $20,000 on a Capital One card, over $15,000 in federal income tax, and more than $533,000 on real estate. Their overall monthly minimum payments exceeded $7,000, with finance charges piling up. Their bank account held $55 one minute, $100,000 the next. Then in February, Amanda had to be temporarily hospitalized with what they worried was meningitis, and more charges accumulated.

By April, Eric had a long list from Tuvshin of the dinosaur parts he could soon expect:

1st box: big Hadrosaur head
2nd box: big Hadrosaur neck, big Hadrosaur half tail
3: 2 ovirators [sic], 1 red protoceratops
4: big Hadro left feet
5: big Hadro right feet…

There were seventeen boxes in all. Seeing that he needed a larger work space, Eric ordered a prefabricated warehouse and installed it beyond the swimming pool. The building came with a fifty-year guarantee— Eric would be eighty-seven when the warranty ran out. He sent another $90,000 to Tuvshin for this latest batch of dinosaurs, and was told the fossils were being prepared for shipment.

Weeks later, on June 24, Eric opened a new email from Tuvshin.

Only it wasn't from Tuvshin.

CHAPTER 15

THE LAST DINOSAUR

"I FEEL VERY SORRY TO CONVEY YOU THIS MESSAGE THAT MY husband Mr. Tuvshinjargal Maam has passed away."

Eric reread the email.

Tuvshin was *dead*?

The email was from Bobo, who said Tuvshin had died the previous night, of "sudden lung failure," after a monthlong illness. The email provided no further information but read, "I am very thankful for your friendship, hard-work [*sic*], and support in the past. I am looking forward to contact you. I pray to god that, 'May his soul rest in peace.'" Eric immediately forwarded the email to Chris Moore in England, writing, "Very Bad news!"

The latest $90,000 that Eric had paid Tuvshin had cleared, but the new batch of bones hadn't shipped. After a couple of weeks passed without word from Mongolia, Eric emailed Bobo, saying, "I hope you are doing ok with the circumstances. I am eager to get our business sorted out."

But the only thing to do was to go to Mongolia in person. Eric and Moore booked flights, Eric letting Bobo know he would arrive on the night of July 22. Hoping to meet the next morning, he told her, "Chris Moore will be coming also. He is my partner in this business and has sent money also." As Eric understood it, five crates were already packed and waiting in a warehouse, and another order was outstanding. Mentioning customs, he said, "Maybe you can use the same person as before."

Three days later, Eric was back in Ulaanbaatar, sharing a room with Moore at the Guide House Hotel, where the presidential suite rented for $95 a night. At Tuvshin's apartment they expressed their condolences to the family. Eric found the whole situation awkward; he didn't know Tuvshin well enough to grieve him, and he didn't know Tuvshin's family well enough to gauge the depth of everyone's suffering. He found it odd that no one wanted to talk about the illness and death, but attributed it to Mongolian custom. What was "sudden lung failure," anyway? A heart attack? Lung cancer? Later, when Eric told his friend Tony about it, they made a macabre joke that, given Tuvshin's various interests, sudden lung failure might mean a pillow to the face. The suddenness of the death seemed odd considering Tuvshin's age—he had just turned forty-three. Eric couldn't make out whether he'd been hospitalized or buried or given a funeral or what. He asked no questions, though. He wanted to pay his respects, get the dinosaurs he'd paid for, and go home.

After Eric and Moore were assured of their shipment, the family took everyone sightseeing. Moore had never been to Mongolia, and before long he was being driven out to the big Chinggis statue in the countryside and learning falconry. Otgo, the chief prepper at the national paleontology lab, joined them at a park for a picnic, as did a customs agent. Within five days, Eric had come and gone from Ulaanbaatar for the third, and last, time.

None of this felt very interesting or exciting anymore, and not just because Tuvshin was dead. The stress had been getting to him for some time. Amanda had noticed that lately, Eric had been drinking too much and acting silly at parties—flirting with her girlfriends, throwing people into the pool. It wasn't like him. Life had become too little about what had drawn Eric to fossils in the first place. He imagined prepping out the rest of the Tuvshin inventory and being done with Mongolia. Maybe he would find some ranch land out West and hunt dinosaurs there. Maybe he would go back to rivers.

For years, lawmakers' attempts to regulate fossil collecting had failed, one after another, but in January 2009 a legislative package called the Omnibus Public Land Management Act had passed the Senate. An omnibus bill is sort of a catchall for a range of legislation. This package consisted of one hundred and fifty-nine bills touching on everything from improvements in rehab treatments for paralyzed Americans to extra lab space at the Smithsonian. Most of it, though, involved nature—the monitoring of ocean acidification, the reimbursement of Native American ranchers whose livestock got eaten by wolves. The Act sought to advance conservation in nearly every state by protecting thousands of miles of rivers and trails; designating millions of federal acres as wilderness; establishing a new national monument in southern New Mexico, an important area for Paleozoic footprints; and providing for better care of national lands that already existed, such as Idaho's Owyhee Canyons, Utah's Zion National Park, and Virginia's "wilderness-quality" forests.

The omnibus package included the Paleontological Resources Preservation Act, which provided for the first uniform protection of fossils on public lands. Anyone convicted of selling fossils found on federal property could be imprisoned for up to five years, and after two or more convictions the penalties could be doubled. Moreover, everyone would need a permit from the Department of the Interior in order to collect vertebrate fossils and other materials deemed "scientifically important" on public property; permits would be issued only to accredited scientists and affiliates of a museum or school, and never to anyone collecting for commercial purposes. Casual collecting of "surface" fossils without a permit was deemed okay in some instances and areas. Some thought the legislation brought necessary and overdue toughness, and others felt it went too far. Commercial hunters had never been allowed to collect vertebrates on federal property, yet bristled at seeing the restrictions so strongly codified. Tracie Bennitt, president of AAPS, the commercial trade group, had written an open letter to Congress on behalf of her colleagues, arguing against the legislation. Noting that the Act called for the management and protection of paleontological resources using "scientific principles and expertise," she had pointed out that some hunters already abided by such practices, writing, "Amateurs are the foot soldiers of paleontology and their activities are to be encouraged." The bill neglected to address the idea that commercial collecting on public lands could benefit paleontology and society if only everyone would work together, she had complained: "Wouldn't this be a better alternative than fossils disappearing from the world forever?"

One aspect of the new law had angered both sides of the debate between paleontologists and commercial hunters. Without the interior secretary's written permission, scientists working on federal property weren't allowed to publish the locations of their finds. The measure was intended to thwart poachers but ran counter to the idea of open science. Paleontologists' research had long included site specifics, down to latitude and longitude; banning the release of that information was seen as tantamount to hiding data.

It had appeared impossible to please everyone. Policymakers had struggled to find a place for reputable commercial hunters without empowering a side-flood of venal yahoos who only wanted to rip at the earth like gold diggers. *That* was the gap that no one had figured out—and that some never wanted to figure out—even as President Barack Obama had signed the PRPA into law in the spring of 2009.

For now, Eric intended to concentrate on the fossils already in his possession, including the *T. bataar* that he shared with Moore. It wasn't the largest skeleton in stock, but it was one of the most visually promising. A *T. rex* requires twenty thousand hours of prep, it's been said, but by the time Eric started working on the *bataar*, it was late 2011 and Tucson was fast approaching. To help with the workload, he and Amanda hired a couple of assistants, including Tyler Guynn, a chipper anthropology student and former roofing-company executive in her thirties, with dark eyes, an olive complexion, and short, chestnut hair lightly streaked with gold. She and her husband had a teenage daughter, Hannah. A military kid, Guynn had grown up partly in Germany but her accent and colloquialisms—"dang it" and "darlin'" and "kiss my grits" and "bless her bones"—were all Florida by way of Georgia. Her interest in anthropology had led to her job with the Prokopis. Eric had set her up at a table in the workshop and showed her how to prep. They worked eight, ten hours a day together, often well into the night, with the bays open and the industrial fans whirring. Tyler had been getting tattoos since she was a teenager—an all-seeing ankle eye, lower-back butterflies, a compass, a peacock feather, a phoenix. As the days passed, she chose a bare spot on her left pelvis and added the large skull of *Tarbosaurus bataar*.

The bones were crumbly, and as the matrix fell away Eric strengthened them with PaleoBOND, a liquid hardener. If a piece was missing, he dug around in his inventory and filled in with random fossils from other specimens. Otherwise he sculpted the replacement with epoxy, or bought cast *Gorgosaurus* parts from the Black Hills Institute in South Dakota. Once all the bones were ready, he placed them in the correct anatomical shape on the workshop floor, like a puzzle, the dinosaur curled loosely at his feet like a sleeping dog. Eric's friend Tony climbed a ladder and shot a photo of Eric kneeling at the snout, hands on hips, smiling up at the camera. "Here is our Tyrannosaurus skeleton. Almost finished!" Amanda posted on Facebook. A friend asked, "If the crab went for 12,500 whats the starting price of a tyrannosaurus????"

Eric didn't have time to mount the skeleton, so he displayed parts of it on a table at Tucson. Other dealers came around to see it, thinking Eric Prokopi was either the bravest or most reckless son of a bitch they had ever seen. The trade in Mongolian dinosaurs so far had been "up to and including skulls," Mike Triebold, a well-respected commercial hunter from Colorado, later said. But this was "a big, whole *T. rex*–y skeleton. This is a big, sexy dinosaur."

The market appeared primed for big, sexy dinosaurs, to paleontologists' continuing dismay. "Luxury Market for Dinosaur Remains Thrives," the *Huffington Post* announced, declaring dinosaurs the "newest hot art objects." Yet despite Eric's hopes of selling the *T. bataar* to a foreign museum, the skeleton failed to sell at Tucson at all.

Natural history now went to auction every spring and fall in the United States. Houses had embraced the category as a promising new revenue stream. Eric, who had been selling items through I.M. Chait and other auction companies, decided to mount the *T. bataar* and consign it to one of the upcoming events of spring 2012.

A giant ground sloth was the largest and most complicated skeleton he had ever assembled, and the only major dinosaur piece he had worked on start to finish was the tarbosaur skull he had installed in Los Angeles at the house he believed belonged to Leonardo DiCaprio. There was no blueprint he could download, no YouTube channel that offered joint-by-bone instructions on assembling a Cretaceous dinosaur. Doing his own research, he downloaded scientific sketches along with photos of Tyrannosaurus Stan, a Harding County specimen found by Stan Sacrison, and illustrations of AMNH 973, the *T. rex* holotype, which the American Museum of Natural History had sold to the Carnegie Museum in Pittsburgh in the 1940s. "Tarbofootpathological," he slugged an image of tyrannosaur feet. "SUEskeletonSide," he slugged an image of Tyrannosaurus Sue.

A thousand things could go wrong on a mount this large. A pelvis alone can weigh over 3,000 pounds. Simply moving a fossil can fracture it. Skulls have been dropped and shattered. Bolts can produce stress fractures. Metallurgical errors can corrupt the integrity of the welds. If the engineering is off, a skeleton can fall with enough mass and force to crush whatever—or whomever—happens to be in its way.

Eric bought the best steel rebar he could find and fired up his welder. For the base he made three steel floor plates, each about 16 inches square, and a fourth plate a little smaller. In the center of each of the three big plates, he added an upright attachment that resembled a large dart sitting on its flight. A steel rod ran straight up from each of the darts, connecting to a frame. Eric shaped each piece of the armature to suggest, overall, a dinosaur in motion, then spray-painted the entire structure a sandy color, to blend in with the Gobi bones.

The skull is the most important piece of any skeleton. The skull helps paleontologists identify a species and provides important clues about traits such as brain size and eating habits. Aesthetically, the skull is the centerpiece. Assembled, this one measured 4 feet long and weighed so much that Eric had to hoist it with a forklift. As he edged the head into place he realized he had miscalculated angle relative to weight—once the hoist fell away, the dinosaur would tip. He made adjustments and reinforcements and tried again. As he withdrew the ropes, the skull held. The ribs held. The vertebrae held. The femurs and tibias held. Snout to tail, all 24 feet of the skeleton held. It has been said that it is most natural to pose *T. rex* hands as if the creature were "holding a basketball—or someone's head—between them," and while Eric had never heard that idea, that's what he'd done. Jaws open, legs in motion, the tarbosaur appeared to be chasing prey. It stood facing the open bays of the workshop as if frozen in the act of dashing out into the Florida sunshine and down the driveway of Serenola.

Eric offered the *T. bataar* to Tom Lindgren at Bonhams. Lindgren had sold Mongolian fossils, and had been the one to connect Eric with Tuvshin, but he and the German dealer Andreas Guhr had gone to court in

California over a co-owned *T. bataar* skull, eventually agreeing to sell it to "Buyer X" through I.M. Chait for $330,000, and now Lindgren wanted nothing more to do with Gobi fossils. Ever since the death of Tuvshin, the very mention of Mongolian dinosaurs made him go quiet and white. So Eric offered the skeleton to David Herskowitz.

Herskowitz now worked for Dallas-based Heritage Auctions, which was poised to sell $800 million worth of collectibles that year. Heritage hosted five hundred auctions annually and claimed 750,000 online bidder-members in nearly two hundred countries. Celebrity clients included Whoopi Goldberg, NASA astronauts, and the estates of Buddy Holly, Ava Gardner, and Malcolm Forbes. Recently, and controversially, Herskowitz had sold a "fighting pair" of Wyoming dinosaurs named Fantasia (a *Stegosaurus*) and Dracula (an *Allosaurus*). Herskowitz's old mentor Henry Galiano had found the skeletons in the Dana Quarry, near the town of Ten Sleep, in the shadow of the Bighorn Mountains. The dinosaurs had been advertised as having died mid-combat, not unlike the famous fighting pair of Mongolia. They had sold for $2.7 million to an undisclosed museum in an undisclosed country in the Middle East and had never been seen since.

The Wyoming dinosaurs had put Herskowitz at the head of a short list of natural history brokers who specialized in major dinosaurs and sales. He wanted to build out this category for Heritage, and he believed the New York City auction he was planning for May 20, with the *T. bataar* as the star attraction, would prove the power of natural history to stand alone. As he promoted the sale in the media, Herskowitz called the skeleton an "impeccably preserved specimen of the sort that is almost never seen on the open market."

Now, though, as the auction proceeded, Mongolia was asking questions, eager to know how an entire Gobi Desert dinosaur had wound up at a New York City sale. Eric paced the beach at St. Augustine, staying in close contact with Herskowitz, Chris Moore, and Heritage's cofounder and CEO, Jim Halperin, about whether to go forward now that the president of Mongolia was getting involved.

By phone and email Heritage continually deferred to the sellers— there was still time for Eric and Moore to back out. Eric liked none of his options. If he pulled the dinosaur from the auction, he stood to lose both the money he had paid Tuvshin and much-needed income. If he proceeded, the situation with Mongolia might get worse. To him, it seemed unlikely that President Elbegdorj had true legal reach, but he decided not to risk it. He told Halperin, "After discussing it with Chris, we have decided to pull the tyrannosaurus from the auction."

Then he spent the next two hours doubting himself. As the auction got under way, he emailed Heritage and said he'd changed his mind—a decision that would become the line of demarcation between a life relatively free of serious drama and his new, increasingly complicated one.

———————

Bolor Minjin, the Mongolian paleontologist who had alerted her native government to the *T. bataar* auction, had returned home every summer while working on her doctorate in New York. At first, she participated in joint AMNH-Mongolian expeditions, delighted to conduct fieldwork in her own country, on Gobi fossils. Early on, she met a young paleontologist named Jonathan Geisler, a South Carolinian pursuing a doctorate at Columbia University, and they got married. Geisler worked on cetaceans—whales, porpoises, dolphins—and Bolor on Paleogene multituberculates, early mammals no larger than a rat. Bolor impressed her American and Mongolian colleagues as a skilled collector and passionate paleontology advocate as she sought to prove herself as a scientist.

In May 2000, as she pursued her PhD, a major exhibition went up at the American Museum, called "Fighting Dinosaurs: New Discoveries from Mongolia." The exhibition featured the stunning "fighting pair" found by Zofia Kielan-Jaworowska, one of Bolor's heroes. The crowds were so large that the museum had to admit people by timed entry. "More original, brand new fossils per square inch than I've ever seen before," one dinosaur enthusiast told members of a vertebrate paleontology email list.

Rinchen Barsbold, the elder statesman of Mongolian paleontology, flew to New York for the occasion. Bolor was thrilled to be invited to a luncheon honoring him, and even more thrilled to find herself seated next to him. Here was a Gobi theropod expert who had advanced the understanding of dinosaur evolution in Eurasia; who had early on ascribed to the idea that birds descended from dinosaurs; and who had described a host of dinosaurs, including *Gallimimus* and the family Oviraptoridae. Bolor anticipated a meaningful conversation, hoping the older scientist would offer career advice and support. She had chosen mammals only because her father had waved her off dinosaurs, just as he had chosen invertebrates because he had been waved off vertebrates, as if there wasn't room in the world for more than one Mongolian expert on Gobi Desert dinosaurs. Bolor wasn't surprised to hear Barsbold discourage her interest in dinosaurs, but it shocked her to hear him discourage her interest in paleontology by calling it a field with "no future." She wondered if scientists of Barsbold's generation felt embittered by delayed recognition and a lack of public interest—Mongolia had a National Geologists Day, but no National Paleontologists Day.

Undeterred, Bolor went on to participate in Gobi projects with her husband and her father and, at one point, Jack Horner, a burly native Montanan who had dropped out of college because of severe dyslexia but whose dinosaur research had earned him a MacArthur "genius" grant. Few names in paleontology were more widely known than Horner's. He was credited with finding the first dinosaur eggs of North America, a discovery that happened because a Montana hunter and rock-shop owner named Marion Brandvold had shown him some tiny bones she had been keeping in a coffee can. Horner had published books and served as a consultant on Steven Spielberg's *Jurassic Park* movies, and by the time Bolor met him, he was teaching in the honors college at Montana State University and curating the school's Museum of the Rockies. Pursuing projects on developmental biology, Horner sought as many specimens of a single dinosaur species as he could get, and the Gobi appeared to be the place to find them.

Bolor joined Horner in the field in 2005, and again the next year, when the team reported finding sixty-seven dinosaurs in one week. As Horner saw it, Mongolian paleontology had two key problems. One, the country needed to develop its next generation of paleontologists and facilities. "Dinosaurs are the one thing to see when you come to Mongolia"—the Natural History Museum should be the best show in town, he told *Discover* magazine. Two, Gobi dinosaur poaching had reached a crisis point. "Smugglers have gotten so bold that they've taken specimens right off the dig sites while paleontologists break for lunch," he said, adding, "There's no control here, and no regulation for digging here."

In 2008, the year she finished her PhD, Bolor accepted a postdoc position with Horner and founded an NGO called the Institute for the Study of Mongolian Dinosaurs, hoping to address some of the concerns she and her new mentor shared. She wanted to build a state-of-the-art dinosaur museum in Mongolia, use her Western education to help the next generation of Mongolian paleontologists, and fight poaching. First, though, she and a Museum of the Rockies colleague developed a Museum in a Box outreach program for Mongolian schoolchildren, wherein Bolor traveled the Gobi with "discovery trunks" teaching the importance of fossils.

Usually, when Mongolian paleontologists finished their education, they went to work for the Mongolian Academy of Sciences, but Bolor didn't want to do that. The institution seemed to care little about public outreach and education. The second time she met Barsbold, she told him about her NGO hoping they could collaborate, but the very idea seemed to agitate him. She would remember him saying, "You have to take 'dinosaur' out of your name," and "You just doing this for business, to make a benefit for yourself."

Bolor believed her achievements sufficient to have earned the respect of her Mongolian and American colleagues, but her résumé, which included grants from Harvard, Berkeley's Museum of Paleontology, and the Society of Vertebrate Paleontology, appeared to impress them little, if at all. Because she worked on mammals, they saw her as unqualified to call herself a dinosaur expert, and they disliked that she tried to circumvent the Mongolian Academy of Sciences to do whatever she wanted. Bolor knew she couldn't look to AMNH scientists for backup because, though they had trained her, they couldn't align themselves with her for fear of alienating the Academy, which granted permissions to work in the Gobi. Bolor had encountered Mongolians who told her they had heard negative gossip about her from other paleontologists, stuff like, "She married an American—she is not even Mongolian anymore."

Bolor had spent most of her adult life in the United States and had married an American, but in fact she wasn't a U.S. citizen. To become naturalized, she'd have had to renounce her home country—Mongolia, ever mindful of its location between Russia and China, doesn't allow dual citizenship. Bolor had never planned to stay in the United States in the first place. She loved certain parts of America—the national parks, Utah, Wyoming—but they weren't home. The in-betweenness was the expatriate's affliction. "If we leave, people think we've abandoned our country, but it's the opposite," she once said. "We're *more* patriotic—because we're trying to make change."

Bolor had started her NGO not to disavow the Mongolian paleontological community but rather to improve it by making her own contributions within a system where she otherwise saw no opportunity for herself. Horner urged her to find a way to work with her Mongolian peers, but Bolor couldn't get past the idea that the paleontological center operated "like a service company" for foreign scientists: "Expedition comes. Provide the cars, go out with them. If they find something, stick their name on the paper." It irritated her that Mongolian paleontologists seemed to expect little of the foreign scientists and institutions that had made their names largely on Gobi fossils. "The whole country's heritage is being handled by just a handful of people," she said. "I am trying to break that thing."

Also, she wanted the bones back. It bothered her that many Mongolian fossils, including type specimens, often weren't returned by the outside paleontologists who had borrowed them for study. Thousands of fossils were still in collections in the United States, Russia, Japan, and elsewhere. Whenever she argued aloud that the bones should come home, including those collected by Roy Chapman Andrews in the 1920s, some of her peers worried that she sounded nationalistic. Even those with no stake in the debate gently questioned her reasoning. Given Mongolia's poor museum conditions and entrenched concerns about government corruption, wouldn't the fossils be safer and more accessible in a place like the American Museum of Natural History?

Bolor decided that if she couldn't bring back the bones, she could at least try to protect the fossils that were still in the field, and that the best way to do it was through a public figure. She wrote to various politicians, with no luck. Then, in the spring of 2011, as the market was flooded with Mongolian fossils, she heard that Oyungerel Tsedevdamba, a popular author and activist, would be in Chicago giving a speech.

Oyuna had attended university in Russia and graduate school at Stanford, on a Fulbright scholarship, and she had spent time at Yale on a fellowship. Now in her late forties, she spoke to women's rights audiences at home and abroad. Youthful and slim, she wore tasteful suits, small earrings, a strand of small pearls, and dark red or plum lipstick that complemented her sleek black hair. One of her editors once described her as a woman who carried "her well-earned success lightly, and with much charm." An aide once watched Oyuna amicably settle a bureaucratic dispute and said, "*Now* I understand what politicians are supposed to do." Bolor showed up at her Chicago event to say, "Sister Oyuna, please mobilize politicians to do something about Mongolian fossils."

Oyuna's areas of expertise involved water quality, law enforcement, and electing women to public office. Like most Mongolians, she had never given one thought to dinosaurs. Bolor pressed her: Mongolia sacrifices too much to outside paleontologists and to poachers. "We are losing our heritage without even knowing what we have."

Bolor struck Oyuna as a passionate science advocate who wanted to raise awareness but lacked the standing to do it. "Give me something to read," Oyuna told her. "Give me something to learn." Bolor assembled a packet of news articles and books, and gave Oyuna one year to read them and to publish a "nice article" on the importance of Mongolian fossils.

Oyuna's already full to-do list suddenly included reading *Dragon Hunter*, the Gallenkamp biography of Andrews; *The Fossil Dealers*, by Australian paleontologist John Long, a former president of the Society of Vertebrate Paleontology; *Tyrannosaurus Sue*, by Steve Fiffer, about the Black Hills case; and Mark Jaffe's *The Gilded Dinosaur*, about the late-1800s bone wars between Marsh and Cope. Reading about Andrews, Oyuna became fascinated. Reading *Dealers*, she became alarmed. In news articles she noticed that critics often blamed Mongolia for not taking action on "our stolen materials."

A year later it came time to write her "nice article." Oyuna knew that as a non-scientist, she couldn't write like a paleontologist. Nor was she comfortable writing as a politician, "preaching others what to do." So she decided to write her newspaper article from the point of view of the dinosaurs in Mongolia's natural history museum.

She chose as her lead narrator *Tarbosaurus bataar*, the fiercest creature ever to inhabit Mongolia, other than Genghis Khan. "Dinosaur's Dream" began: "For some time, I've been talking to the dinosaurs of the Mongolian Gobi in my heart." The dinosaur was bothered that the Natural History Museum's plaques were written in a way that "only a paleontology student can understand." The dinosaur eggs dreamed of not being poached: "Please, dear Mongolians, stop stealing eggs and selling us to foreigners." Mongolia's "fighting pair" spoke up to say, "We need a law that prohibits people from taking us out of Mongolia"—if foreign scientists were banned from borrowing fossils, they would be forced to conduct their research in Mongolia, bringing Mongolia acclaim. It wasn't fair that "American, Canadian, and Chinese scientists are doing research on our dinosaurs and becoming stars in famous science magazines."

Oyuna wrote her article for *Ardchilal*, a newspaper founded in the early 1990s by her boss, President Elbegdorj. It was published on May 17, 2012, to coincide, Oyuna later said, with International Museum Day. At one point the *T. bataar* character said, "The truth is, I'm not a star yet, but I do know that I have all the potential to be a star."

That potential was already on display in New York City: Oyuna's article appeared as the Heritage auction opened for previews.

"Dinosaurs!" Oyuna said when Bolor beseeched her to do something about the *T. bataar* sale. She was through with dinosaurs!

But Oyuna took the matter to President Elbegdorj. "You're talking to me about *dinosaurs?*" he asked when she called him that morning. "I didn't appoint you science adviser."

"This dinosaur is getting sold and somebody needs to complain," she told him. "Could you please just give me half an hour?"

Oyuna was soon in the president's office, spreading out the dinosaur books and articles Bolor had assigned her as homework. "Boss, I'm not a paleontologist, but I read all of this stuff," she told Elbegdorj. She then showed him images from the Heritage catalog and said, "This dinosaur is Mongolian. Please claim it for Mongolia."

But what can we do? Elbegdorj asked. When Oyuna explained temporary restraining orders, Elbegdorj asked, "What if we lose?" Oyuna told him, "If we lose, you're still going to be the first president who tried to claim Mongolian property. If we win, you will be the first president to ever bring Mongolian property back. Nobody has ever claimed Mongolian properties from international auctions or sales before."

An "urgent appeal" then posted to Elbegdorj's presidential website: "President of Mongolia Is Concerned That T-Rex Skeleton May Belong to Mongolia." The species was inaccurate and the release credited Mongolia's education minister, not Oyuna, with alerting Elbegdorj, but the point was made. Mongolia wanted to know who was selling the dinosaur and how the seller or sellers had acquired it. News outlets were picking up the story. David Herskowitz told Live Science the dinosaur had entered the United States legally and that its consigners warranted in writing that they "held the title to the fossils." He said, "No one knows where exactly it was dug up. They'd have to find the hole and match up the matrix," which any paleontologist could tell you is impossible.

Heritage's history defense, meanwhile, was that laws didn't apply because the dinosaur's existence predated them, a stance that was deemed absurd. "That's like saying the Saudis aren't entitled to their oil," said Oyuna's husband.

Once President Elbegdorj agreed to pursue a legal remedy, Oyuna worked on hiring a lawyer. In Houston, Texas, Robert Painter was on the verge of leaving for a work trip to Singapore when he stopped to check his messages one last time.

CHAPTER 16

THE PRESIDENT'S PREDICAMENT

PEOPLE HEARD ABOUT PRESIDENT ELBEGDORJ'S NEW LAWYER and thought, *Whoa, what? How does a Texas attorney in a small family firm wind up representing the leader of Mongolia?* To understand it, you had to go back to even before the first days of the democratic revolution.

The eighth son of a herder, Tsakhia Elbegdorj came from Khovd, a province of lakes and valleys in the far southwest, on the Chinese border, at the foot of the Altai Mountains. Born in 1963, he "proudly" wore the distinctive red kerchief of a Communist Young Pioneer as a child. When he was sixteen, his family moved way up north, to Erdenet, a city four hours south of the Russian border and 250 miles northwest of Ulaanbaatar. Erdenet had recently been built to tap the world's fourth-largest copper deposit. After high school, Elbegdorj went to work in the mine as a machinist, joining a labor force of some eight thousand people, many of whom commuted from Russia. A year later, in 1982, he was drafted into the military, where his leadership of a Revolutionary Youth Army unit and his fondness for submitting poetry to the army newspaper led the government to suggest that he study military journalism in Lviv, Ukraine.

In Lviv, Elbegdorj listened closely to Gorbachev's speeches about change, and later exulted at the televised sight of President Ronald Reagan saying, "Mr. Gorbachev, tear down this wall!" After college, Elbegdorj moved to Ulaanbaatar to work for an army newspaper, *Ulaan Od*, "Red Star." By 1989, he was thinking about starting Mongolia's first independent newspaper and advocating for democratic reforms alongside other young revolutionaries, calling perestroika a "timely and brave step." Mongolia had quietly become the first Eastern bloc satellite to break away.

Elbegdorj was a compact, soft-spoken young man with wavy hair and a full face and, later, eyeglasses. He married his college sweetheart, Bolormaa Khajidsuren, who at one point studied at business school in Roanoke, Virginia. They would have four sons together and adopt a daughter from an orphanage, as Elbegdorj was poised to start *Ardchilal* and to claim a seat in parliament as a member of the nation's newly formed Democratic Party.

A total of six Americans lived in Ulaanbaatar as Mongolia prepared for its first multiparty elections in late July 1990. Ambassador Joseph Lake and his wife became numbers seven and eight, discovering "this incredible clicking between Mongolians and Americans." Mongolians were the nicest people Lake said he had ever worked with, and the country was one of the few Soviet satellites that had "truly progressed" under Communism. "My perspective was that if the Mongolian people voted for a democratic process we could influence the process of change in Mongolia," he said. The United States suddenly had the chance to influence the course of events at a time when Mongolia's leaders were looking for help, a deputy secretary of state, Desaix Anderson, had told the House Foreign Affairs Committee several years earlier, as the United States and Mongolia established diplomatic ties. A push for free elections and economic freedoms in Mongolia was a sandbag against the spread of authoritarianism.

Independent observers declared the first multiparty election fair and democratic. The Democrats grabbed a handful of seats in parliament and slowly gathered strength. The prospect of a positive, violence-free transition to democracy appeared so promising, Secretary of State James Baker visited Mongolia with a tantalizing offer: the United States would serve as Mongolia's "third neighbor," helping the country broaden its contacts and influences beyond Russia and China. When Mongolia drafted a new constitution in 1992, Americans were among the outsiders invited to participate. "Mongolians seem remarkably free of false pride," one participant told *The New Yorker*. "Did you ever hear of a country asking foreigners to help draw up its constitution?"

Among the consultants were twenty-three Christians from Sioux Falls, South Dakota. They already happened to be in Ulaanbaatar, having "adopted" Mongolia at the urging of Dr. Bill Bright, founder of Campus Crusade for Christ. As part of Bright's mission to "teach Christianity to everyone on the planet by the year 2000," the South Dakotans had traveled to UB to show the film *Jesus*. Then, as now, the capital was a small town kind of city—people mysteriously seem to know your business almost before you do. It didn't take long for men in uniform to show up at the South Dakotans' hotel and ask what they were doing in Mongolia. Jesus sent us, said one of the South Dakotans. When government leaders heard this, they supposedly said, in a line almost certainly too good to be true, "Well, do you think this Jesus could help us write our constitution?" And so it was that one of the South Dakotans helped write the religious freedoms passage. Other sections were "lifted verbatim from the election law manual of Texas" and the U.S. Constitution. The Mongolian constitution's preamble began, "We, the people..."

The South Dakotans met a parliament member named B. Batbayar, who, during college in Poland, had lived with a missionary from Milwaukee. Batbayar thought Mongolia needed an independent TV news station, to counter state programming, and offered the South Dakotans the opportunity to start it. "His motive was to bring free and objective news coverage to the nation, believing that without a free press their young democracy would collapse," one of the South Dakotans later explained. As Batbayar secured a broadcast license for channel 8, Eagle TV, the South Dakotans founded the AMONG Foundation, a nonprofit based in Sioux Falls, to oversee the project. AMONG would provide a newsroom, studio, equipment, and staff, introducing CNN and other American programming like *The Waltons* and *Touched by an Angel*. The foundation's stated mission was to "further a better understanding of the democratic form of government, promote and foster the free market system and the values upon which the democratic society is built;" but, to be clear, AMONG was also there to proselytize. One employee later said the foundation helped write a constitution "guaranteeing religious freedom in the most Buddhist nation on earth," as if the ancient faith were some intractable scourge.

It was hard to know precisely who else was behind Eagle TV or the rest of the media, for that matter. Then, as now, Mongolian laws made it almost impossible to know the names of the owners and players, or to understand their intentions. Many outlets were said to be owned by politicians or political actors who used the media as a secret extension of their public agenda. Bribes—"squeeze," as Roy Chapman Andrews put it in the 1920s—were "epidemic across the media spectrum," one Eagle TV station manager, an American, later wrote. "Bribery was almost cultural." It was known that AMONG owned half of Eagle TV, while the other half was owned by an entity identified only as the Mongolian Media Corporation; together they comprised the Mongolian Broadcast Company. The individual Mongolian owners were generally identified only as "Democrats," but some press reports mentioned founders by name, including Elbegdorj. Eagle TV went on the air on September 28, 1995, less than a year ahead of the pivotal parliamentary elections of the summer of 1996, where the Democrats hoped to unseat the Communist Party for the first time.

Now that Mongolia was a burgeoning democracy and had declared itself a nuclear weapons–free zone, President George H. W. Bush authorized financial aid, which zoomed from zero to $30 million. Secretary of State Baker had already returned to Mongolia to build on his third-neighbor offer: a concept initially seen as "a rhetorical gesture" was now codified into foreign policy and law. Baker, a former Houston lawyer who served as the Reagan White House's chief of staff, worked with the International Republican Institute, and now the IRI began working in Mongolia. In the days leading up to the crucial elections of June 1996, they used an unlikely strategy, deploying Congressman Newt Gingrich's "Contract with America" as a model. Two years earlier, the Georgia Republican's national campaign blueprint for a "conservative revolution" had triggered a GOP sweep in the House of Representatives, resulting in the first majority since 1952. The IRI now offered to teach Mongolian party leaders how to forge coalitions, draft position papers, gauge popular opinion, and mount campaigns. The Democrats, the only party to accept the American Republicans' consulting offer, adopted a "Contract with the Mongolian Voter." Drafted in part by Elbegdorj, the platform called for cutting taxes and social services, and privatizing most of state property. A herder with fifty cows and sheep told the *Washington Post*, "I read the Contract with the Voter closely. Everybody did."

The Republican-trained Democrats dominated the 1996 elections, knocking the Communist holdovers out of power for the first time in nearly a century. Elbegdorj, who was thirty-five, was reelected to parliament and named prime minister. Within two years he blundered politically by trying to sell a state-owned bank to a large private bank owned by Democrats. "The merger, illegal and corrupt, made several Democrats and their friends extremely rich," wrote Michael Kohn, an American who edited a state newspaper, the *Mongol Messenger*. After the MPRP staged a walkout of parliament in protest, Elbegdorj was forced to resign. He later said, "Okay, the bank merger was a bad idea."

Two years after that, as Mongolia flailed economically, a particularly devastating *dzud* killed 2 million livestock over the winter, and another 300,000 animals died weekly during the spring, exhausted from fighting starvation and the catastrophic cold. Depressed herders were committing suicide. Elbegdorj soon decamped to the United States. He stopped first at the University of Colorado's Economic Institute, in Boulder, for English-language immersion, then attended the John F. Kennedy School of Government at Harvard, in 2002, as an Edward S. Mason Fellow, a mid-career program for leaders of developing nations and transitional democracies. He enrolled in courses with star professors like the leadership guru Ronald Heifetz and Jeffrey Sachs, whom the *New York Times* called "probably the most important economist in the world." Sachs had spent time in Russia a decade earlier during the breakup of the Soviet Union, where everything was a "mess of every kind." There he had encountered what others were experiencing across the border in Mongolia: "One could go for literally miles...with utterly one hundred percent empty shops. And if you wanted something you had to go to a black market. You had to make secret deals."

If the West wanted to groom someone for the Mongolian presidency, Elbegdorj was an ideal candidate. National security experts observed that he expressed positive views of the United States; he favored English to replace Russian as the primary foreign language in public schools; the traits that he revered in Genghis Khan he praised about America, such as a society built upon rule of law. In the States, he tended a growing network of American contacts, including Heritage Foundation leaders, Senator Bob Dole, and the Nobel-winning economist Milton Friedman, who argued that nations and individuals survive only via free markets. Friedman's critics argued that the economist and his powerful followers had perfected the strategy of "waiting for a major crisis, then selling off pieces of the state to private players while citizens were still reeling from the shock, then quickly making the 'reforms' permanent," Naomi Klein wrote in *The Shock Doctrine: The Rise of Disaster Capitalism*. Friedmanesque policies were ideological smokescreens for multinational corporations to acquire wealth, she wrote; what inevitably emerged was a "powerful ruling alliance between a few very large corporations and a class of mostly wealthy politicians—with hazy and ever-shifting lines between the two groups." Some of the fiercest fans of Friedman's policies were the "Communists turned capitalists" in China and the Soviet Union, whose collapse left nothing standing in the way. Klein had published her book as a challenge to the idea that "unfettered free markets go hand in hand with democracy." Elbegdorj admired Friedman's doctrine so much that his NGO, the Liberty Institute, translated his work into Mongolian.

In the spring of 2003, toward the end of his time in the United States, Elbegdorj attended a leadership conference at Michigan's Mackinac Center for Public Policy, once described as a "conservative think tank school." Then he traveled to New Orleans for a Heritage Foundation conference, where he met a young lawyer named Robert Painter.

Painter had lived in Texas for years, but he came from Beckley, West Virginia, from a family with a political pedigree but little involvement in current politics. As a young boy in a state more blue than red at the time, Robert somehow became such a fan of Ronald Reagan that he taped the news coverage of Reagan's assassination attempt. At West Virginia University, he became active in College Republicans. He volunteered for the Bush-Quayle campaign and wrote a manual on how to politically organize teenage Republicans. When the handbook went national, he and his cohort became known to high-profile conservatives like Rush Limbaugh, who was busy "sweeping college campuses around America."

Painter, who was six foot three, had the resonant voice of a radio announcer, the gregariousness of a networker, and the neatly combed hair of a church deacon. At first he planned to become a doctor, but in the late 1990s, he enrolled at Baylor Law. Baylor, a private university founded in the mid-1800s, was the world's largest Baptist campus. *The Princeton Review* had named it a "Best Western College" and, recently, the most bigoted school in America. Boosters bristled at the characterization, though the *Baylor Lariat*, the student newspaper, agreed with the assessment, writing in an editorial, "There is seemingly a good number of Baylor students who just aren't willing to accept those whose beliefs or skin color are different from their own."

Serving on Bob Dole's 1996 presidential campaign, Painter had made a lot of contacts, developing unlikely acquaintanceships with both former president Reagan and Margaret Thatcher, then the prime minister of the United Kingdom. While in law school he started a nonprofit called Liberty Forum, backed by a Virginia-based conservative youth group called the Leadership Institute, an organization that had helped him and a friend found a conservative newspaper at Marshall University, where Painter had briefly attended medical school and his friend had hosted a Rush Limbaugh fan club in the cafeteria. Painter named the Texas publication *Libertas*. Once a month, he and his friends distributed the tabloid on campus. Articles referred to Martin Luther King Jr. as a "Communist-sympathizer and plagiarist" and called out faculty who supported liberal political candidates. Headlines included "AIDS Quilt: is it laced with more than just compassion?" wherein the author characterized the disease as the "medically documented consequences to sinful behavior." Baylor's student newspaper wasted no time responding to the publication's arrival: "*Libertas* looks like a newspaper and reads like a newspaper, but that doesn't make it a newspaper," read one *Lariat* editorial. A Baylor senior, Alyson Ward, published an unequivocal column on *Libertas*'s abuse of journalistic standards, acknowledging that there was "something admirable about being controversial and hard-hitting" but noting that *Libertas* had utterly failed in that regard by publishing rumors and name-calling to a degree "reminiscent of McCarthyism." She wrote, "It's not that they're veering too far to the right but that they're moving in some unclear direction under the guidance of people with suspicious motives and methods."

In late January 1999, Painter and a handful of friends headed to Washington, DC, for the Conservative Political Action Conference, or CPAC, the world's largest annual gathering of conservatives. The Leadership

Institute paid their way. The students planned to meet with Republican figures like Dan Quayle and Oliver North. Taunya McLarty, a friend of Painter's who worked for Senator John Ashcroft, arranged for the group to observe a day of President Bill Clinton's impeachment trial. Less than two weeks after the Washington trip, *Libertas* announced plans to broaden its distribution base to include "conservative donors across the nation." Twenty-five thousand copies of the latest issue were mailed to "high-dollar conservatives." Painter told the school newspaper, "Many of the conservative donors are very interested in what's going on at college campuses."

Liberty Forum dissolved later that year, after Painter graduated from law school. He joined Fulbright & Jaworski, a pedigreed megafirm in Houston, working in the area of medical malpractice. A few years later he married McLarty, a fellow lawyer who had joined the legal staff at Walmart. Around 2005, as the Painters started having children, they opened their own practice, Painter Law Firm, in a two-story brick office building near their upscale Houston neighborhood, Champion Forest.

By that point Painter had founded a nonprofit called Bellwether Forum, for young professionals who wanted to learn about government and politics. Directors, members, and advisers included or would include senators and judges; an appointee in George W. Bush's executive office for Homeland Security; and Jay Weimer, who would become an assistant U.S. attorney in the Northern District of Texas. Painter planned a speaker series echoing a similar project he'd started while at Baylor, whose guests had included the Watergate burglar Bernard Barker. Bellwether acknowledged the challenge of "selling politics to the average person," but noted that people "secretly yearn to be in the inner-circle of influence in government." Members attended events with titles like Cocktail Party Capitalist and You Can Change Your Local Government; they could join the "The Washington Experience," a DC field trip to CPAC, where they might meet privately with insiders like Supreme Court justice Antonin Scalia and Texas congressman Tom DeLay.

In April 2003, Bellwether members traveled to New Orleans for Heritage Foundation Resource Bank, an annual gathering of conservatives who wanted "to share the lessons they've learned in the battles for freedom." John Ashcroft, by now the U.S. attorney general, was keynoting, and the Painters wanted to say hi. As Robert waited to speak to Elaine Chao, then the labor secretary, he noticed an Asian man wearing a name tag that read "Former PM, Mongolia." Intrigued, Painter approached him and said, "What's *your* story?" Instantly, he and Elbegdorj—"E. B."— were friends. Two months later, the former prime minister addressed Bellwether members in Houston and spoke at a continuing education event for attorneys, telling his audience of two hundred, "You know, even though Mongolia is seven thousand miles away from here, my country is becoming much closer to this part of the world than ever."

Whenever anyone asked Elbegdorj about Mongolia, he explained his homeland in three parts. One, Mongolia was a great empire in the twelfth, thirteenth, and fourteenth centuries. Two, Mongolia was the second country in the world, after Russia, to become Communist, a system that he believed created criminals by prohibiting free enterprise and private ownership of property and businesses. Third, Mongolia was in transition as the only Asian country pursuing economic and political freedoms and reforms. People assumed that such reforms "could not be achieved in Asia," Elbegdorj told his audience. "Mongolia actually broke that stereotype, showing that if these reforms and freedom could be introduced in Mongolia, they could work anywhere in the world." He added, "Making this kind of transition, considering Mongolia's location between Russia and China, is a huge and challenging task."

By autumn, the former prime minister was home in Ulaanbaatar, sporting a seat on Bellwether Forum's advisory board and a Harvard class ring, which his wife let him wear on his wedding finger.

Elbegdorj became prime minister again in June 2004, this time in a power-sharing agreement with former president Nambaryn Enkhbayar, a member of the MPRP, or the old Communist Party. Right away Elbegdorj reached out to the United States, announcing his new position in a letter to Senate majority whip Mitch McConnell, and saying the peaceful transition represented "Mongolians' continuing commitment to democracy and human rights." When Eagle TV, the station he had reportedly helped found, returned to the air on October 22, 2005, after a hiatus, the network spent the first day, all day, on live coverage of the inaugural visit to Mongolia by a U.S. secretary of defense, Donald Rumsfeld.

Soon, George W. Bush became the first sitting U.S. president to visit the country that was being called a "poster child for democracy in Eurasia" and a "desirable security interest." In *Far Eastern Economic Review*, John Tkacik, a retired foreign service officer, wrote that "the American stake in Mongolia is not insignificant," noting, "Although rarely recognized, Mongolia is of critical geopolitical importance.... Its 1.5 million square kilometers of real estate is a stabilizing element in Eurasia that keeps border frictions between its two giant neighbors, Russia and China, from reaching a critical mass of conflict." While Mongolians had "never dared think of themselves as anything but real estate over which Russians and Chinese had fought for centuries" they were now "very self-conscious about their importance in the global scheme of things," and it thrilled some to see a U.S. defense secretary and a president visit back to back. "Americans and Mongolians have much in common," Bush told an audience of eight hundred on November 21, 2005. "Both our nations were settled by pioneers on horseback who tamed the rugged plains. Both our nations shook the yoke of colonial rule and built successful free societies. And both our nations know that our responsibilities in freedom's cause do not end at our borders and that survival of liberty in our own lands increasingly depends on the success of liberty in other lands." Thanking his host country for backing the United States in Iraq and Afghanistan after the terrorist attacks of 9/11, Bush pledged $11 million to help improve Mongolia's military.

The next several years were tumultuous. Elbegdorj was in and out of power as the young democracy came to life, stumbled, and then rose again. When he and MPRP leaders accused each other of malfeasance, well-crafted press releases and news items went out, calling the old Communists "the godfathers of corruption" and Elbegdorj "Mongolia's Thomas Jefferson." (Elbegdorj's speechwriters included Bellwether Forum board member and public relations executive Robert Bauer, who also wrote speeches for George W. Bush and Afghanistan's former president, Hamid Karzai.) At one point, Elbegdorj warned that the disintegrating government "immediately created a dangerous situation in our country." In a phone interview with the *Harvard Crimson*, he said, "It's kind of a coup d'état."

To regroup, in May 2007, Elbegdorj invited Painter and a Bellwether team to Mongolia to train the Democratic Party in American-style politics. Instructors included a National Rifle Association executive and a fund-raiser for the American Legislative Exchange Council, or ALEC, a lesser-known organization of right-wing activists that a *New York Times* editorial once described as a generator of "voter ID laws that marginalize minorities and the elderly, anti-union bills that hurt the middle class, and the dismantling of protective environmental regulations." ALEC had been instrumental in the passage of the "Stand Your Ground" laws that gave armed civilians more legal leeway to shoot to kill. Its major funders included foundations with ties to Koch Industries' Charles and David Koch, brothers who contributed heavily to conservative and libertarian think tanks, campaigns, and causes. While Mongolia's opposition was called the Democratic Party, by U.S. standards its politics were decidedly right-wing.

By the time Elbegdorj ran for president in May 2009, he had dropped visa requirements for American tourists and pushed to abolish the death penalty. He had announced plans to create a tourism ministry and privatized the nation's yaks. He had removed the Soviet-era mausoleums from Sukhbaatar Square and brought a massive bronze of Genghis Khan to the front steps of Government House. He had been the target of a suspected assassination attempt and accused of inciting deadly riots after the parliamentary elections of 2008 by questioning the validity of the vote.

Mongolia faced an ever-widening wealth gap, the privatization of the past two decades having laid a foundation for a "system whereby a small elite controls disproportionate resources and a large population of poor are without basic services," as one Asia analyst put it in a report for the Brookings Institution. President Enkhbayar signed for a whopping $285 million in relief from the Millennium Challenge Corporation, which President George W. Bush created to "reinvent" foreign aid, but the aid package didn't help him hold on to the presidency. When Elbegdorj defeated Enkhbayar by 41,770 votes in 2009, everyone worried that the close count would trigger more violence. But nothing much happened. To everyone's surprise, Enkhbayar conceded. On May 24, 2009, Elbegdorj became the first Mongolian president educated in the West, and the first never to have been a member of the Communist Party. In his presidential suite, he hung a picture of Margaret Thatcher and one of Ronald Reagan.

To those invested in seeing democracy develop a stronger foothold in East Asia, it was essential that Elbegdorj stay in power for two terms—eight years—and that the Democratic Party take control of parliament in the 2012 election. The MPRP had been in charge for much of the past six decades, and the next four-year parliament would, among other things, establish mining policies and decide how to distribute the "tens of billions" of dollars expected to "flood" the country in a bid for natural resources. If Mongolia behaved fairly and transparently, it could parlay its staggering mineral wealth into "a stable, middle-class society like Qatar," analysts were saying, while continued corruption would trigger something similar to Nigeria, "where an oil boom in the 1970s led to environmental degradation and conflict, the country's wealth frittered away by corrupt officials, its average citizens mired in poverty," the *Guardian* reported.

The government had finally worked out a deal to develop Oyu Tolgoi, an enormous copper and gold mine in the southern Gobi with an estimated life span of over a hundred years. A Canadian company, Ivanhoe Mines, had landed a $6 billion deal to develop the project with its parent, the Anglo-Australian mining giant Rio Tinto, after much negotiation with the Mongolians. The ultimate agreement called for Ivanhoe to receive 66 percent of Oyu Tolgoi and Mongolia 34 percent, but some parliament members had decided the ratio was unfair and that Mongolians deserved better. The mining companies, which had already invested over $2 billion, expected the government to honor the contract, and Mongolia's sudden reneging made international investors nervous.

Elbegdorj went on pitching his country as a stable, pro-America democracy "deserving of more attention." For the past twenty years, the United States had been sending Mongolia among the largest, if not *the* largest, aid packages per capita in the world, but Elbegdorj pushed for more. He briefly considered making the nation a player in the global uranium fuel business, then "abruptly pulled the plug," Marin Katusa wrote in *The Colder War: How the Global Energy Trade Slipped from America's Grasp.* What Elbegdorj appeared to want was a trade deal with the United States. He was sipping from an American flag mug in June 2011 when he told the *Washington Post*, "Maybe if we caused problems—if we hid bin Laden or atom bombs—America would pay more attention."

By now, Robert Painter, the Houston lawyer, represented various Western business interests in Mongolia. He set up another company, Mongolia Research Group LLC, then changed the name to Bellwether Mongolia, echoing that of his nonprofit. The company's website consisted of a static single page that offered the address and phone number of the Painter Law Firm, and read only, "Bellwether Mongolia invests in mining and infrastructure projects in Mongolia." The United States was also lobbying for mining rights, pushing for Peabody Energy, a private company based in St. Louis, to win the right to develop Tavan Tolgoi, a southern Gobi deposit that held enough coking coal to fuel China's insatiable industry for the next half century. To prevent the Chinese from buying a majority interest in Tavan Tolgoi, the government hurriedly passed the Strategic Foreign Investment Law, what the *New York Times* called "the latest step in the complicated political choreography aimed at appeasing a rising nationalist fervor while encouraging foreign investment." In short, mining was a mess. But many citizens, young urbanites in particular, saw Elbegdorj as the politician who could navigate the boom in natural resources. To his supporters he represented stability and moral courage in the face of rampant corruption because he had launched investigations into the national airline and the customs agency, among other entities. Elbegdorj's critics, meanwhile, argued that he had failed; some still blamed him for the election riots of 2008, in which at least five people were killed.

On the day Elbegdorj captured the presidency, Painter was there to see it. He watched the victory speech as a VIP and afterward joined the president-elect for lunch at Bellagio, a Vietnamese restaurant at the Bayangol Hotel. At one point Elbegdorj turned to him and asked, "What's next?"

"For you? For me?" Painter said.

"For us," Elbegdorj said.

Voting machines, they decided.

"Mr. President, if you go forward with this it'll cut out most of the fraud and we'll see the Democratic Party sweep the [next] elections," Painter said he told Elbegdorj. "You'll already have the presidency. You'll have a working plurality of parliament. You'll even win mayor of UB."

Over the next several years, Painter brokered a deal for Mongolia to buy voting machines. The contract ultimately went to Dominion Voting Systems, a private company based in Toronto. Dominion's premiere system featured a hybrid of optical scanner and paper, providing a hard-copy backup of the digital tally. Dominion's founder and CEO, John Poulos, said publicly that it would be extremely difficult to rig the company's machines, but there were skeptics of electronic voting in general. Machines were known to misplace or erase votes, attribute votes to the wrong party, and tally "yes" for "no." It was happening across the United States alone. "Our entire governing system is based on the sanctity of the vote," noted Bev Harris, a Seattle writer who had become known for documenting voting-machine snafus. The vote, she said, "is the underpinning for our authorization of every law, every government expenditure, every tax, every elected person. But if we don't *trust* the voting system, we will never accept that those votes represent our voice."

Mongolia ordered what today would be $10 million worth of Dominion machines. Painter later said he was paid for his assistance by a U.S.-based nonprofit, which he declined to name. He never registered under the U.S. Foreign Agents Registration Act (FARA), which requires people acting on behalf of foreign principals in a "political or quasi-political capacity" to disclose the relationship along with activities, receipts, and disbursements. He said he thought he didn't have to, though FARA experts disagree. Despite his friendship with the president, and the president's position on Bellwether Forum's advisory board, and the president's intentions to run for reelection, Painter had no ethical concerns about helping Mongolia buy voting machines. "I thought it would be good for the country—I even thought it would be good for democracy—to have a new system," he said. Painter made nearly a dozen trips to Mongolia to help work out the details as everyone hurried to get the new voting system in place before the crucial parliamentary elections of June 2012.

Elbegdorj's most vocal critic was his predecessor, Enkhbayar, who planned to run for parliament and possibly again for president. So some thought it curious when Enkhbayar was unexpectedly charged with corruption in the early spring of 2012, accused of wrongly privatizing a hotel, misappropriating TV equipment intended for a Buddhist monastery, and illegally shipping eight copies of his autobiography to South Korea. Calling the charges bogus, he refused to acknowledge a court summons. On the night of April 12, 2012, officers from the Independent Agency Against

Corruption (IAAC), which Elbegdorj had recently created, tried to arrest Enkhbayar as he arrived at his apartment. After he fled inside, riot police converged on the building. At dawn, officers forced their way inside and hauled the former president out of his home, shoeless, with a black bag over his head, live on national television.

Enkhbayar's allies were quick to point to Elbegdorj as the culprit of a "naked attempt" to "vanquish a political foe." On the day of the raid, "Mr. Enkhbayar had released internal government documents finding Mr. Elbegdorj responsible for inciting deadly violence" by alleging voter fraud after the last parliamentary elections, *The Economist* reported. The *Guardian* noted that some of the complaints against him seemed "rather picayune." The *New York Times* wrote, "Mongolia...has been widely lauded by American officials as a bastion of democracy in a region where the rule of law and due process are rare. That reputation is now being called into question." The arrest appeared to be a "dangerous regression to practices honed during the Soviet era." Enkhbayar's lawyer told the press, "Everyone had hoped Mongolia had broken away from the Soviet chain. But this is Soviet stuff." Some noted that the arrest had been carried out by an intelligence officer named Bat Khurts. Elbegdorj had recently appointed Khurts deputy director of the new anticorruption agency, despite Khurts having once been wanted in Germany on charges related to the kidnapping, torture, and forced extradition of a Mongolian suspect in the murder of Zorig, the famous democratic revolution leader, who had been stabbed to death by masked assailants as he rose in national politics.

When Enkhbayar began a hunger strike and his health deteriorated rapidly, human rights watchers in the United States, Australia, and Britain lobbied for fair treatment, a worrying development that suggested the "demise of democracy in Mongolia." After appeals from UN Secretary-General Ban Ki-moon and others, Enkhbayar was released on bail and hospitalized. Western reporters found him fragile and gaunt after ten days without food or water. The courts, historically among the most corrupt of Mongolia's institutions, set his trial for June 4, effectively blocking him from running for office. "Their task was to make me a criminal before parliamentary elections next month and the presidential election in 2013," Enkhbayar told the *New York Times,* saying, "Mining is the reason they're so cruel and antidemocratic in trying to prosecute me. Copper and gold have made people crazy." Mongolia was now a "pawn in a global game involving China, the United States and Russia."

Enkhbayar's opponents alleged that it was he who wanted to make his own grab for Mongolia's mineral fortune. One of Elbegdorj's defenders was his aide Oyuna, Bolor Minjin's ally in the fight to preserve Mongolian dinosaurs. "This is really a case of [Enkhbayar] finally being brought to justice after years of the people being too afraid to file complaints," she told the *New York Times*, calling the manner of Enkhbayar's arrest a positive sign: "When they came to arrest him, nobody was beaten, and the press could broadcast openly. Those are marks of democracy in action."

By mid-May, the matter was attracting criticism from high-profile leaders of the country Elbegdorj valued most. U.S. Senator Dianne Feinstein, a Democrat from California, issued a statement about Enkhbayar's health and the importance of due process, saying, "It has been deeply troubling to follow the way in which he has been treated." Elbegdorj, in turn, put out a press release saying Enkhbayar had been treated fairly and humanely, and reminding everyone that Secretary of State Hillary Clinton had recently commended Mongolia on its commitment to human rights.

All of this happened just before the *T. bataar* went to auction—Enkhbayar's scandalous arrest, U.S. admonishments about human rights, the entrenched corruption, the mining controversy, the rising inflation, the pressure for Elbegdorj's party to both win the parliamentary elections in June and retain the presidency in 2013, and the growing concern among some Western diplomats that Mongolia appeared to be losing its tenuous hold on democracy. Elbegdorj needed a victory.

Oyuna went to him with an opportunity that in the moment barely made sense: *dinosaur...auction...New York*. She told President Elbegdorj, "Win *T. bataar* for Mongolia and *you* will be the hero."

Like most lawyers, Robert Painter had never handled a case involving a dinosaur. But he approached the matter as he had learned to do at Baylor: "Assess the big picture, formulate a strategy, prepare as much as humanly possible, then swing with the punches." A temporary restraining order seemed like a solid first move. As evidence, he used the letters to Heritage from Mark Norell and Bolor Minjin; the auction listing; an English translation of Mongolian laws on natural history; and a May 16 article from a U.K. newspaper that flat out said the skeleton had been "found in Mongolia," was "owned by a fossil collector in Dorset," and had been shipped to Florida for preparation and sale. Painter then scrambled to find a judge who would sign the order, forbidding the *T. bataar* sale until the question of ownership could be resolved. Carlos Cortez, a state district judge in Dallas, agreed to do it. Painter flew to Dallas and got the signature Saturday morning, the day before the auction, then notified Heritage and jumped on a plane to New York.

On Sunday, he made his way into the auction venue, past Bolor Minjin and a few other Mongolians protesting the sale with a banner reading RETURN OUR STOLEN TREASURE. Surely Heritage would stop him entering, Painter thought, but no one approached him. As he waited for the auction to begin, he browsed the merchandise, at one point bidding on and winning a wristwatch made of meteorite.

After a German sea-lily fossil sold for $40,000, Heritage's president, Greg Rohan, who had been standing near the lectern murmuring into his cell phone, handed the auctioneer a note. The auctioneer scanned it and announced, "The sale of this next lot will be contingent upon a satisfactory resolution of a court proceeding." Largely intact dinosaur skeletons of any species are not easily found, and this one had been advertised as 75 percent complete, the auctioneer told the audience, saying, "It can fit in all rooms ten feet high, so it's also a great decorative piece."

Painter couldn't believe Heritage would defy a court order and go through with the sale. As the bidding opened at $875,000, he dialed Judge Cortez in Dallas, then held his BlackBerry aloft and announced to the room, "I hate to interrupt this, but I have the judge on the phone."

The auctioneer eyed Painter but never broke patter. He called for $900,000.

As Painter started up the aisle toward the auctioneer, Rohan met him halfway, and for five seconds they squared off in a silent little dance, four arms waving, Painter trying to stop the sale and Rohan trying to stop Painter.

"Come with me, just come with me," Rohan said.

A security guard stepped in and told Painter, "Sir? Sir, you need to walk—"

"I have the judge on the phone," Painter told her.

"Okay, well, you need to walk outside," the guard said, pointing him toward the back of the room.

Heritage's attorney materialized and Painter handed him his Black-Berry. As the attorney apologized to the Texas judge, the dinosaur skeleton sold.

CHAPTER 17

UNITED STATES OF AMERICA V. ONE TYRANNOSAURUS BATAAR SKELETON

As FAR BACK AS MEDIEVAL TIMES, LAW ENFORCEMENT authorities seized property believed to be a tool or the fruit of wrongdoing. America's approach dates to English maritime law of the seventeenth century, when "the English Crown issued 'writs of assistance' that permitted customs officials to enter homes or vessels and seize whatever they deemed contraband," Sarah Stillman reported in *The New Yorker*. The writs, she wrote, "were 'among the key grievances that triggered the American Revolution.'" The Bill of Rights protected citizens against "unreasonable searches and seizures" and promised that they wouldn't be deprived of life, liberty, or property without due process, fair treatment in the judicial system. But "Congress soon authorized the use of civil-forfeiture actions against pirates and smugglers," Stillman wrote. "It was easier to prosecute a vessel and seize its cargo than try to prosecute its owner, who might be an ocean away."

After a somewhat dormant period, the practice returned as actions known as *in rem* cases, Latin for "against a thing." Because the process required prosecutors to pursue an object instead of a person, the cases tended to have curious names, like *United States of America v. Approximately 64,695 Pounds of Shark Fins*. Asset forfeiture laws gave police agencies the power to take a person's property without filing criminal charges, and to keep it indefinitely on the hunch that the seized goods were connected to a crime. Almost anything could be confiscated: cash, cars, tractors, homes, kegs of Coca-Cola. In the 1970s, the federal government began using the law to prosecute criminal organizations, such as drug rings, expanding asset forfeiture's reach. The pace really quickened after 1984, when Congress created the Assets Forfeiture Fund under the Justice Department; the government could sell seized property and keep the proceeds if there were no identifiable victims of the crime in which the property had allegedly been involved. The fund grew rapidly as the government confiscated items believed to have been procured with tainted money, and cooperating states received portions of the proceeds. As many states amended their laws to give local police seizure powers similar to those of the federal government, critics complained that innocent people, who were often too poor or too daunted by their immigration status to fight back, were being systematically and unjustly relieved of their belongings. Congress attempted reform in 2000, but state statutes continued to proliferate and revenues to climb. The Justice Department's forfeiture-related revenues spiked from $27 million in 1985 to $556 million in 1993. By the summer of 2012, as the *T. bataar* skeleton went to auction, the DOJ was claiming almost $4.2 billion a year in forfeitures, much of which reportedly had been returned to victims such as swindled investors.

The Southern District of New York's asset forfeiture unit was, by far, the most successful at what then U.S. Attorney Preet Bharara called "taking the profit out of crime." Bharara, whose staff worked out of the Daniel Patrick Moynihan United States Courthouse in downtown Manhattan, once told the *New York Times*, "Asset forfeiture is an important part of the culture here and an example of the government being efficient and bringing home the bacon." In 2012 alone, the unit was on its way to seizing $3 billion in tainted property, a record amount for a single U.S. attorney's office, and 68 percent of the national take. Usually, the "defendant" object was surrendered, and the suspect associated with it walked away rather than risk a protracted, expensive legal battle, or prison.

The assistant U.S. attorney who ran asset forfeiture at the Southern District of New York (SDNY), or "Southern," was Sharon Cohen Levin. A Chicago native in her early fifties, she had worked in the DOJ's civil division in Washington before moving to New York. Since 1996, she had run Southern's anti–money laundering and asset forfeiture unit, overseeing billions in seizures and restitution. Her favorite case had happened only recently, involving the return of a 1912 painting by Egon Schiele, *Portrait of Wally*, which the Nazis had looted in 1939. The painting had wound up in an Austrian museum and then on exhibit at the Museum of Modern Art in New York, where SDNY seized it after its ownership came into question. The case ultimately yielded a $19 million settlement for heirs of the Vienna art dealer who had left *Wally* behind when fleeing the war. The lawsuit "rocked the foundations of the art world," *Variety* reported. "How could they function, museum officials tried to argue, if stolen art was going to be seized by American authorities?" Levin pointed out that she had used the law to "correct a historical injustice."

The *National Law Journal* had named Levin one of the seventy-five "most accomplished female attorneys in the legal profession today." Of all the federal prosecutors in all the jurisdictions of the United States, she was an officer of the court whose attention you did not want to attract if you hoped to get away with something. Hearing about the contested dinosaur in the news, Levin called Robert Painter, reaching him on the tarmac at LaGuardia Airport as he headed home to Texas.

"Want some help?" she said.

"Yes," he told her. "Because I have no idea what to do next."

Levin said, "Oh, I do."

Painter already had issued a press release praising Elbegdorj's wisdom in fighting for the *T. bataar*, saying, "This is a victory not only for the people of Mongolia, who are one step closer to proving the true ownership of this important dinosaur, but also for the important friendship between the people of the United States and Mongolia." Now he suggested that Elbegdorj formally write to Levin, asking the U.S. government to step in. The move might prompt the seller—whose name Robert Painter and the Mongolians still did not know—to give up the dinosaur. Elbegdorj responded immediately, requesting "legal action."

A summons soon arrived by fax at Heritage's Park Avenue office in New York. Homeland Security Investigations, a division of the Department of Homeland Security, sought "any and all records relating to Lot 49315 'SUPERB TYRANNOSAURUS SKELETON,'" including the name and location of the "owner, seller, consignor, shipper, importer, exporter" and buyer. Heritage complied, sending federal investigators what few papers they had on file: a commercial invoice, a UPS Air waybill, a UPS Supply Chain Solutions invoice, and customs forms, which revealed the names Eric Prokopi and Christopher Moore.

Eric had been in near constant contact with Heritage, but now Heritage and the Mongolians appeared increasingly aligned, issuing joint press releases announcing their cooperation in investigating the dinosaur's path to market. They agreed that an international team of paleontologists would inspect the skeleton and formally confirm its species and, if possible, its origin.

A couple of days into the negotiations, Heritage received a surprising email from someone named Don Lessem, who identified himself as "Steven Spielberg's dinosaur advisor" and said he worked "closely with the Mongolian government." Lessem was a former journalist from Philadelphia and a promoter of exhibits involving natural history and cultural artifacts. He had worked with the Mongolian government on a Genghis Khan exhibit at the Field Museum, coauthored a book with Jack Horner, and produced a *Giganotosaurus* cast for the Academy of Natural Sciences. Mongolia's Ministry of Culture had now asked him "unofficially" if he might resolve the *T. bataar* situation quietly, he told Heritage, saying the whole matter "reflected badly" on all: "Chris Moore is viewed as a commercializer of stolen goods, and has already felt harassed by Homeland Security; Heritage is viewed as knowingly dealing in objects of highly questionable provenance; the Mongolian government is seen as lax in protecting its cultural heritage." With luck, they could put the situation to rest "without court appearances or negative press attention." Lessem felt certain that the Mongolians considered Heritage's decision to pause the *bataar* transaction a "great service to the nation" and said company officials would probably be invited to Ulaanbaatar to "receive an award from the President." That "same possibility of honor and award" might be extended to Chris Moore if Heritage cooperated. Lessem didn't know Moore, but liked him; he didn't know Eric Prokopi, either, but considered him foolhardy and arrogant.

Then he made a suggestion: perhaps Mongolia could reimburse Moore for the no doubt staggering expenses involved in shipping and prepping the *T. bataar*. In Lessem's scenario, the Mongolian government would pay

Moore $150,000. Heritage would be seen as having killed the sale once questions of scientific integrity were raised, and Mongolia would receive a nice skeleton to display at its natural history museum. "The alternative— further questioning by the press, and legal wrangling—is far more distasteful to all involved," Lessem told Heritage. If the plan worked, "Heritage gets off the hook for selling a controversial specimen, and the Mongolian government can say it came to the rescue," he added. "There's a LOT of politics involved in this of course. Yucky business, huh?"

Eric did not appreciate the idea of being cut out of such a deal. If anyone deserved the $150,000 it was he. *He* had worked around the clock prepping the skeleton. *He* had seen it to market. He tried to get a piece of the proposal, telling Heritage, "Although I am sure that everything with this specimen is legal as far back as I can tell, I do know just about all of the people involved in the business of [Central Asian] fossils, and could offer ideas and help to make permanent changes that would nearly eliminate the black market and benefit all sides. If the Mongolian president is indeed only interested in getting to the bottom of the sources and wants to look good for his people, I think I can help him do that if he is willing to cooperate and compromise. If he only wants to take the skeleton and try to put an end to the black market, he will have a fight and will only drive the black market deeper underground."

Eric's response reflected the opportunistic aspect of the murkier realms of the fossil trade. The black market was a myth; the black market was real. Permits existed; permits probably didn't exist. Eric wasn't saying he *ran* a black market business in Mongolian dinosaurs, but the aggressive email sure made it look like he did.

Robert Painter decided he wanted nothing to do with Don Lessem's proposal. In his opinion, everything needed to go through President Elbegdorj's office rather than involve outsiders or even Mongolian ministry officials. The opportunity for a payoff deal vanished almost as quickly as it had appeared.

Now that the temporary restraining order was active, the skeleton was locked up at a warehouse in Sunnyside, Queens, belonging to Cadogan Tate, a fine-arts storage company. Heritage was still waiting for Eric to provide the papers he claimed to have, proving that he and Moore had had permission to export the *T. bataar* in the first place. "We do have documents," Eric blustered to Jim Halperin, the Heritage CEO. "But obviously it depends on how much the Mongolians are willing to work with us as to whether we are willing to provide all the info they may want."

With uncharacteristic somberness, Halperin told him, "That would be a very foolish position to take, IMO."

The dinosaur inspection was scheduled for June 5 at the warehouse in Queens. When Eric insisted on uncrating the dinosaur himself, the government of Mongolia agreed to cover his travel expenses to New York and asked, through Heritage, where he wanted to stay. Eric said the Shelburne, where he and Amanda had stayed when they sold the "Nic Cage skull." The hotel had no rooms available, and Eric ultimately wound up at a La Quinta Inn in Queens. He woke up there on the morning of the inspection and walked over to the warehouse before the paleontologists arrived. Piece by piece, he laid out the *T. bataar* on long tables, leaving the larger bones in their crates, nestled in padding and Bubble Wrap. Then he returned to the hotel and waited, staying in touch with a Heritage employee by text.

The inspection team consisted of Tsogtbaatar, the Mongolian paleontologist who ran the national paleontological center under the Mongolian Academy of Sciences and one of the scientists named in the 1993 *GEO* article about commercial fossil-hunting in Mongolia; Philip Currie, the tyrannosaur expert from Canada; and Bolor Minjin. Mark Norell couldn't make it, but agreed to coauthor a report with Currie—collectively the two scientists had logged forty field seasons in the Gobi. The other members of the delegation were President Elbegdorj's chief of staff, P. Tsagaan, an exceedingly discreet former government minister, and B. Naranzun of the Mongolian Ministry of Education, Culture, and Science.

The team arrived to find the crates and tables filled with femurs as fat as clubs, vertebrae the size of carburetors, ribs curved like a display of unstrung bows. Fossil by fossil, the paleontologists studied the coloring and made measurements. Noting the anatomical features—the disproportionately small forelimbs, the number of tooth sockets, the domed shape of an eye bone, the narrow skull—they agreed the dinosaur was *Tarbosaurus bataar*. Which meant the dinosaur was Mongolian. Because except for isolated scraps found in Kazakhstan and China, significant specimens of *T. bataar*, especially largely complete skeletons, had been collected only in or around the Nemegt Basin of Mongolia. The "fairly light" coloring with "ivory staining" matched the other fossils from that part of the Gobi. The scientists suspected the skeleton had been constructed with parts from different specimens, though it was hard to know for sure. They knew this, though: the skeleton was so nicely prepared, the work had been done abroad.

The paleontologists' reports were forwarded to the U.S. attorney's office. On June 19, Sharon Levin filed a new asset forfeiture case, *United States of America v. One Tyrannosaurus Bataar Skeleton*. The complaint publicly named Eric Prokopi and alleged that the dinosaur had been smuggled from Mongolia into the United States via "several misstatements" on customs documents. To have listed Great Britain as the country of origin surely signaled intent to avoid mentioning Mongolia. The listed value of $15,000 clearly conflicted with Heritage's estimate and subsequent million-dollar sale price. The contents' description—"2 large rough (unprepared) fossil reptile heads" and so on—was surely an attempt to circumvent calling the fossils what they were: the remains of a Cretaceous dinosaur.

An arrest warrant was issued for the *T. bataar*. Federal agents backed up a truck to the Queens warehouse, loaded the crates, and took them to a government storage facility. Eric soon received a letter from Levin, notifying him that he had thirty-five days to either surrender the dinosaur or legally claim it in court.

———

Eric had been relying on Heritage's attorneys—"You couldn't be in better hands, in my opinion," David Herskowitz told him—but now it seemed clear that he should hire his own legal representation. Eric had never needed a lawyer in his life and didn't know the first thing about finding one, especially one who specialized in natural history. As he thought about what to do, a lawyer named Peter Tompa from the firm of Bailey & Ehrenberg, in Washington, DC, reached out to him.

Tompa handled contracts and employment cases but also matters involving antiquities and cultural artifacts. For this case he partnered with Michael McCullough, a New York lawyer who had worked as vice president for compliance at Sotheby's, specializing in U.S. customs law and international trade. Neither McCullough nor Tompa had handled a case involving a dinosaur, but the niche opportunity seemed to be presenting itself now that fine-arts auction houses sold fossils. The partners would represent Eric pro bono and, if successful, take a cut of the dinosaur sale on the back end. They told their new client that they intended to fight the forfeiture while pushing for a settlement, hoping a "white knight" would buy the skeleton and donate it to Mongolia.

Eric now staked an official claim to the *T. bataar*, saying he'd spent a year of his life and "considerable expense identifying, restoring, mounting, and preparing it" for auction. He instructed Heritage to release any and all relevant documents to the government. Chris Moore lawyered up, too: he cooperated with investigators and his name was barely mentioned again.

The Heritage brand also faded from the story. On the long July 4 holiday, Greg Rohan, the company president, went on vacation to the Hamptons. One afternoon he stopped to reflect on the auction, saying Heritage would now have to rethink its natural history category. What a shame! People loved natural history! His wife, for instance, had once seen a river stone inlaid with fossil fish at auction: "In her mind she goes, 'We could have the corners polished and put a piece of glass over the top and make a base, and it would be a perfect table.' And then she sees some ammonites that are just sitting there and she goes, 'Oh my God, you could...make a lamp.' Well, now we have ammonite lamps in our guest bedroom!" Rohan was fifty-one and a lean six feet one. In khakis, a navy Izod, blue leather loafers, and a canvas belt embroidered with sailboats, he appeared ready to tack. "If you've got a large space for a dinosaur skeleton—I'll tell ya," he said, then shook his head. "The successful bidder for the *Tarbosaurus*"— he pronounced it *Ta-bor-a-saurus*—"he was gonna display it in a public building that he owns. Then all this hoopla happened, so." Rohan was still trying to process what had happened. "The person who consigned it is not a novice—he's been a paleontologist for decades. There's a thriving trade in black market things that have nefarious pasts and origins and title issues, but I would certainly think that if he thought there were any title issues the last thing he would do is consign it to a major auction in New York City that was advertised worldwide."

If the *T. bataar* sale had gone through, the Heritage auction altogether would have yielded $1.7 million—nowhere near a dominant chunk of the company's overall business, but also nothing to snub. "Had there not been a lawsuit filed right before the auction, there would've been many more buyers and it would've brought a much, much higher price," Rohan said. *That* was his big regret. That and the lost commission and the legal expenses. "But that's the cost of doing business," he said, adding, obliviously, "I don't think that anybody thought Heritage did anything wrong."

Anyway, he said, "We're out."

And so they were.

Eric had not uttered a word in the press, but now that his name was public and he was fighting the forfeiture, he issued a statement. He admitted to acquiring the dinosaur without knowing "for sure" where it came from. He said he had bought the skeleton from someone in Japan and sold it to Moore, then took it back after his partner ran out of time to prep it; the Japan bit was thrown in partly to scare the hell out of Hollis Butts. Commercial dealers and private collectors were a "vital part of bringing some of nature's most precious treasures to museums worldwide," he went on. "The commercial paleontology business is full of intelligent, passionate people who love paleontology, not bone smugglers just looking to steal from important scientific research. If it weren't for people like me, some of these bones would just turn to dust and none of us would ever get to see or study them."

The case threatened to destroy him, he went on. "Imagine watching your house burn down with everything you have in it and knowing you have no insurance." Later, in court, he would also argue that "fossil collecting is well established, and has been intertwined with paleontology in the United States at least since Thomas Jefferson converted the entry room of his home at Monticello into a natural history museum," and that for years, fossils from China, Mongolia, Russia, and Kazakhstan had been "openly sold on the international markets."

The case went before U.S. District Judge Kevin Castel on September 7. Castel had heard matters involving mobsters (John Gotti Jr.) and rappers (Kanye West), but never a party from the late Cretaceous. "I stand to be educated," he told the courtroom. "I'm not going to claim that I have dinosaur arrests presented to me with any frequency."

Eric skipped the hearing because its sole purpose was to determine whether the case would go to trial. Also, he had no desire to meet the press. For the moment he let his one public statement stand: "I'm just a guy . . . trying to support my family, not some international bone smuggler like I have been portrayed by some."

The judge wanted to know who Eric Prokopi was. The name was always a bit of a tongue tripper. It was not *Pro-COPE-y* or *Pro-SKOP-y* or *Pro-COPES-y* or *Ko-PROSK-y*. It was *Pro-COPY*. Prokopi. He never corrected anyone.

"He collects fossils and builds dinosaurs out of them," McCullough, one of his lawyers, told the judge.

"How did he acquire it?" Judge Castel asked.

"He purchased it."

"Where did he purchase it?"

"Multiple sources."

"What sources?"

"From dealers who sell dinosaur parts."

Learning that the skeleton had been assembled from the bones of more than one dinosaur, the judge said, "You're telling me that what was being auctioned did not come from one once-upon-a-time living creature?"

"That's exactly what I'm saying," McCullough said. Most of the skeleton was real bone; half of that was from one specimen; Eric had filled in the rest with matching bones from his inventory, casts bought from the Black Hills Institute, and other bits sculpted to match. Suddenly, and somewhat misleadingly, the *T. bataar* was being called a "Frankenstein" dinosaur, even though museums are known to construct display skeletons with parts from different animals of the same or even different species.

The judge turned to whether prosecutors could prove that the skeleton was even Mongolian. Wasn't it possible that *bataar* lived elsewhere on Earth? Five esteemed paleontologists had provided letters to the court saying the answer was no—the fossils in question almost certainly had come from Mongolia.

Judge Castel appeared unconvinced, but he moved on. "Any idea how large this dinosaur is, when fully assembled?" he asked at one point.

About 24 feet long and 8 feet high, said Martin Bell, the assistant U.S. attorney handling the case.

The judge said, "So it would fit nicely in my courtroom."

PART III

CHAPTER 18

RAID!

WALK AWAY, FRIENDS TOLD ERIC. SURRENDER THE DINOSAUR, get out, start over. Others wanted him to fight. Unable to tolerate the idea of losing so much money or of giving in, he decided to fight.

Their income stalled, the Prokopis started offloading belongings. The Florida Fossils eBay account came alive with available merchandise: a fuel tank, a forklift crane, the gooseneck Continental Cargo race car trailer, an E-Z-Go golf cart. The Ford F550 pickup sold for $30,000, but it wasn't enough. Their mortgage holders were poised to foreclose on Serenola because the house note had gone unpaid since December.

At night, there wasn't a crevice of the internet that Eric failed to Google to gauge the depth of public venom being leveled at him. "Grave robber," "greedy slimeball," and "destroyer of science" were only some of the names strangers called him. A few suggested that he die or "ROT IN HELL." To Amanda, the whole thing was baffling. "It's not like we murdered someone!" she kept saying. "People might think we were living this luxurious, wealthy life, but we made that life for ourselves. We didn't have a Ford550 with a fifteen-hundred-dollar payment every month because we wanted to show off; we had to pull a heavy trailer. I like nice things but I'd rather eat off the appetizers list because it's cheaper. We're not crazy throw-money-in-the-air people. We had a few luxuries, sure—like taking my friends to the Bahamas. What are you working toward if you don't eventually enjoy what you have? A lot of the successful people we know come from family money or family businesses. I was always proud of Eric that he hadn't; he came from nothing, and he was scrappy and surviving. When he wanted a car, when he wanted to go to college, he had to do it on his own. It was like a dream come true, to be able to turn what you love into a real life. It was never our goal to get rich. When I met Eric, he was doing well but he wasn't killing it. We killed it *together*."

Eric's most recent tax return showed reported income of slightly more than $103,000. He and Amanda owed $450,000 in mortgage loans, well over $70,000 on credit cards—Home Depot, Tires Plus, Best Buy, Lowe's—and more than $41,000 to the Black Hills Institute. They were behind on their taxes. Overall, they owed more than $11,000 a month. The seizure of the dinosaur was a setback no one had expected. "Like, America just got *involved* all of a sudden," Amanda said. "I didn't realize that could happen."

Eric still had all these other Mongolian dinosaurs, and he intended to sell them. Instead of backing off, he went on prepping, partly because he thought he'd win his case and be cleared for business. Even with the federal crackdown—or maybe *because* of the federal crackdown, given that notoriety often fuels demand—other dealers were still openly selling Gobi bones on eBay and independent websites. One seller in France was offering a *Protoceratops andrewsi*, a *Conchoraptor gracilis*, an *Oviraptor philoceratops*, and a *T. bataar*, all outright labeled Mongolian. After Eric prepped out an *Oviraptor* that David Herskowitz had placed with a private client, that sale fell through, too, the client claiming that Eric had missed the deadline. The panic was now palpable, as was Eric's tendency to cast blame. "This couldn't come at a worse time for me and it is unacceptable for your client to back out," he told Herskowitz, furious. "There is actually a good chance that this is what puts me over the edge and I will have to file bankruptcy."

Eric's civil lawyers soon referred him to a criminal defense attorney in New York, just in case. Eric took countermeasures. He put his passport in a Ziploc bag and hid it in a crawl space beneath the house. Leaving his cell phone at home, in case anyone was tracking him, he paid cash for a U-Haul and stored plastic crates filled with dinosaurs in a pole barn at the home of his old log-pulling buddy Joe. Eric figured the feds wouldn't connect them; he hardly ever talked to Joe on the phone and they'd never exchanged a single email—as far as Eric knew, Joe didn't even *have* email. Tyler, Eric's assistant, rented a self-storage unit in her name and they stashed more bones there, then quickly thought better of it and moved the fossils elsewhere.

In Mongolia, police carried out their own investigation. The government wanted to know Eric Prokopi's connection to Mongolia and whether he had ever been there. A search of customs records showed that he had visited the country three times. The first time, he had arrived by air on the night of October 23, 2008, departing four days later at around four in the afternoon. The second time, he arrived on the night of June 23, 2009, departing on July 8. The third time, he arrived late at night on July 22, 2011, departing five days later at six in the morning. A search of hotel records showed that on his last trip, he had shared a room with Chris Moore.

In each instance, customs had required him to list the purpose of his visit. "Tourism," he had written on his second and third visits. But on his initial arrival he had written "private" and provided a local phone number that Tuvshin had told him to use. Investigators traced the number to a former Natural History Museum employee, which led them to the name Tuvshinjargal Maam, which led them to the mint-green building on Peace Avenue. The suspect was many months dead, but his computer still existed, and while it had been scrubbed of data, investigators were able to restore deleted material and discover emails and photos—photos of dinosaur skeletons and of Tuvshin posing with dinosaur skeletons, and emails to Eric and other dealers about the sale of dinosaur skeletons. Bobo, Tuvshin's wife, was arrested on September 13 and held for at least nine days, but was never charged.

A draft letter also turned up on the computer. It appeared to have been written in the spring of 2008. This would have been just before Eric's first trip to Mongolia and just before that summer's deadly post-election riots; it was also the year Mongolia amended its criminal code to toughen punishments for illegally crossing the border with "restricted goods, rare animals... minerals and natural elements." The letter began, "Dear Butts." It read, "You told me that you would buy a head of tarbosaur, meat eating dinosaur of any size whether its [sic] small or big as long as it is whole for 40000USD while we were sitting in the coffee room on the first floor of the hotel Chichibi [sic]."

At the meeting, in Japan, on July 12, 2007, "Butts" supposedly had told the author he would buy "hand claw of tarbosaur" for $5,000, as well as three tarbosaur skulls, *Gallimimus* claws, and a *Velociraptor* skull. "Following my offer many expeditions went to countryside," read the letter. They had talked prices: $65,000 for a *Velociraptor* skeleton, $70,000 for tarbosaur, plus another $20,000 for "ger (tent), custom problem, transportation, and container shipment charge." When the payments from "Butts" slowed, the Mongolians wanted either the rest of their money or the return of a skull. The letter read, "They assaulted me several times and hurt."

By now "Butts" owed Tuvshin $37,600, the letter went on. If he wanted to buy more dinosaurs, he would have to come to Mongolia in person and pay half up front. The diggers would bring "their goods from the gobi" only after getting paid. "I don't have any money either to pay them in advance or to make shipment," the letter said, adding that a shipping container alone costs $20,000. "I 100% guarantee this business to you. You will never lose," the letter said. But "Butts" should hurry. "People are getting more and more cautions [*sic*] here. In Mongolia the price everything is rising." A three-bedroom apartment now costs the equivalent of $200 a month. The global economic crisis of late 2008 and early 2009 had "savaged Mongolia's currency, capital, and equity markets," according to one U.S. embassy report. The *tugrik* fell some 40 percent against the dollar.

As investigators continued their inquiry into Tuvshin's activities, Oyuna went on television and pleaded for more information. A key witness soon came forward.

Otgo, the chief fossil prepper at the national paleontology lab, sat in a chair at police headquarters wearing a beige-plaid shirt and a windbreaker, his hands crossed in his lap. It was September 20, four months to the day after the Heritage auction. Snapshots of four different white men lay in a row on a table. The first showed a sunburned man in a hoodie, with a Ferris wheel in the background. The second was a cell phone photo of Eric Prokopi that a member of the Mongolian delegation to New York had surreptitiously shot in the La Quinta Inn lobby on the day of the *T. bataar* inspection: it showed him sitting on a sofa in jeans and a brown knit shirt, talking on his cell phone. The third showed a mohawked guy perched on the stone wall of a restaurant patio. The fourth was a selfie of a buzz-cut young man in a Jack Wills T-shirt.

Otgo pointed right away to photo number two. That is Eric, he told the investigators. Eric came to Mongolia with "Vadim"—for some reason, that's what he called Tony—in the summer of 2009. Otgo admitted that he had served as the paleontologist of the group, which had traveled by Land 80 and Land 260, following a road map highlighted in yellow. During the excursion, Otgo had understood only small parts of the conversation but recalled that one of the Americans introduced himself as Ukrainian. They had met a Gobi family that helped "collect dinosaurs." They had found a dinosaur egg at Ergiin Tsav and "small bones" at Bugiin Tsav.

Among the photos investigators found on Tuvshin's computer was one of Eric and Tony climbing a stony embankment; they had scaled the rock to look at a falcon's nest, but within the context of black market fossils they looked every bit like poachers. Another photo showed Eric in the desert, holding what U.S. authorities would later call a clipboard; in fact, he was holding a roll of plastic wrap used for collecting bone scraps. Otgo was pictured in the second photo, a fact he now didn't deny.

The cops soon brought Otgo in for a second interview. This time he revealed details about 2011, when Eric returned hastily to Mongolia after Tuvshin's death. Tuvshin's wife had called and said "Eric is here, come meet him," Otgo said. "I went to the green building and Eric was there with a tall Englishman with gray hair, named Chris. Tuvshin's children were there, too. I used my limited English. Bobo's youngest son translated. We went to the Natural History Museum and to a Christian church. We went to Terelj for a picnic. Several young people joined us in eating and drinking. We recalled Tuvshinjargal fondly. Before they left, the two foreigners asked me to join them on their next expedition to the Gobi. They gave me their name cards and I gave them my son's email address, but I lost the cards."

This was the first admission that a Mongolian paleontologist was connected to the *T. bataar* case. The information wasn't publicized widely, if at all. Despite Otgo's acknowledgment that he had acted as the Gobi guide, the Mongolian public was assured that the *T. bataar* case had nothing to do with Mongolian scientists.

"Isn't it because Mongolian paleontologists are involved in this that so many dinosaur skeletons are smuggled out of the country?" one reporter asked Oyuna.

"I *also* had the fear that recognized paleontologists of Mongolia may have participated in the smuggling," Oyuna responded, adding, "Their participation would have influenced and reflected negatively [on] the reputation of the Mongolian paleontology sector." The public should trust that Mongolia's scientists were innocent, she said.

In fact, more paleontologists than Otgo had been involved.

In the 1990s, Tuvshin called himself the head of the Mongolian Private Museums Union, which, as his acquaintances understood it, was another of the NGOs that proliferated after the fall of Communism. Once the law allowed for the creation of antiques shops and private museums, Tuvshin started contracting with scientists and academics, hoping to start his own. In the spring of 1999, he signed a contract titled "Financing Geology, Paleontology, and Natural Science Study," under which Tuvshin would finance paleontological expeditions for university students and researchers; in exchange, the museum association could keep or sell the findings, splitting the proceeds with the university. Three people signed the contract. One was Tuvshin. Another was Boldsukh, a vice president of the National University of Mongolia. The third was the paleontologist Altangerel Perle, who had participated in early joint expeditions with the American Museum of Natural History before moving on to other work.

Perle had died of a stroke. Boldsukh either had died or would soon die in what Western scientists later heard described as a fishing accident. At one point, Oyuna fleetingly mentioned their names to the Mongolian media in connection with Tuvshin, but shut her mouth after one of the men's families threatened to sue her. In Mongolia, this was an unusually harsh threat: the country's libel laws were among the worst in the world. Anyone found guilty could be fined and imprisoned for up to six months. For years, the American embassy had urged the Mongolian government to decriminalize libel because the law was so often used to intimidate people, and to stop them from telling the truth, but nothing changed. The flicker of information about Boldsukh and Perle—confirmation of a connection between scientists and commercial interests—quickly disappeared.

Whenever Tuvshin had talked about "permits" that authorized him to sell fossils, Eric had never really probed what he meant. His simple concern was losing money, "getting ripped off," he said later. "Tuvshin wouldn't sell in small quantities. You had to risk a lot of money to deal with him. He wanted to deal with only a few people and sell big loads. That deterred a lot of buyers." All Eric knew was that there were supposedly fifty diggers in the Gobi and Tuvshin worked with two of them. If ever called upon to produce papers, Eric had figured he could ask Tuvshin to email documents. He had never counted on the guy dying.

As *United States of America v. One Tyrannosaurus Bataar Skeleton* proceeded in the United States, Homeland Security Investigations, working in tandem with the Southern District of New York, quietly built evidence for a different kind of case. The federal prosecutor, Martin Bell, and the lead HSI agent, Daniel Brazier, wanted to know how and where Eric did business, and with whom. Brazier obtained a search warrant for the AOL account Eric had been using since he was a teenager in Land O' Lakes. The online service harvested emails for the period between April 2010 and August 2012, downloading them onto a searchable disk.

On the morning of Wednesday, October 17, Amanda got up early to get the kids ready for school. She'd slept downstairs on the couch; as she walked out of the guest bathroom and headed for the stairs, she was startled at the sight of a man looking through her front window. Opening the door, she didn't bother asking the Homeland Security agents why they were there. One of the few details she would remember was that one asked, "Is Eric Koproki here?" and that she said, "Do you mean 'Prokopi'?" and one of the agents said, "Oh. We've been saying his name wrong for months."

Eric was asleep upstairs, sprawled out with the kids in the master suite's big black Shaker-style bed. Amanda scooted the children to their rooms and got them ready for school, saying, "These nice men are here to see your Halloween costumes!" After she scrounged up some clothes for Eric to wear, the agents put him in handcuffs and leg chains and took him away. Amanda drove the kids to school, trying not to panic. The agents had taken her phone but they hadn't said not to call anyone, so she borrowed another mom's cell in the carpool lane and called Georges Lederman, Eric's criminal defense attorney in New York. Stay calm and cooperate, Lederman told her.

But it was hard to think. The agents were taking stuff out of Serenola; Eric was going before a judge at noon; the kids would get out of school at like, two; friends were texting, asking, "Why are there TV news trucks lined up outside your house?"; she needed to find real estate documents, to "do a bail."

The HSI agents gathered evidence by labeling each room with a letter of the alphabet for inventory. They packed up phones, fossils, casts, reproductions, the family iMac, two Gateway laptops, a MacBook Pro, two SanDisk storage drives, Eric's passport (plucked from beneath the house), tax documents (which happened to be sitting beside the front door), and an empty shipping container with another dealer's name on it. When Amanda started crying, an agent told her, "Don't worry, it's just stuff, you'll get it all back." Amanda could admit that none of this looked good, and that she and Eric hadn't "always been angels," but she wanted to scream, "It's not just *stuff*—it's my whole life."

When it didn't seem possible that the situation could get worse, a delivery truck entered the front gate. The 400-pound shipment was addressed to Eric and had come from I.M. Chait, the Beverly Hills auction house. One of the federal agents called Bell, the prosecutor in New York, and said, "You're not gonna believe this."

Georges Lederman told Amanda not to open the box, so the agents got another search warrant and opened it themselves, finding Lot 291 of a Chait auction from December 5, 2010: "Graveyard of the Oviraptors, *Oviraptor philoceratops*, Cretaceous, Central Asia." Given the discovery of more Mongolian dinosaurs in Eric's possession, Bell, the prosecutor, requested bail of $100,000, telling the magistrate, "This fresh conduct, in light of the known legal issues concerning the defendant's business, is alarming."

In Manhattan, the federal authorities announced the arrest. Whereas the civil action had been handled by the Southern District of New York's anti–money laundering and asset forfeiture team, Bell would prosecute the criminal case within the Complex Frauds unit, on charges related to smuggling. James Hayes, special agent in charge of Homeland Security Investigations, declared to the press, presumably with a straight face, "We want to make this illegal business practice extinct in the U.S." Preet Bharara, the U.S. attorney, called the *T. bataar* "merely the tip of the iceberg" and said, "Our investigation uncovered a one-man black market in prehistoric fossils."

It stunned Amanda to see Eric in handcuffs, "like he was some bad guy," and wearing shackles, "like he'd done something wrong." Florida's laws were such that Eric, after being booked in the federal criminal justice system, was released on his own recognizance, placed under home arrest, and ordered to appear in court on Monday in Manhattan, where he would have to post bond in order to be free until trial. This time he'd have to pay for his own travel.

That afternoon, Amanda picked Eric up from jail and drove him home, where the raid was ongoing. Each was panicked about a hundred different things: what to tell the kids; what their parents would think; what their friends would say; how to process the fact that Eric was now an accused felon and that they'd have to put up Serenola as collateral. How much money was in the bank? Would anyone ever do business with Florida Fossils again? Was Eric now "radioactive," as his new attorney put it?

Eric had sat in jail that morning, frantic that the cops would tell Amanda the secret he'd been keeping. If investigators had been monitoring him, they knew that for months he had been having an affair with Tyler, the woman who'd been working as his assistant. The Prokopis had hired her in December and the affair had started in March. Eric and Tyler had found that they liked the same things—both were "treasure hunters by nature." They could talk for hours. "He's sometimes such a painful introvert but I can't shut his ass up," Tyler later said. "I can look at him and finish his sentences. There's a wavelength with us. He likes that I'm independent, and I'm smart, and I'm as fearless as he is—we'll sneak into quarries together. We're both attracted to that fearless part of one another." She later said, "This is going to sound terrible—I'm a very moral, guided, compassed kind of gal—but we couldn't live without each other."

As Eric agonized over how to tell Amanda about the relationship, he acted so weird, she worried that he was either planning to kill himself or run. Finally she confronted him: Are you having an affair? Yes. Is it Tyler? Yes. Are you in love with her? Yes. Eric kept saying, "We didn't *mean* for this to happen," which to Amanda was like saying, "I didn't *mean* to kill you, the gun just went off." It infuriated her that Eric refused to give details. He kept living at Serenola, but after a while Amanda finally said, "I can't go through my mourning if you're here," so Eric moved in with Tyler, everyone navigating the delicate task of how to explain all of this to the children.

Now that Amanda knew the secret, Eric felt not especially unburdened. Guilt was wearing him down. The guy who once could handle everything now couldn't handle anything. Every passing day zombified him more. To Amanda, he had become a different person almost overnight: paranoid, angry, immobilized. Their bank accounts were depleted. Friends and family stepped in to help without being asked. Amanda's brother sent a $500 grocery card. Girlfriends tucked cash into her purse when she wasn't looking. People cooked for them and offered to keep the kids, like you do when a loved one is sick. They had been eBaying certain items, to generate income, but now decided to sell everything they owned—cars, furniture, almost all of it. Serenola's front parlor suddenly looked like one of Amanda's Bizarre Bazaar booths: twinkly, earthy, For Sale. In a series of tag sales, they sold their possessions right out of the rooms, right off the walls, telling the children, "Don't worry. These are just things."

On October 22, a cold, overcast Monday, Eric climbed the front steps of the federal courthouse in lower Manhattan and cleared security alongside Georges Lederman, his attorney. He wore a dark suit and a red-print tie, and was carrying a laptop bag as if it were a briefcase. At 3:07 p.m. his case was called.

The criminal charges in *United States of America v. Eric Prokopi* were altogether unprecedented in American jurisprudence. Count 1 accused him of lying on customs forms in order to import that Chinese *Microraptor* in 2010. Count 2 alleged that he had imported "multiple containers of dinosaur bones of Mongolian origin, by way of Great Britain" between 2011 and 2012 by misrepresenting their contents and value. Count 3 accused him of importing the fossils knowing them to have been "taken from Mongolia by fraud or conversion." This charge alone carried a maximum prison sentence of ten years.

Not guilty, Eric pleaded.

Should the defendant be jailed while awaiting trial?

Well, he's a flight risk, the prosecution argued: after all, Prokopi had international contacts and the means to make quick, off-the-books money.

He's got fifteen hundred dollars in the bank and seven hundred in cash, the judge responded, to which Martin Bell, the prosecutor, said, Yeah, but he's sitting on "about half a million dollars' worth of dinosaurs."

After Bell argued for a stricter bond agreement, the judge raised the amount to $250,000, and required that at least one financially responsible person of "moral suasion" cosign for it. Eric, who was using Serenola as security, chose his father, dreading the words "I told you so" but knowing that, if called upon, his parents would put up their house in Land O' Lakes.

As Eric left the courthouse that day, the commercial fossil world was talking about his rapidly worsening legal problems. Few people knew him well, but the dealers who considered him a friend tried to assure the rest of the trade that, no, really—good guy, bad mistake. The Heritage auction "was like rubbing the tarbo right into the American Museum's face, and so they had to react," Kirby Siber, a longtime hunter and natural history museum owner in Switzerland, told George Winters, then president of the Association of Applied Paleontological Sciences, the trade group for commercial fossil hunters. "The conflict had to erupt." Commercial dealers already felt misunderstood and maligned because of "irreconceivable views (fossils are such pure cultural items/fossils are commercial items like other resources)," Siber said. He wondered if perhaps there wasn't some other issue driving the tension. Maybe the American Museum of Natural History had protested the auction for fear of *T. bataar* overshadowing *T. rex*, America's signature dinosaur. Or maybe AMNH scientists were protecting their Gobi turf. "I have seen this over and over again in many countries, that some official institution makes a claim to all the fossil resources and only uses them partially," Siber said. "This is what really causes the 'black market.' Is an unfair monopoly of a cultural and commercial resource."

As the weeks passed, Eric began to realize the enormity of the odds against him in criminal court. If there was one place to avoid standing trial it was the Southern District of New York: those lawyers rarely lost. Also, juries are a notorious wild card—there's no predicting what they'll do. His best chance at getting back to his life rested with a plea.

Two days after Christmas, Eric returned to federal court in New York. He wore his peacoat, no sunglasses, no gloves, and a bemused grimace. When the magistrate asked if he understood the counts against him and Eric said yes, the magistrate then said, "Tell me what you did." This might've been a moment for revelation, but true to character Eric basically just repeated the charges as they'd been read to him, admitting he'd done what they accused him of but offering little more. The magistrate started to ask how Eric had obtained a permit to export Mongolian fossils in the first place but interrupted himself. The interrogatory turned, and the question went unanswered.

Now that he had pleaded guilty, Eric faced a possible million dollars or more in fines and up to seventeen years in prison. If sentenced to the full ride, he'd be in his late fifties when he got out. His children would be in their twenties. In private, he wept at the thought of it.

CHAPTER 19

VERDICT

Amanda filed for divorce on May 16, 2013, almost a year to the day that the dinosaur went to auction. By June 11, the Prokopis were done: twelve years of marriage, gone in twenty days.

In the settlement, Eric got the johnboat and the flat-bottom boat, plus an old Toyota truck, and the box truck. Amanda got the 2011 Toyota Camry, which they'd bought after selling the Lexus. They hoped to sell Serenola for over a million dollars. Under the terms of the agreement, Amanda would receive 60 percent of the proceeds and Eric the rest after their legal expenses were paid; the house ultimately fetched $575,000; the court ordered Eric to pay $2,000 a month in child support and five hundred in alimony, plus 75 percent of the children's medical expenses. Custody of the children would be shared.

Now that the Prokopis' time in Gainesville was coming to an end, their friend Jill Hennessy Shea assembled and bound a 240-page photo book called *Never a Dull Moment*, celebrating the Prokopis' twelve years in town. "Never a dull moment" was another of Amanda's mottos, deployed cheerfully whenever something went wrong. The book's photos told an ongoing story of birthday parties, Easter egg hunts, home renovations, Junior League soirees, and *T. bataar*, including a shot of Eric and Amanda captioned, "Installing a dinosaur skull in Leonardo DiCaprio's foyer." Amanda hoped the next family would love Serenola as much as hers had, but then she put Gainesville behind her and pointed the car north toward Williamsburg. It took less than an hour to move herself and the kids back into her childhood home, beyond the guard gates of Kingsmill. For the first time since SeaWorld, Amanda started looking for a full-time job.

Before long, Eric and Tyler found a house together, half an hour east of Williamsburg, in Hayes, on the other side of the George P. Coleman Memorial Bridge, where ospreys and peregrine falcons flew sorties to and from their nests with live fish clutched in their claws. At the end of a gravel lane stood a white house with red shutters and terrible pink carpet. Eric had found the property via Craigslist as a land barter, a sort of rent-to-own arrangement. There were three bedrooms and two baths downstairs, and two small bedrooms and one bath upstairs, plus a sunporch, a garage, and two outbuildings, the smaller one as red as a Vermont barn. The house's previous owners had tacked an incongruous vaulted family room onto the kitchen—the place looked like a cottage that had sprouted a ski lodge. The sliding back door didn't lock—Eric stuffed a rag where the handle used to be, to keep the mosquitoes out. The pitch of the staircase was such that you had to lean into it going up and lean back going down. The upstairs bath's wallpaper (ducks) was peeling, as was the linoleum. There were holes in the kitchen wall and the ceiling of the master bedroom. The air vents were jacked, the floorboards soft. Someone told Eric it must be depressing to live this way. Eric, who was waiting to learn whether he was going to prison, replied, "This is the least of what's depressing."

Even though he now faced a possible long sentence and crushing fines, he and Tyler hoped to renovate the house, whose chief attraction was its location on the York River, not far from where George Washington's army and their French allies won the battle that secured America's victory in the Revolutionary War. Beneath the back deck lived a woodchuck. A parabolic sloop of a clothesline hung between pines. Down past the tall grasses lay a white knuckle of beach where turtles hatched in the sand, leaving bits of shell in the seaweed. The property once had a boardwalk, but the planking had disintegrated somewhere around the third piling, disconnecting the pier from shore.

To earn money, Eric prepped fossils for the dealers who still spoke to him, or sold leftover Pleistocene items from his early inventory, from back when he hunted in rivers. He and Tyler spent hours at low tide, searching the shoreline for shell and bone. A nice *Chesapecten jeffersonius*—the extinct scallop that is the state fossil of Virginia—alone could fetch up to twenty bucks. For steady pay, Tyler took a bartending job. Eric found a used van on Craigslist for next to nothing, with side panels that bore the peeling remnants of cartoonish decals that once advertised a preschool. On weekends, he and Tyler prowled country auctions, collecting inexpensive vintage objects that they could refurbish and sell at those same auctions or on eBay. They found that they could buy a load of scrap metal, take it directly to a scrapyard, and drive off with cash. Whenever they saw an obviously abandoned house, they ventured inside, "treasure hunting."

Amanda, meanwhile, found a full-time job at an interior design company, helping rich people redecorate their yachts. One day she worked up her nerve and went to the county welfare office to see about getting health insurance for herself and the kids, but was told that in order to qualify, she'd need to prosecute Eric for being behind on child support, which she wasn't willing to do. His inflamed eye was getting worse, and they briefly worried that he had Crohn's disease, a painful bowel condition. She didn't want to add one more stress to her children's father's life, because he still seemed close to breaking.

Eric's best chance at avoiding lengthy incarceration rested with acting as a government "cooperator." Cooperators help prosecutors understand an industry and in exchange may avoid a long prison term and high fines (think Frank Abagnale Jr., the prolific con artist, forger, and FBI dodger whose crimes inspired the movie *Catch Me If You Can*). Prosecutors could recommend a lenient sentence for a nonviolent cooperator, but ultimately, sentencing judges could do whatever they wanted, within the federal guidelines. The consensus among case lawyers was that as a cooperator Eric wouldn't see time.

To start the process, he'd had to make a proffer, which, within the context of criminal law, is a written understanding between the government and a defendant who volunteers information hoping the court will go lighter on him. A proffer is technically different from turning state's evidence, in which a defendant flips after being guaranteed a certain outcome, such as immunity or witness protection. A major protection that's built into a proffer is that prosecutors can't use a defendant's revelations against him unless he lies about something. The one outcome Eric could probably count on, if he *didn't* cooperate, was prison. "Ultimately we're the house," one prosecutor later explained. "And the rules are set up in such a way that if you're not candid with us in that setting, we've got ways to come back and bite you."

Eric's enemies were fine with the thought of him heading to lockup. Others, including at least one somewhat remorseful scientist involved in the *T. bataar* case, felt probation was probably a fair enough punishment, even though they considered Eric reckless and a liar. One of the government officials thought of him as "one of the more sympathetic defendants," saying he appeared "driven by the bottom line but also by a genuine fascination with fossils themselves and a joy and pride in having learned how to put them together into art in a way that other people can't do." Sharon Levin, the assistant U.S. attorney who had brought the civil asset forfeiture case, felt less sanguine about it: "In the scheme of things, in terms of crime, no one died; he didn't defraud people of millions of dollars; he was no Osama bin Laden. That said, he participated in a conspiracy to steal from the people of Mongolia and to make money off of that. Maybe he thought this was a minor evil versus supporting his family, but at the end of the day the consequences of his actions are that he's going to be punished. He took a risk and it didn't work out for him. No one hated him, he wasn't evil; but he was caught doing something illegal."

Prosecutors and federal agents were used to all kinds of defendant behavior—minimization to mania to emotional distress signals like anger, shame, and fear. But Prokopi, man—he was a sphinx. "It did take some work...speaking with him and understanding exactly the precise series of events," Martin Bell, the assistant U.S. attorney on the criminal case, later told one judge in court. Talkers are exhausting, but at least they provide insights into why they do what they do, why they are the way they are. John Laroche, the Florida plant poacher Susan Orlean wrote about in *The Orchid Thief*, once said, "People look at what I did and think, Is that moral? Is that right? Well, isn't every great thing the result of that kind of struggle? Look at something like atomic energy. It can be diabolic or it can be a blessing. Evil or good. Well, that's where the give is—at the edge of ethics. And that's *exactly* where I like to live." You weren't about to get that kind of sentiment out of Eric. The authorities realized it was simply in his nature to say little. It was a personality trait that probably would have worked against him at trial. "He's such a difficult person for the average person to believe, just because of the way he communicates," said someone close to the case.

The heart of his proffer involved surrendering the *T. bataar* and acknowledging that there were more Mongolian dinosaurs in circulation. Eric wanted to cooperate without ratting anybody out, figuring that investigators already knew plenty from having read his emails, but the prosecutors walked away with what they wanted to know. Federal agents were soon knocking on doors in several states.

When all this started, it simply never occurred to Eric that he might get caught, or that there were serious consequences to getting caught. To him, the first trip to Mongolia was equal parts income and adventure, not unlike the setup he had seen described in the *GEO* cover story about the German dealer Andreas Guhr. He had never been arrested and didn't think to be afraid of it.

If he had read up on Mongolia's criminal justice system alone, he might have decided to stay home. Despite the nation's impressive democratic strides, a Mongolian detention center was an awful place to be. One of the sights Roy Chapman Andrews never forgot, albeit in a much earlier and different era, was that of the prison at Urga. "There the unfortunate prisoners, some of the condemned on charges of petty larceny or the like, were placed in coffins and covered with earth except for the face," he once said. Others were held in a position in which they could neither stand up nor sit down. "The most fiendish tortures were practiced on the inmates, tortures in which the savagery of the Russians was combined with the fiendish and refined cunning of the Chinese."

Mongolian detention centers and prisons were still so bad that the government had asked the United Nations for an independent assessment in order to meet international human rights standards, which partly dictated foreign aid. "Torture persists, particularly in police stations and pre-trial detention facilities," Manfred Nowak, the UN's special rapporteur on torture, found as recently as 2005. Two prisoners had been tortured to death.

The torturers got away with it partly because while Mongolia's constitution provided that "no one shall be subjected to torture, inhuman, cruel, or degrading treatment," the criminal code failed to address torture at all—didn't even *define* it. Even Mongolia's legal and judicial community appeared fundamentally unaware of what constituted torture. Nowak was, however, able to document methods: needles under the fingernails; cigarette burns; beatings; "flying to space," wherein a person stood on a stool and had it kicked out from under him; and electroshock via a light-bulb socket, wires, and a puddle of water. A suspect could be detained by an ever-changing confusion of government agencies. Pretrial confinement could range from fourteen days to thirty months. Suspects under eighteen could be held for up to a year and a half without being charged. Foreign citizens who hadn't been charged with a crime could be denied exit visas and banned indefinitely from leaving the country, even for vague and unsubstantiated business disagreements. In one high-profile case, a U.S. businessman was prevented from leaving the country for two years. "Mere complaint by an aggrieved party is sufficient to deny exit," the U.S. embassy reported.

When Nowak asked about capital punishment, the government refused to talk about it. Every facet of the death penalty was a "state secret." In fact, in Mongolia almost *anything* could be declared a state secret. For years the U.S. embassy had urged the government to either amend or abolish the state secrets law, which the State Department called "among the most restrictive and punitive in any post-communist country." As a pony caught between two elephants, as President Elbegdorj once put it, Mongolia had good reason to be wary of Russia and China, but this law exceeded even the usual paranoias. "Authorities remain fearful of information and, thus, reticent to comply with citizens, media, or civil society organizations' requests for information," the U.S. embassy wrote, quoting one USAID-funded report. A person could be imprisoned for up to eight years for revealing even a banal state "secret."

Nowak finally learned that Mongolia executed twenty to thirty people every year. Death row inmates had been held in "complete isolation, hand-cuffed and shackled, and denied adequate food" before they were killed. The public was never informed who, when, or how. Even the families of the condemned were not told the date or location of their loved one's execution, and afterward they weren't allowed to collect the body.

———

Eric's mom thought he had made a mistake by admitting guilt—she wanted him to fight the criminal charges. His father advised taking a deal. "I feel that he did something wrong, but I don't think he did it intentionally," Bill said one day over lunch at Chili's in Land O' Lakes. "He got involved with the fossils and that was more important to him than—"

"A *lot* of people are into fossils and they never get caught," Doris said.

"What do you mean, 'get caught'?" Eric asked. He was eating barbecue ribs and drinking Diet Coke. "That's implying that everybody who's interested in fossils is hunting illegally."

"I mean *you* didn't get caught," his mom said. "It's just—you did everything in the open."

"He didn't go through the proper channels, that's the thing," Bill said.

"He didn't go through the proper channels," Doris said.

"There *are* no proper channels," Eric said, now that it was almost all over.

"That's why they dragging it out so long," Doris said. "They didn't know what to do. In the fossil clubs we met lots of fossil people. They were all really into fossils but they all had regular jobs. Eric was the only one who went really into it."

"That's not true!" Eric said. "There's a lot of dealers that do it for a living."

"There's a few, yeah," Doris said.

"There's *hundreds* of people," Eric said. "Look at Tucson. Most of those people, that's *all* they do."

Doris changed the subject. "When I go swimming I forget everything," she said. "I forget to eat sometimes."

Later, as the date of Eric's sentencing hearing approached, Doris stuffed a big brown envelope with mementos from his youth—newspaper clippings about his school honors, fossils, and swimming; photocopies of his dive cards and fossil club memberships. On a piece of loose-leaf notebook paper she wrote her son a note, apologizing for her spelling mistakes—

I am doing thomthing crazy...I send you a lot of paper clippings. I want you send some of the paper story from you, that the lawer got to know you a little better. I help [hope] that he sees you as a yung good man. What you did all your years. I hope you going to do this. Just send it to his office. If I had his adress I would have send it. But Dady sad you do it. If you have your Trial we want both be there. You must fite for it.

She added, "Well I wisch you good luck with what you doing. You are my boy and I love you very much."

She folded the note, put it in the packet, and mailed it.

On June 3, 2014, Eric stood before the last judge in the last courtroom he hoped to ever see in his life. Two months shy of forty, he wore the one good suit he owned. Eric had always had to YouTube how to tie a tie, but, for court, his father had tied it for him. Strands of gray now laced his hair. His Florida tan had faded, but only slightly, the way a baseball glove loses its brown. The dinosaur, the Gobi, Tuvshin, Serenola—all of that seemed distant now.

The judge wanted to know what the confessed smuggler had to say for himself. Alvin K. Hellerstein, sixteen years on the bench in the federal courts of the Southern District of New York, appeared dissatisfied with the level of detail that had been elicited from the defendant by the magistrate who had taken Eric's guilty plea—he wanted more. The lawyers did the talking. *Jurassic Park* came up, as did Rodin's *The Burghers of Calais*, lava flow, English law, ankylosaurs, Mongolian tourism, Wyoming, the phrases "law enforcement renaissance," "sophisticated ringleader," "dark of night," "expensive boats, expensive cars, expensive planes," "media frenzy," and "public indignity, shame, and humiliation." Doris and Bill sat in the courtroom; they hated to drive farther from home than the supermarket or church, but they had driven to Virginia and Tyler had chauffeured them the rest of the way to New York. Amanda had traveled separately from Williamsburg. Eric's friend and fellow fossil dealer Andreas Kerner had driven in from New Jersey.

Judge Hellerstein had before him the pre-sentence investigation from the office of probation and parole:

Offender's citizenship: U.S.

Dependents: two.

Aliases: none.

Criminal history: none.

Offender's family: father, eighty-three, retired schoolteacher; mother, seventy-eight, homemaker; one maternal half-brother, truck driver. "He described his childhood as 'great' and noted no problems in the household." Started scuba diving at age ten in the local rivers, collecting fossils. Good relationship with parents.

Divorced: "He said that they were not getting along and there was a great deal of stress placed on them, which included financial issues. He said that his ex-wife always sought financial stability and his business has more peaks and valleys.... His court case was the final straw."

One son: Greyson, age six. One daughter: Rivers, age four. "He stated that his kids and family are the most important thing to him in the world and his biggest worry is being taken away from them."

Girlfriend: Tyler, thirty-six. "She has remained very supportive."

Drugs? No.

Drinking? Occasionally.

Drug test results: negative.

Offender employment record: Florida Fossils since 1992, dinosaurs since 2005.

Current job: "He still collects fossils on his own for prep and resale. In addition, he has been purchasing antiques from estate sales and restoring the items for resale."

Offender's physical condition: six feet tall, 210 pounds, no tattoos, no scars. Having some eye problems, requires medicinal drops. "The defendant indicated that post-seizure of the Tyrannosaurs [sic] bataar, he attended one or two treatment sessions with a mental health specialist. He indicated that he was prescribed an antidepressant, however he stopped taking the medication after about a week because he did not like the side effects. He said that he is handling things as well as he can. He remarked that he spends more quality time with his children now than he did previously, as he was always working."

Nothing in the investigative report promised to undo Eric. It was the million-dollar Heritage sale that threatened him. By going forward with the *T. bataar* sale Eric had unwittingly contributed to his own prosecution:

the auction results sealed the dinosaur's commercial value, allowing the court to impose a higher sentence.

Eric's lawyer wished to make clear that his client admitted the smuggling but maintained that he never did anything wrong in Mongolia. The judge replied, "My concern has to do with the extent to which there is acceptance of responsibility. Someone who is fudging what happened, in my mind, doesn't have a clear sense of his own responsibility for his crimes."

Martin Bell, the prosecutor, told the judge the U.S. government had no reason to believe that Eric was "explicitly aware" of Mongolia's laws but that the constitution rendered "all of the natural things under the ground the property of Mongolia."

"Look, that is all you need, to convict," Judge Hellerstein told him.

"As black market enterprises go, this one is a little unusual, and not simply because it involves *Tyrannosaurus* bones," Bell said. "It is a little unusual because for the most part it is a black market that has thrived in plain sight, owing among other things to lack of enforcement." Federal law enforcement was "only recently realizing the contours of that black market," he went on. Whereas other illicit trafficking carried the "standard sort of earmarks" such as a "high level of secrecy," fossils were sold at auction and large trade fairs.

"Everyone averts his eyes," Judge Hellerstein said.

Pretty much, Bell said. Under scrutiny, dealers might resist selling questionable materials, "But that scrutiny hasn't existed for some time."

Lederman offered that Eric wasn't some criminal mastermind, saying, "There has been some confusion about the law abroad, in Mongolia, and to what extent he understood that."

"I have to feel, Mr. Lederman, that if Mr. Prokopi lied on the customs declaration form, he lied to cover up something, some uncomfortable knowledge, and that uncomfortable knowledge was that he was importing something against the law," the judge replied.

Bell had written the all-important "501" letter to the court, saying that the defendant had been a good cooperator and recommending leniency "in recognition and consideration for Mr. Prokopi's substantial assistance to the government." The *T. bataar* case, in fact, had "proven to be something of a net gain for the overall cause of law enforcement." Federal investigators now knew of two dozen Mongolian dinosaurs in circulation, and Mongolia was getting them back. Also, law enforcement now had more insight into the murkier side of the fossil trade, an area Bell said had been "sufficiently ignored." In fact, the case had prompted "something of a law enforcement renaissance. There is probably not an active fossil investigation at this point that doesn't owe on some level to information that Mr. Prokopi has furnished law enforcement with, at least indirectly." The case had altogether resulted in "frankly getting federal law enforcement's act together with respect to the policing of this admittedly obscure area."

Bell emphasized that by recommending leniency he didn't mean to suggest the crime wasn't "a serious one." He said, "These are natural resources that literally cannot be duplicated. You would need, frankly, the sort of [resurrection] power that the movie *Jurassic Park* suggested but which doesn't exist in reality." At one point he mentioned *Oviraptor*. "I think a number of them stampeded in the 1996 movie *Jurassic Park*. It might have been 1992. I was young and awestruck in any event, Your Honor."

"I missed the movie," Judge Hellerstein said. "Maybe I should go back to see it."

Bell said, "Every now and then it airs on TNT."

The probation department had recommended prison—a sentence of two and a half years. Hellerstein wanted to think about it for a few minutes, and to "size up Mr. Prokopi and how he feels about what he did and what are the chances of a rehabilitated life." Lederman asked the judge to consider the degree to which Eric had "already been punished indirectly." He was divorced; he'd lost his home "to effective foreclosure" and his business "to fear, whether actual or perceived, by others who no longer will work with him for fear that they will suffer reprisals by the government." When Eric was offered a chance to speak, he stood and read a printed statement: "I would like to apologize to this court and the government for my actions. What I did was wrong, and I failed to appreciate the gravity...My life has been devastated by these mistakes, but I have not lost my love for paleontology and hope to rebuild my business with more emphasis on proper documentation. I sincerely love fossils, and...I am remorseful for the damage my actions may have caused to hurt the relationship between commercial and academic paleontology."

Judge Hellerstein was in his early eighties, the same age as Eric's father. He had presided over cases involving 9/11 insurance claims, and he had ordered the federal government to make public certain evidence of detainee abuse at Abu Ghraib prison. Just recently, he had passed sentence on Kareem Serageldin, one of the few Wall Street executives convicted in the massive mortgage fraud that led to the economic collapse of 2008. In that case, Hellerstein had handed down a relatively lenient punishment, partly because Serageldin had worked "in a place where there was a climate for him to do what he did." The judge had confessed to feeling baffled by the banker's actions, telling the courtroom, "This is a deepening mystery in my work. Why do so many good people do bad things?"

Hellerstein soon returned to the courtroom and announced, "Mr. Prokopi is an unusual person. He is following a discipline that not very many people follow. The fact that he's following it and helping to create a market for it is important in the study of fossils, and the study of fossils is important in our understanding of life on earth and where we came [from] as men and women. So he's to be commended for that."

But in society, trust and honesty are important, the judge said. "That is particularly important in relationship to the discipline that Mr. Prokopi has engaged in all his life, because he in effect has made a living on the scarcity of history, and in engaging in this reputation he in fact has committed himself to respect the history and the patrimony of countries offering that history."

The sentence needed to be a deterrent, Judge Hellerstein went on. Then he surprised everyone by sentencing Eric to six months in prison.

In the courtroom, Doris and Bill looked stricken. Tyler sobbed. Amanda thought, "Okay, we can deal with this." Eric could think only of being away from the children.

Under the agreement negotiated by Lederman, he would spend three months in a federal prison, followed by three months in a halfway house. There would be no fine. Lederman asked that Eric be allowed to serve his time near Williamsburg and that his incarceration start in the fall, so that he could spend the summer with Greyson and Rivers. It was so ordered. Eric would report to prison on September 9, 2014, at two in the afternoon, at a facility to be determined.

A few days later, Seth Meyers, the late-night talk show host, mentioned the resolution of *United States of America v. Eric Prokopi* in his opening monologue: "Just a word of prison advice: Don't tell the other inmates you're a bone smuggler." Eric posted the clip on Facebook. "Well, at least you're famous," a friend commented. When Eric changed his profile photo to a courtroom artist's chalk rendering of his face, another said, "Your nose isn't that pointy." Another said, "Too jowly." A third said, "Really, who IS that person??"

CHAPTER 20

TARBOMANIA

In Mongolia, people could talk of little other than the dinosaur smuggler, delighted by the thought of their underdog country recapturing, Genghis Khan–style, a national treasure from a superpower like the United States. Mongolians who had never so much as uttered the word *dinosaur* enjoyed speculating about how Eric Prokopi got the skeleton out of the country. Stolen gold framed their understanding of smuggling—thieves hid nuggets in their intestines or anus. How big *is* this dinosaur? they wanted to know.

Oyuna had become the face of Mongolian dinosaurs, and more. The Heritage auction had happened mere days before the deadline to declare one's candidacy for parliament. "You're in the media ten times a day. How about running?" President Elbegdorj had asked her as the filing date approached. Oyuna had already run for parliament twice and lost. Mounting a strong campaign was expensive, but Elbegdorj told her she needn't worry—*T. bataar* provided plenty of publicity for free. Oyuna had won her seat in parliament, and Elbegdorj had appointed her minister of culture, sports, and tourism. "Thank you, dinosaur!" she'd said.

As Robert Painter had predicted, the Democrats had swept Mongolia's 2012 elections. For the first time in the country's history, the four most important political seats belonged to the opposition. The Democratic Party now had the collective power to make appointments and affect all manner of policy, including laws and regulations involving mining, transportation, and the judiciary. Segments of the public appeared skeptical. The government had hired a local software company, Interactive LLC, to help rush the new voting machines into service mere weeks before the elections. The machines stood about belly high and resembled smallish photocopiers. The Mongolians had taken one look at them and called them *khar khairtsag*— "black boxes." The machines represented "something of a (literal) black box to voters," wrote Julian Dierkes, an international election observer who worked in Mongolia. He said it remained to be seen "to what extent voters will trust a counting mechanism that neither they nor anyone else can observe directly." Nine political parties had called for a recount, but the elections commission refused.

Voters' questions persisted as the nation prepared for the presidential election of 2013. Who controlled the black boxes? Who paid for them? Who programmed them? Who oversaw the programmers? The machines became such a prevalent concern and subject of conspiracy chatter that two University College London anthropologists, Bumochir Dulam and Rebecca Empson, eventually wrote about the constant rumors. "Despite their alleged 'impartiality,' speculation has continued to circulate about the possibility that the *khar mashin*s are rigged or can be 'hacked,'" they wrote. "The General Election Committee maintains and runs the black boxes and its IT officers are in charge of installing and updating programs on the machines. According to some, this means that powerful politicians could very well influence and manipulate the General Election Committee, making sure the machines were used for their own means."

Oyuna, meanwhile, focused on cultural heritage and tourism. Half a million foreigners were expected to visit Mongolia by the end of 2013. Special-interest tourism appeared an ideal way to attract bird-watchers, skiers, golfers, and spiritual pilgrims. A new international airport was being built in Khushigiin Valley, 30 miles south of Ulaanbaatar, with plans to develop a city around it. Oyuna targeted 2015 as the year for "significant promotion of Mongolia to the world." The government began testing new tourism slogans; "Go nomadic, experience Mongolia" was followed by "Mongolia—Nomadic by Nature."

Nothing attracted tourists like the Gobi. Oyuna envisioned dinosaurs as central to Mongolia's new age of tourism. Dinosaurs could lead the nation out of a recent tourism slump and its outdated ways. Mongolians loved their festivals—Ice, Camel, Eagle, Snow—but one could never be sure when the events would happen or how long they'd last. A noncommittal attitude toward scheduling made it difficult for tour companies to advertise and for travelers to plan. At a recent Mongolia Economic Forum event, everyone had agreed that dinosaurs held "vast potential," and Oyuna was already thinking about a dinosaur-centric tourism route capitalizing on President Elbegdorj's "retrieval operation" of the *T. bataar.* "Mongolia's been criticized by world paleontologists and many science lovers that government didn't do anything when smuggling was going on," she told Al Jazeera. "But there was no particular case from which government could grab, and start stopping the illegal fossil dealing. So *T. bataar* was an ideal situation."

It was Oyuna's job to decide where to put the dinosaur once the United States repatriated it. The Natural History Museum wasn't an option because it was in such bad shape. Oyuna decided to develop a new museum, with *T. bataar* as its centerpiece. When fellow cabinet members teased her about creating a whole museum around a single dinosaur, she told them, "Roy Chapman Andrews's expedition took fifteen hundred pieces from Mongolia, and after that many other expeditions from many other countries took things. If you bring it all together it's going to fill *ten* houses." The government didn't have the money for a new building, so Oyuna decided to repurpose an old one.

The Lenin Museum, honoring the founder of Russia's Communist Party, opened in 1980, a two-story square monolith. Inside loomed a minor mountain of a pedestaled bust of the goateed old Bolshevik with his eyes closed like a death mask magnified. ("If they decided to use it as a permanent [dinosaur] museum, I would think Lenin's head would need to be removed, because in terms of content it doesn't really go," Bolor Minjin told the press.) Other than the giant Lenin head the place was perfect: large, air-conditioned, with decent lighting and an excellent center-city location on Independence Square, where the first stages of the democratic revolution had unfolded barely twenty years earlier. The building had since housed high-rent restaurants, a karaoke bar, a florist, and a billiards hall, with the Mongolian People's Republic Party as landlords. Previous cultural ministers had tried to claim the property for the state and always lost in court, but Oyuna believed she could win.

She did win. After renovations, the Lenin Museum would become the Central Museum of Mongolian Dinosaurs and, in her vision, nationwide satellite museums would follow. "We have a wonderful dinosaur heritage but people are not aware of it," she told the media in early 2013. In the meantime, she worked on finding the *T. bataar* skeleton a temporary home.

On May 6, at eleven a.m. a small crowd—Oyuna, Bolor, Philip Currie, the Painters—gathered in the Manhattan Room of 1 United Nations Plaza in New York. As news cameras rolled, John Morton, then the director of Immigration Customs and Enforcement, took the lectern. Customs had repatriated lots of crazy stuff over the years—an 18-karat-gold bookmark supposedly given to Hitler by Eva Braun; a chrome-plated AK-47 with Saddam Hussein's image on it—but the dinosaur was something else. "In this business I hesitate to identify any particular return as the most unusual but this is clearly one of the most exceptional, if not the most exceptional, we've ever returned," Morton had told the *New York Times*. Now, he said, "We simply cannot allow the greed of a few looters and schemers to trump the cultural interests of an entire nation."

The crates of *T. bataar* bones were moved out of government storage and into a jetliner belonging to Korean Air, which had offered to fly the dinosaur home for free.

———

Ulaanbaatar's main city square has been called Sukhbaatar Square, Central Square, Parliament Square, and, most recently, Chinggis Khaan Square, the latter honoring "every Mongolian's pride and idol." The limestone heart of the ancient city used to be all temple and palace. Where distinguished clergy and nobility once gathered to watch dancing and wrestling, children now zip around in electric toy cars. Chanting lamas in robes of red or gold still gather to pray. On auspicious days for weddings, brides and grooms materialize on the picturesque steps of the colonnaded Government House, at the gargantuan bronzed feet of Genghis Khan. The most distinctive, incongruous building on the square is the futuristic Blue Sky Hotel and Tower. Eighteen stories tall, the tower is a monolith of azure glass and steel whose shape has been compared to a shark fin, a sail, and, from the vantage point of Government House, female genitalia. Whereas Genghis once oversaw an empire that came to rule half the known world, he now enjoys an eternal view of the architectural equivalent of blue labia.

At nearly 82,000 square feet, the square is large enough to accommodate military parades (1920s–1950s), a democratic revolution (1990), small-plane landings (a theory; it could happen), or, as transpired in late May of 2013, the frenzied construction of a pop-up dinosaur museum. The dinosaur would go on display for three months, then move temporarily to Darkhan Province, 75 miles south of the Siberian border.

Oyuna's new cabinet position appeared to bode well for Bolor, who still struggled to find her place within paleontology. "She's mostly castrated while she's in Mongolia," Oyuna later said. "The paleontological community is very envious or negative about her because her primary subject is geology. They'll say, 'Okay, we will give her permission to observe the area geologically but not permission to dig.' When Bolor got credit for *T. bataar* the paleontologists were envious, saying, 'Well, she didn't actually do anything.' We said, 'Bolor's input was crucial.' They said, 'No, it was you and Robert Painter.' They tried to limit Bolor all the time. If you ask Mongolians they will say, 'Oh, I admire Bolor.' But if you ask a scientist they'll say, 'Bolor did nothing.' It's a club that never let Bolor enter." Even though Bolor lived on Long Island, Oyuna put her in charge of the pop-up museum.

The dinosaur arrived in Ulaanbaatar at midnight on May 17, a year to the day since the publication of "Dinosaur's Dreams," Oyuna's "nice article" that ran in the newspaper President Elbegdorj had founded. In a khaki trench coat and a bright yellow scarf, Oyuna met the plane on the tarmac. As the plastic-wrapped crates of dinosaur bones were rolled out of cargo, TV cameras followed along live, as if the shipment contained the mortal remains of a head of state.

Bolor arrived a few days later, facing the daunting prospect of erecting the pop-up exhibit within her mandate of two weeks. The building went up before the fixed eyes of Genghis: a large, prefabricated rectangle with a high ceiling, a platform, lighted display cabinets, and LED screens. As an exterior touch, Oyuna commissioned her own brother to paint a mural—it showed *T. bataar* as a grinning cartoon figure making tracks from the Statue of Liberty to the green hills and *ger*s of Mongolia, holding balloons and a Mongolian flag. The museum's theme was "I am home."

Oyuna planned an unveiling befitting the prime minister. But the day before the event, he backed out. Mongolia's death-related sayings involved bones, so to attend the *T. bataar* ceremony loosely meant that you were "dying on the square." Oyuna's critics already called her Bone Oyungerel or the Dinosaur Minister, but she kept explaining, "It's not bone. It's just stone." She told the naysayers, "If you hate bones, all of Sukhbaatar Square should be peeled off, because it's also bones—limestone." Turning to the leaders of the largest lamaseries she said, "Please come and contribute to science by your appearance and admiration. People have very wrong understanding about bones. Please come and say that you admire this paleontological finding."

On the morning of June 8, hundreds of spectators queued in Sukhbaatar Square, including the lamas. White-gloved dignitaries cut a green ribbon. Oyuna made opening remarks, but instead of giving a bureautic speech, she told a fairy tale, made up on the spot, about *T. bataar* reuniting with his mother.

Inside the pop-up museum, visitors found an exhibit more efficient and informative than in any museum in Mongolia's history. The *T. bataar* stood on a bed of Gobi sand. Spectators circled the skeleton, their cameras and phones held aloft. Onto one wall, a video loop projected the U.S. government's repatriation ceremony and the dinosaur's arrival by plane. Displays held photos of the New York City auction, including Robert Painter with his BlackBerry, squaring off with Heritage's Greg Rohan. When Painter arrived from Texas and donated the phone to the exhibit, a news station interviewed him live on national TV for eighteen uninterrupted minutes. Persistent rumors about vote-rigging and public corruption had followed Mongolian officials all the way to the dinosaur exhibit—in one version, the American government, backing Elbegdorj for reelection, had engineered the entire *T. bataar* matter, with Robert Painter working as a CIA asset. Until officials saw the dinosaur in person, some assumed it had been "made up."

For the first four weeks, admission to the pop-up exhibit was free. An estimated 267,000 people visited. Oyuna had made the mistake of hiring only two guides, but when she saw the size of the crowd, she told the entire staff, "Okay, you guys, this is a new-era museum, a new kind of exhibit, a new kind of interest. You have to change your management. *Everybody* has to be guide, *everybody* has to be curator." She asked a guard, "Have you heard about *Protoceratops*?"

"I can't even say *Protoceratops*," he said.

"*Pro-toe-SERRA-tops*," Oyuna said. She turned to a janitor: "Okay, you, Cleaner. Say 'Tarbosaurus bataar.'"

"Tarbosaurus bataar."

"You. Say 'Oviraptor.'"

"Oviraptor."

"Learn to explain things in simple words," Oyuna told them. "Learn this word and this word and this word, because starting tomorrow, everybody is an educator."

When Oyuna returned to the pop-up, she barely recognized the employees. The staff was busy explaining what is *T. bataar*, what is dinosaur, what is paleontology. A tent bazaar had sprung up in the square, where vendors sold *T. bataar* trinkets, teacups, and pillows. People were wearing *T. bataar* T-shirts and baseball caps. Thirty mothers were raising money for their physically disabled children by selling handmade *T. bataar* toys. Drafting on the pop-up's popularity, the city museum of Ulaanbaatar created an exhibit called "Indiana Jones: Roy Chapman Andrews in Mongolia," showcasing photos from the American Museum of Natural History's Central Asiatic Expeditions of the 1920s. In an accompanying photo book, the curator, B. Tungalag, wrote, "Those expeditions proved that Mongolia was a scientifically unique and important region with large amounts of ornithology, geology, paleontology, paleobotany, and archeological resources."

By September, the pop-up dinosaur exhibit had taken in 200 million *tugrik* (about $80,000). By the end of 2013, the government claimed some 750,000 people had seen the *T. bataar*. Watching videos of the event in Canada, Philip Currie said, "This is the first time, really, that Mongolians have even been aware of their own [paleontological] heritage." When the paleontologist Rinchen Barsbold saw the throngs of people lined up to see the dinosaur skeleton, witnesses remembered him tearing up.

There was still no national Paleontology Day, but there was a national *T. bataar* Day. It fell on October 17, the anniversary of Eric Prokopi's arrest.

———

The day after the United States repatriated the *T. bataar*, the Democratic Party of Mongolia unanimously nominated Elbegdorj as their candidate in the 2013 presidential election. Ten days after the dinosaur went on display at Sukhbaatar Square, where the exhibit featured documents prominently showing Elbegdorj's name, the president's chief of staff said the return of *T. bataar* "enriched the true nature and spirit of the cooperation between the Mongolian and American peoples....America has once again shown to the world that it is a beacon of justice." Thirty-nine days after that, Elbegdorj was reelected president.

Over four hundred outside election observers, most on behalf of the Organization for Security and Co-operation in Europe (OSCE), had fanned out across Mongolia to monitor the election. While the election overall was deemed open and fair, the OSCE reported concerns. The General Election Commission had displayed an "overall lack of transparency in its decision-making," structure, and procedure; and during a pre-election systems check of the new voting machines, testers had found a "programming error" that prevented the machines from "correctly calculating the number of invalid ballots." Dominion Voting Systems confirmed the error but failed to provide the source code that would have allowed a fix: the company said the error had been discovered too late to reprogram the machines. Dominion issued assurances that the errors "would not affect the results," but the OSCE reported that since Mongolian law required a winning candidate to receive the majority of all votes cast, it was "essential for the [vote-counting equipment] to accurately establish and report the number of invalid ballots cast." The OSCE also had a problem with the media environment: the overwhelming majority of news outlets were reportedly "directly or indirectly owned by political actors." Mongolian journalists told the OSCE that two big ethical no-no's were common: people paid for news coverage and media owners interfered with editorial autonomy and attempted to discredit opponents with "black PR." The OSCE reported, "A lack of transparency in media ownership leaves the public unable to fully evaluate the information disseminated by the media."

Elbegdorj had been re-elected by only about twenty thousand votes. Yet there were no public protests. The peace held. On July 10, he was sworn in for his second and final term. Wearing full traditional regalia at Sukhbaatar Square, he bowed to the statue of Genghis Khan.

Recently, Hillary Clinton then secretary of state, had visited Mongolia for the first time in some twenty years. She had last traveled there as

First Lady. The place was different now that a surging Chinese economy increasingly put its northern neighbor's natural resources in demand. The fitful Mongolian economy had grown a staggering 17 percent in 2011, and an influx of foreign mining investments had given Ulaanbaatar a boom-town vibe. The potential windfall threatened to either bring Mongolia great riches or worsen corruption and the wealth gap, putting Mongolia at what Clinton saw as a "crossroads." She later wrote that Mongolia would either "continue down the democratic path and use its new riches to raise the standard of living of all its people, or it was going to be pulled into Beijing's orbit and experience the worst excesses of the 'resource curse.'"

Some thought democracy wasn't "perfectly at home in Asia," but Clinton said Mongolia proved otherwise. She praised the fact that nine women had recently been elected to parliament, and in a public speech she obliquely criticized China for its human rights violations. Her support of Mongolia's democratic struggles offered "hope that the U.S. pivot to Asia will go beyond simple muscle-flexing and become a multilayered approach to match the complexity of China's rise as a modern superpower," the *Washington Post* editorialized.

While in Ulaanbaatar, Clinton had met with Peabody Energy executives and Mongolian leaders to speak privately about Tavan Tolgoi, the Gobi Desert coal deposit, whose development contract Peabody ultimately did not win. She and Elbegdorj spent some time talking in the ceremonial *ger* inside Government House—"our Oval Office," as the president put it. Later, Eric wondered if they talked about the dinosaur. In the one and only statement he released to the public, he had complained that federal authorities had had "political" reasons for bringing actions against him.

The idea seemed farfetched. Why would any U.S. government entity beyond law enforcement care about a dinosaur case? There was no evidence that anyone other than Homeland Security and the Department of Justice had taken an interest in the Mongolian matter. Yet less than twenty-four hours before the Heritage auction, Homeland Security Investigations had received "information" that "the auction of a *Tyrannosaurus bataar* skeleton from Mongolia" was to take place the following day. The tip, noted in case officers' internal files, came from the State Department.

The State Department's mission is diplomacy, a crucial adjunct to military strategy, peacekeeping, and intelligence. The agency exists to "shape and sustain a peaceful, prosperous, just, and democratic world and foster conditions for stability and progress." If Mongolia was to withstand pressure at the borders, and if the United States wanted a continued presence in a difficult region, the two countries needed to maintain their "friendly relations." After all, it wasn't easy pushing toward direct democracy when your neighbors were Russia and China, Elbegdorj explained weeks later during a public appearance at Harvard.

Standing before a small audience at the John F. Kennedy School of Government, he recalled telling Clinton a story during her visit to Ulaanbaatar. The story involved "dinosaur *bataar*," which had "slipped away from Mongolia" and ended up in the United States. "That dinosaur actually lived seventy million years ago," Elbegdorj said he told Clinton. "Now, in these days, in these years, we got one dinosaur, called corruption."

That year, a study commissioned by USAID and the Asia Foundation found that corruption in Mongolia had worsened since 2005, not improved. "Opportunities for corruption have increased at both the 'petty' or administrative and 'grand' or elite levels," the report noted. "Both types of corruption should concern Mongolians and investors, but grand corruption should be considered a more serious threat because it solidifies linkages between economic and political power that could negatively affect or ultimately derail or delay democracy and development."

CHAPTER 21

PETERSBURG LOW

GOING TO PRISON FELT A LITTLE LIKE KNOWING YOU WERE about to die. Arrangements had to be made, loose ends secured. Several weeks before he had to "self-surrender," as it's called, Eric made a big decision and called off his relationship with Tyler, whose own marriage, by now, was ending in divorce. His life was too confused to include anyone other than the children and Amanda. As Tyler climbed into her old Saab and left Virginia for Florida, they both crumpled, even though both could admit feeling relieved to see "some of the bricks drop," as Tyler put it.

Eric was losing the river house in Hayes because they had fallen behind on the payments. His last chore before vacating the property was getting rid of the thousands of auction items he and Tyler had amassed in hopes of reselling them: oyster tongs, a pump organ, a coal bucket, Pyrex, pewter, maps, model sailboats, cameras, hooks, pulleys, levels, meters, a porthole, a Victrola, a cow skull, milk cans, canteens, compasses, antique fish knives, clocks, bottles, doorknobs, radios, linens, clothes, toys, corals, ammonites, whale vertebrae, and clamshells. The antique wedding dress cabinet that had once stood in the front hall at Serenola sat empty and listing. Eric had contracted with a live-auction company to help him dispose of everything. They were coming on Saturday morning to group the items into 467 lots.

The only way on and off Gloucester Point by land ran through the toll booth at the foot of the Coleman Bridge. To avoid the $2 toll, Eric tried not to go back and forth unless he had to, but at five o'clock on the evening before the auction, he headed toward Yorktown, passing the late-summer fields and the boatyards and Rising Sun Baptist Church, whose marquee read, "God has subpoenaed all of us as witnesses." Amanda met him in the parking lot of the Watermen's Museum, where, thirteen years earlier, they had held their wedding rehearsal dinner.

Eric buckled the kids into the van and drove back over the river, past the osprey nests, past their favorite Mexican restaurant, Juan's. The instant he turned onto the gravel driveway, the kids burst out of their seatbelts like they always did. "Get up here, Grey, I need you to drive," Eric said. Grey clambered into his father's lap and steered; then Rivers took a turn, passing the little red workshop and the giant magnolia tree where she had once mistaken a snake for a garden hose.

Later, watching his father sort auction items, Grey said, "Are you moving back home, Daddy?"

"Sort of."

Eric's plan was to move into the garage apartment at Amanda's mom's house. At Betty's, he and the kids could see each other every day until he went to wherever the Bureau of Prisons was sending him. Technically, he could be sent to any federal penitentiary in the country, but Eric had petitioned the court to serve his time in Petersburg, about an hour away from Williamsburg.

He heated up a frozen pizza and the kids turned on *Peppa Pig*. As they ate, Rivers said, "Daddy, do we have to sell everything?"

He said they did.

"But this is my art desk," she said. She was sitting in a small school chair, spreading her hands out over the wooden surface.

Eric thought for a second and said, "You'll be fine."

Saturday morning, the tenth of August. The river throbbed with jellyfish. Last night's moon had been fuzzy and almost full. By the time the auction preview started at ten, the weather was stifling. Up walked a stocky guy in amber-tinted eyeglasses with thick gold bracelets halfway to both elbows, and men in trucker hats and tank tops, and women in capris and sun visors, holding sweaty, napkin-wrapped glasses of iced tea. Eric had returned the children to Amanda and had been working nonstop ever since, sorting what was left of his material life. Throughout the day he ghosted around, talking to hardly anyone as the auctioneer pattered.

He had hoped to bank $20,000 from the auction, but the total came to about ten, and the auctioneers took 40 percent of that. After everyone left,

Bennie, Eric's next-door neighbor, came over to pick up the Harbor Freight welder he had bought, the one Eric had used to make mounts, including the *T. bataar* armature, which had been sent to Ulaanbaatar along with the skeleton. As they stood talking in the driveway another neighbor, Jerry, an old guy with a glass eye, came riding up on a John Deere lawnmower, driving it like it was a car. When Jerry cut the engine, Bennie said brightly, "I been meaning to come over there all day and kick your ass."

"Well, you better find you a stick!" Jerry said.

For the first time that day, Eric laughed.

On Wednesday, he turned forty. All day he ferried loads of possessions to the garage apartment at Kingsmill and put the rest in the box truck or took it to the scrapyard and traded it for cash. At one point, he arrived at the river house to find an eviction notice taped to the front door. He stood looking at it for a minute without touching it. Then, without saying a word, he unpeeled it from the door and went inside.

That night he met Amanda and the kids for dinner in Williamsburg. Amanda had reserved the wine room at a nice restaurant called Waypoint and had invited Betty and her boyfriend, Ben, and a few of their friends. She wanted Eric to eat good food and feel loved. Rivers had on a sundress patterned with seahorses and crabs. Grey wore an orange-checkered Brooks Brothers shirt and white pants. Eric sat at the head of the table, beneath a cluster of helium balloons, looking pleased but distracted. Just before the food came, he got an email from his lawyer. Instead of assigning him to a prison near Williamsburg, the feds were sending him to a prison *called* Williamsburg—in Salters, South Carolina.

"It'll be okay," Amanda told him.

"No, it won't!" Eric said. "That's like six hours away!"

"They can move you," she said.

"They don't give a shit."

The kids sang "Happy Birthday" and they all ate Harris Teeter sheet cake. Rivers gave her father a book of hand-drawn flowers and sea monsters and stick-daddies holding hands with stick-daughters. Grey gave him a small box; inside lay a tiny crab he'd found on the beach at the Outer Banks. One of the crab's arms had broken. "That's okay," Eric said. "We can glue it back together. We'll put it in a frame."

The next day at the river house, Eric scooped all the fish out of the kids' aquarium, along with Denton, a turtle they had found in the driveway, and drove them in water-filled buckets to Williamsburg. At Betty's, he reassembled the aquarium in the narrow stairwell that led to the garage apartment, filled it up, and put the fish back in. The turtle did fine, but within hours the fish started floating to the surface, and by day's end, they were all dead. When Amanda found Eric bawling, she said, "Aren't you sick of being sad? Aren't you ready to just get on with it?"

In a flower bed at the corner of the garage, they gave the fish a family memorial service. Grey and Rivers played "Taps" on Amanda's iPhone and arranged stones in the shape of a heart. Their construction-paper tombstone read FISH DIED HERE A LONG TIME AGO.

The prison mistake was quickly corrected, and Eric got assigned to the satellite camp of the medium-security facility in Petersburg. He took one last trip to Florida in the box truck, driving the whole stretch in one fourteen-hour all-night haul. The truck was neither registered nor insured. It wouldn't crank with a key, so he basically hot-wired it. The speedometer was broken, so he used an iPhone app. The gas gauge didn't work, so he kept track of the fuel level with a stick that he poked into the tank every now and then. Whenever the busted glove compartment flopped open, he reached over and wedged it shut with a scrap of cardboard. Wires dangled from the dashboard. The truck occasionally went *pop-POP*—the sound of a tire shredding.

Dealer friends had hired him to prep fossils for a few days at thirty dollars an hour, for pocket money. At his friend Tony's welding shop in Venice, he cleaned the remnants of a fossil camel, saying, to no one in particular, "It was the size of a small dog. Or like a newborn deer. The horses of that age weren't much bigger."

At one point he and Tony looked at old Mongolia photos on Tony's computer. Here was the place where they saw all those dead hedgehogs. Here was the frozen waterfall. Tuvshin's basement. Otgo, sleeping. "Oh, this is one of the many tea breaks," Tony said. "They were always stopping for tea, and it was the hottest milk tea—the hottest liquid you could ever imagine short of molten lava."

"Horse milk," Eric said. "But it's not the fermented—"

"Let's put it this way," Tony said. "They drink fermented mare's milk and it's awesome to them. Tuvshin was disgusted with our hard cheese. Like, cheddar: to him, that was gross." He clicked. "There's Tire Kingdom."

"There's a pile of horse and camel bones," Eric said. "There's the fossil market, out in the middle of the desert."

"There was *nobody* there, two minutes prior," Tony said. "Then suddenly they came out of the ground and ran to their tables, trying to sell us rocks."

"It was *mostly* rocks," Eric said. "But there were nice toe bones and vertebrae and stuff."

There was the military base, the Natural History Museum, the picnic. Eric and Ulzii wrestling. "I got my ass kicked," Eric said.

Tony said, "That's what happens when grown men get together and drink."

Eric drove out to Joe and Charlene's house in the country for an early supper. He hadn't been out there since the day he hid the dinosaurs in the pole barn. Charlene fixed spaghetti. They talked about old times pulling logs. The TV was on. Ferguson was happening. "You loot, we shoot," Joe said. He disappeared and came back to the table with a huge pistol, then sat down and wagged it around, like he couldn't decide where to point it. Charlene promised to visit Eric in prison. "What do the kids think?" she asked. They don't know, he told her. He and Amanda had decided to tell them Daddy was going on a business trip.

They walked around the yard at twilight and stopped at a table where Joe was drying some zinnias for their seeds. He pulled out a box of bullets he'd found in a river. Then he fired up a table saw and started slicing pine. Joe was "not Braille blind, but legally blind," as Amanda put it, and every push toward the whining blade was like watching cars speed toward each other and just miss. It was getting dark out there, and the overhead lights didn't work, and bats were flapping around in the eaves, and blind Joe went on sawing. He thought of Eric like a son, *that's* why he hid the dinosaurs. "Fuck the government!" he said. "Fuck 'em!"

In one building, Joe kept river treasures in the kind of glass display cases you might see in a jewelry store. Inside one was a rusted stake, an alligator skull, and beaver-gnawed wood as sharp as a spear. Pointing to an arrowhead, Joe said, "That's Prokopi right there," the way someone might say, "That's Van Gogh."

———————

Surrender day was warm and damp. The sidewalks of Kingsmill gleamed with humid sweat. Autumn leaves lay pasted to the streets, the gutters banked with gold. In the backseat, on the way to school, the kids yammered about nothing, oblivious to the fact that today, the ninth of September, was the day their father would become federal inmate number 21746-017. When they arrived at the school, Eric couldn't make himself get out of the car to hug them, so he stayed in his seat and said, "Be good" and "I love you" and "See you later."

As Amanda drove him to Petersburg, he stared at a commissary form, which was all he knew to do. He had on cargo shorts and an O'Neill T-shirt, and his "good" flip-flops. "Toothpaste, toothbrush—," he said. On the form he made a mark beside "Sony SRF-M35."

"Eric, do you really need a radio for forty-one dollars?" Amanda said. "If you don't have a radio you might get more done. I'm just saying. You might get more letter-writing done."

"I can only write so many letters."

"You'll be the new guy. They'll steal your radio."

"They're not stealing my radio."

They wondered how Petersburg Low, the low-security satellite camp, would look. "Maybe it's in a pretty tent," Amanda said, trying to keep things light. "Maybe it's *like* camp. Maybe they're wearing sweaters over their shoulders and playing tennis. That's what I'm envisioning."

They were 40 miles away, obeying the GPS. Eric wasn't due until two o'clock, but he wanted to be there by twelve thirty, just in case. Amanda said, "One person I know said, 'I think what they have the prisoners do is clean the beach.' I was like, Eric will love that!"

They crossed the James River. Before long, Amanda said, "Seven point five miles away. I'm hungry." When Eric didn't say anything she asked, "You okay?"

He didn't answer.

"Can I do something?"

He didn't answer.

"It's kind of like going on a trip to a place you've never been before," Amanda said as she exited toward McDonald's.

The GPS said, "Recalculating."

At McDonald's Eric ate a Quarter Pounder, then borrowed Amanda's phone and went out to the patio to call his mother. As he talked, a cicada landed on the extended index finger of his free hand and stayed there, like a pet parakeet. A man inside McDonald's saw Amanda watching and said, "Ain't that ugly? They call that a *ci-ca-dia*."

When they were back on the road, Eric checked the time. "We'll be there at one fifteen," Amanda told him. They passed a boatyard where yachts were being washed and wrapped for the winter. "Turn right on West Broadway Avenue," said the GPS.

Amanda flicked on the windshield wipers as it started to drizzle. They stutter-dragged across the glass.

"What do they do with people who show up with no money?" Eric wondered, still staring at the commissary form. "They can't get a toothbrush or—"

"Well, that's when you do stuff to earn money from your friends," Amanda said. "Like, tattoo or hair-braid. You'll figure out your trade."

Eric said, "Yeah."

They were on a two-lane road, in a residential area. "In three-tenths of a mile, turn right on River Road," said the GPS. Grabbing Eric's hand, Amanda said, "Well, that's a cool road to be on. River Road in Newport News is where all the fancy houses are. And you have Rivers. Rivers Road."

Fields came into view, then a water tower, then security lights overlooking a cluster of buildings. The GPS said, "Approaching your destination."

Looking at the expanse of fields, Amanda said, "Well, you can't run."

"My area's not even fenced in," he said. "I could walk off at any time."

"Are you for real?"

"Why would I be escaping when I'm self-surrendering?" he said irritably. "I could escape *now*."

Amanda was reading the signs. " 'Administration, parking, warehouse.' What the fuck?" In the parking lot she spotted a uniformed employee with a giant set of keys dangling from a belt loop. Pulling alongside him with her window down, she said, "Hey! I've got a question. For the Low satellite camp, is this where you go to surrender?"

"No, Low is over there," the guy said. He hollered over at another employee, who told Amanda, "Follow me." Amanda said, "See, look how nice these people are, Eric! He's escorting us in his family van."

The medium-security unit was surrounded by high fences topped with concertina wire, but Petersburg Low looked like dorms. Inmates were outside lifting weights; one was driving a golf cart. "Look, they're having soda and walking around!" Amanda said. "They're working out! This'll be good!" She had been envisioning Eric using his time to "get ripped."

As she parked the car beneath a guard tower, Eric said hesitantly, "I told Tyler I would call her today." Without a word, Amanda handed him her phone.

She walked over to Administration to see what they needed to do next. Just inside the front door, she saw a correctional officer at a security checkpoint and said, "Hi! My ex-husband is surrendering today and—"

"He needs to be here *now*," said the officer, a woman who wore her hair pulled back tightly from her face. As Amanda tried to explain that Eric wasn't due until two, the officer, without asking for a name or consulting a document, said, "If you don't want him in the hole, he'd best be here *now*. He was supposed to be here at twelve."

Amanda sprinted back to the car. Eric was hanging up the phone, wiping his eyes. Amanda cranked up and zoomed over to Administration. A male officer came out and redirected them to a visitors' lot, down the hill, and told them to wait in the visitors' booth, which looked like a bus stop with plastic strips for doors. The booth had a phone; Eric was supposed to use it to call up to Administration, the building in front of which he now sat. When Eric started to speak the officer said, "Listen at what I'm telling you! You're not listening! You gotta listen when you go in *there*, so you'd better start listening now!"

Eric said quietly, "I know."

As they parked Amanda said, "I think they assert themselves so that you remember that they are the ones that work here and you are not."

The visitors' booth consisted of a metal bench, a table, fluorescent strip lighting, and a wall-mounted phone. A sign gave instructions to call on the red phone, but the only phone in the booth was black. "What'd he say? Pick up the phone?" Eric said.

"I guess." Amanda studied the numbers people had scrawled on the wall. "Four-three-five-seven? Just try one. Do you want me to do it? I'm not afraid."

As they searched for the right number, the phone jangled. Eric answered it, and a firm female voice said, "Hug your family good-bye and come up." "Right now." Amanda hugged Eric and said, "Just think of it as college."

He met a Billy, an Anthony, a Jason, a Terry, a Randy, a Nick, a D.B. One was an ex–Navy SEAL. One was a country boy, in for cocaine trafficking. One was ex–Coast Guard, in for something about taxes. One had embezzled money from the Dave Matthews Band. The one they called Mayday had gotten drunk on a boat and radioed a false call for help. Eric met a guy named Flowers who weighed 450 pounds, and Lefty, who was missing an arm. When they heard Eric was in prison for smuggling a Mongolian dinosaur, they called him "Indiana."

The monotony set in pretty fast. Up at six. Back to bed. Long walk around the track. Lunch. Nap. Volleyball. Watch *Captain Phillips* on DVD in the library. Watch *Superman*. Walk five laps around the track. Draw animal anagrams for the kids—

Giraffe
Rat
Emu
Yak
Snake
Owl
Narwhal

Rooster
Ibex
Viper
Eel
Ray
Spider

On the ninth day, he ran a lap, then a mile, then two. He started drinking hot tea. On the twelfth day, he got his job assignment, working the morning shift in the cafeteria. By now, he knew that dried mackerels, from the commissary, were currency because they were protein, and that one mackerel equaled a dollar. One mackerel paid a guy to do your laundry for a week. On the eighteenth day, Amanda came to visit, but it was awkward; she came twice more and then never came again.

He kept a prison diary—laundry, noisy guards, "great apple muffin." "Beat Carlito in arm wrestling." "Started playing flag football." "Felt sad all day so no exercise." "I can't seem to not eat tons of carbs here." In October he made Amanda and the kids a Halloween card of plain white paper and autumn leaves he'd collected on the prison grounds, laminated with clear packing tape. In a letter he said he was "fantasizing about doing some things as a family." Maybe they could go to Florida after Christmas and see his parents, or to Gainesville to see their old friends. Maybe SeaWorld, to show the kids where Amanda used to work. Maybe Northern California, for spring break. "I loved that trip and want to show our kids so much and take them to see the world," he wrote. He had come across a boating magazine in prison and was amazed by the different kinds of sailboats you could live on. How cool it would be, to "just go."

As Amanda and the kids went on with their autumn, she realized she had been having "Lifetime Movie fantasies" about getting back together with Eric. It wouldn't work. And maybe she didn't want it to. Right now all she wanted was to rebuild her credit, pay bills, and find a "cozy nest" that she could afford on her own. During an otherwise abysmal period of online dating, she had met a sweet army guy and father of four named George Bryan. She loved her job at Jackson Thomas Interiors, where she specialized in staging houses. For the first time in a while, her life felt somewhat stable, and she wanted to see where it would go.

One day, she received a thickish envelope in the mail, with no return address. Folded into a blank sheet of paper were two long, thin pieces of cardboard pressed together. Opening the lengths of cardboard like a book, she found crudely cut, oval indentions. Nestled into each indention was a flat object wrapped in toilet tissue. Peeling back the toilet tissue, she saw Indian arrowheads. Right then she knew that if Eric could manage to find arrowheads while in prison, he would never stop hunting treasure.

CHAPTER 22

THE DINOSAUR BUS

Bolor Minjin wanted to know who had dug the famous *Tarbosaurus bataar* skeleton, and where. "Like, the *hole.*"

Any paleontologist knew it was impossible to match a specific fossil to a specific hole in the ground, even within a geological formation as signature as the Nemegt. If a poacher did admit to being the digger, he would almost certainly be winging it, or lying. A range of clues would surface the discrepancies, not least of which was the fact that the million-dollar dinosaur was not one creature, but rather an amalgam of various skeletons, as had been revealed in court. Still, Bolor wanted details. Then again, she didn't. She had reservations about nosing around in criminal matters, and she certainty did not want her fellow Mongolians wrongly assuming her a snitch. Anyway, she had found another way to combat dinosaur poaching, and in the summer of 2015 she went home to Mongolia to try it.

As July gave way to a blistering August, she settled in with her mother. One Friday morning she dressed in army green capris, a pink T-shirt, and a long-sleeved plaid shirt, a *Spinosaurus* pin affixed to her drab-olive Baggallini purse. Her father had died recently, in Virginia, where Bolor's sister lived, of pancreatic cancer. Bolor was using his cell phone case, his Canon camera, and his car, which he had trimmed with padded seats and a cloth steering wheel, a sky-blue *khadag* scarf threaded around the sun visors. The rearview mirror twinkled with rhinestones; on the dashboard bobbled a plastic turtle. Steering through the snarl of Sukhbaatar Square, Bolor left the city center and came to an unpaved street, where she parked at Wagner Asia, a Denver-based Caterpillar equipment company that had been operating in Mongolia since the early days of capitalism. She walked past a security guard in combat fatigues and toward a service warehouse where workers in hard hats were driving forklifts and whacking at steel bars with large hammers. Nearby sat a city bus encased in a blue vinyl skin. Emblazoned on both sides were the image of a long-necked fat-tailed sauropod and the words DINOSAURS—ANCIENT FOSSILS, NEW DISCOVERIES, MOVABLE MUSEUM.

In the late 1800s, American educators debated how best to teach the natural sciences in public schools, often cross-pollinating disciplines such as biology and geography with other topics. Massachusetts offered "Nature-Study and Literature." California had "Nature-Study and Moral Culture." New York City simply added "Nature Study," allowing public school students to examine ten small cabinets of preserved birds on loan from the newly opened American Museum of Natural History. The program proved so popular that in 1904 it expanded, reaching well over a million students by 1916. The institution's "movable museums," which had "started with a suitcase," progressed to delivery by motorcycle, and in the 1930s, operated out of converted ambulances. In the 1990s, AMNH switched to what were essentially city buses retrofitted with walk-in, interactive exhibits. Each bus featured a different subject, like anthropology and astronomy. The buses lumbered throughout the five boroughs, often drawing three hundred students a day while parked outside public libraries or churches.

The museum eventually discontinued the program, and in 2013, announced plans to donate the buses. Bolor put her hand up for paleontology, with the idea that she'd drive it into the Gobi. The displays were interactive, a concept that did not yet exist in Mongolia's museums, and conveniently, they already featured Gobi dinosaurs, one species of which was now well known because of *USA v. Eric Prokopi*. The museum approved her application, and before long Bolor's NGO, the Institute for the Study of Mongolian Dinosaurs, was in possession of a modified Winnebago Adventurer still marked with AMNH branding and the name of the program's early financial sponsor, Bloomberg.

There arose a logistics problem: a bus measures 40 feet long, weighs over 10 tons, and is located in New York City. It must be moved to the other side of the world, yet it cannot be driven there. How many dollars are needed to relocate the bus the 10,783 nautical miles to the port of Tianjin, China, and then deliver it overland another thousand or so miles to Ulaanbaatar?

After finishing her postdoc, Bolor had stayed close with Jack Horner and the Museum of the Rockies in Montana. Some scientists are good at science, some are better at explaining it to a lay audience, and a few are good at both. Horner had been called "the Tom Sawyer of paleontology" because he "gets people excited about dinosaurs, and then he recruits them to come and work (free!) at his digs." That talent could also be parlayed into funding. The Museum of the Rockies's board of directors included wealthy science lovers such as Nathan Myhrvold, Microsoft's former chief technology officer. A polymath in his fifties with a Princeton PhD in theoretical and mathematical physics, Myhrvold was a trained chef and cookbook author who once teamed up with Microsoft cofounder Paul Allen to donate over $12 million to the SETI Institute, which searches for intelligent extraterrestrial life. Through his investment firm, Intellectual Ventures, whose value has been estimated at $5 billion, Myhrvold irritated the technology world by controlling billions of dollars in patents as "one of Silicon Valley's favorite villains." Dinosaurs were among his many interests. Myhrvold published on dinosaur growth rates and collaborated with Horner, and, teaming up with Philip Currie, he used computer simulations to theorize that a sauropod could flick the tip of its tail like a whip at supersonic speeds. ("To be blunt, the computer simulations are another case of garbage in, garbage out," one paleontologist complained.) Along with Nicolas Cage and Leonardo DiCaprio, the *New York Times* had named Myhrvold as an avid collector of fossils, which suggested that he engaged in the commercial practice that two of his beneficiaries, Horner and Bolor, publicly professed to despise.

The board also included Gerry Ohrstrom, a New York investor with a Mexico-based asset management firm called Epicurus Fund. The Ohrstrom family annals involved references to Virginia horse breeding, the Olympics, Mel Gibson, Securities and Exchange fraud, Viscountess Rothermere, the CIA, the *Paris Review*, a New Jersey gangster named Longy, and a woman called Bubbles. The patriarch was George Ohrstrom Sr., who, in the 1920s, founded a private equity firm, G.L. Ohrstrom & Co., on Wall Street. Four of the companies that he bought in the 1930s and '40s, which manufactured parts for oil rigs and pumps, were later conglomerated into Dover Corp., which made elevators. His son was George Ohrstrom Jr., a Greenwich Country Day School classmate of President

George H. W. Bush and an early investor in President George W. Bush's oil ventures. *His* son was Gerry, a science enthusiast. Gerry Ohrstrom once served on the board of the Property and Environment Research Center, or PERC, a think tank based in Bozeman, Montana. PERC encouraged "free market environmentalism," which the group defined as "an approach to environmental problems that focuses on improving environmental quality using property rights and markets." Ohrstrom served on the board in 2007 when *PERC Reports* published its fall issue. One article, headlined "Fossil Farming Blooms Where Barley Withers," touched on the "huge disservice to science" posed by the fossil trade, but reported that "private landowners on barren stretches of the western plains are glad to have something to take to the bank."

> Once upon a time, an original Picasso on the living room wall was more than adequate proof that the owner had successfully summited the economic peaks. Now, the work of famous artists is no longer proof of stratospheric wealth. Instead, high-end decorators hired by the wealthy are busily tracking down T. Rex teeth on eBay, bidding millions at Christie's Auction House for mounted mammoth skeletons, or competing for foot-high dinosaur eggs. How else do you add interest to the den, drama to the entry hall, and curiosity to the coffee table? Fabergé eggs be gone!

Ohrstrom helped finance Bolor's outreach project, covering the cost of shipping the former AMNH bus from New York to Ulaanbaatar. Bolor rebranded the vehicle, replacing "Bloomberg" with "Ohrstrom/Epicurus," and plastering over "American Museum of Natural History" with the logo for her NGO.

Recently, she had hired a quiet, towering fellow with thick black hair and rosy cheeks named Ganbold to work as the driver/mechanic. He had spent years as a city bus driver in Ulaanbaatar, and Bolor liked his quiet, obedient temperament. As she crossed the parking lot at Wagner Asia, Ganbold stood waiting for her. She handed him a new windshield wiper (mostly for dust, not rain), and they spoke for a moment about a bad tire for which there was no spare. In the Gobi there would be no service stations, no tow trucks, so they needed to fix the tire before going to the desert on Monday. The exhibit's touch screens also needed attention, as did the hydraulic system that lowered the steps so that people could board.

Ganbold brought out his tools, which he kept in a Winnie the Pooh backpack. He opened the bus's side panel, exposing the vehicle's guts. Pulling out cables and hoses, he looked as baffled as a small-animal veterinarian tasked with saving the life of a water buffalo. Curious mechanics from the service warehouse gathered around, their denim coveralls inscribed "Safety First!" As they stared at an electrical diagram, trying to make sense of the engine, they uttered what occasionally sounded like English but wasn't: *Millicent did…a turkey and a ham…my daughter.*

After a while, Bolor walked next door to a company cafeteria, and stood in line with men in coveralls and hard hats, and women in high heels and short skirts. A cook handed her Styrofoam takeout containers of baked chicken in gravy, white rice, Russian potato salad, and shredded beets. Back at the bus, between consults with Ganbold, she ate her lunch in the cab with the windows open, a hot breeze blowing through.

On the dashboard lay her pink leather notebook, which held her jotted schedule and ideas. While she still showed up at the American Museum of Natural History now and then, her official work for the institution had ended. Mark Norell and Mike Novacek and their team had already come and gone for the Gobi field season. Philip Currie was coming soon for a quick trip to the eastern Gobi with Tsogtbaatar and some students, but Bolor wasn't involved in that, either.

She had thought that Oyuna might name her the director of the new Central Museum of Mongolian Dinosaurs, but the job had gone to a woman with experience in museums but not science. Bolor had been hired as the museum's chief paleontologist, but within six months she had quit. The disappointments had been depressing, and at times Bolor wanted to drop everything and focus on science or write children's books about Mongolian paleontology. *I've brought it this far: now somebody else can take over*, she would think, and then something positive would happen and she would keep going. The dinosaur bus appeared to be a positive. She hoped that a smooth trip to the desert would allow her to raise money to continue, or even expand, her educational outreach. If she gave Mongolians information, they could speak for the Gobi bones, and she could stop shouting. The Mongolian Academy of Sciences still wouldn't issue her a collection permit, but she needed no one's permission to drive around in a big blue bus, teaching people about dinosaurs.

On Monday, Ganbold drove the bus south, out of Ulaanbaatar, with Bolor riding shotgun in an "I Dig Dinosaurs" T-shirt. They were trailed by two rented Land Cruisers and a Russian UAZ Bukhanka—a khaki-colored van that resembles a tough cousin of the VW bus—altogether carrying a small scrum of media that had become interested in the project. The halfway point on the two-lane road from Ulaanbaatar to Dalanzadgad, the largest town in the desert, is Mandalgovi. On the way there, the bus stopped at a roadside café for ramen, dumplings, and the restroom, an outhouse that consisted of rickety planks spanning a deep pit in the earth.

The sight of the bus drew a small crowd of children and their parents. The kids ventured toward the bus as if unsure whether it was safe to approach even the illustrations of the enormous creatures depicted on either side. Bolor opened the door and let down the stairs, and the first children of the Gobi Desert followed the memory of the children of New York City inside. There wasn't enough time to hold an impromptu workshop, but for a few minutes the children got to walk in the "footsteps" of dinosaurs and see displays like "Mesozoic Mysteries" and "What's for Dinner?" Those who read English learned that paleontologists "examine fossil jaws and teeth for clues to what dinosaurs ate;" after studying the "What caused the mass extinction" wall, they could watch a video of a re-creation of the asteroid impact that is thought to have killed all the non-avian dinosaurs.

The bus closed up and moved on. Ganbold drove past cattle wading in a pond, a solitary *ger*, and two horses standing side by side at the edge of the pavement like a pair of wisecrackers in a cartoon. When the highway got rough, the caravan went off road, the passengers bouncing like popcorn. At the small city museum in Mandalgovi, Bolor was treated to *airag* and *buuz*, then she moved everyone along to Dalanzadgad, a city of about twenty thousand whose small airport had recently paved its dirt runway. They checked into the Dalanzadgad Hotel, which, depending on who you were and where you came from, was either spartan or the essence of luxury.

The next day, Bolor met with government officials and the local tourism director, Tumendelger, who lived in DZ in a nice brick house with a fence around it. He had on a suit and a white shirt and a dinosaur-printed tie that he had bought at the American Museum of Natural History gift shop. The local media had been summoned. Bolor explained her plans, in Mongolian, using the word *tourism*. Speeches followed, on the front steps of the government center.

In a children's workshop, Bolor and an assistant distributed sheets of paper and colored pencils, and the kids got busy working on an exercise involving a plastic toy mastodon and a dinosaur egg. Next, Bolor handed out flash cards as a sort of puzzle: the students studied laminated drawings of how an animal becomes a fossil and had to put them in the right order. Obviously Death came first and Erosion-and-Discovery came last, but did Transportation happen before or after Decomposition? And when was Weathering-and-Burial? After that, she passed out papers with images of animals on them and asked the children to think about which creatures were dinosaurs. Dinosaurs were built with their legs underneath them, not splayed out like a crocodile's, Bolor explained. Dinosaurs were reptiles, which are different from mammals.

The children worked in earnest silence, handling the small plastic toys that sat before them as reference. The plesiosaur had them stumped. Yes, it ends in "saur," Bolor told them—but it's not a dinosaur! Plesiosaurs were enormous marine reptiles that lived in the age of dinosaurs, but they were flippered and their skulls were built differently. She held up an image of a saber-tooth cat. Is it a dinosaur?

No!

What about this crocodile?

No!

This *T. rex*?

Yes!

Looking at the *rex*, the children were shocked that such an animal ever existed—tall and strong enough to trample a *ger*, with a mouthful of spiky teeth. But Bolor had more shocking news: right here in the Gobi, there once lived a remarkably similar animal called *Tarbosaurus bataar*.

Late that afternoon, leaving the dinosaur bus parked in town, the caravan moved west from DZ, led by a car driven by the museum director from Mandalgovi. She had changed out of her pretty work dress and now wore

Juicy Couture sweats, huge sunglasses, driving gloves, and pearls. She led the caravan through the undulating Gobi for half an hour, through sage and low hills, zigzagging along old trails and making new ones, the landscape a flat-open nothing. Around eight thirty, as the sun spread fire along the unbroken horizon, a *ger* came into view. A couple hundred yards from the *ger*, across a slope, stood two more *ger*s, set side by side.

In the flat gravelly pasture behind the *ger*s, nomads were milking their horses. Horses are a sign of wealth, and this family owned dozens. Mongolian horses are short, stocky, fast, and strong, with long tails and manes. Herder families breed and race them, and consider them kin. A herd of them, tied together along a rope stretched between rough-hewn poles, is an iconic Mongolian image. The dust kicked up by these horses' shuffling hooves hung as a gauzy filter, backlit by the setting sun. A young girl hustled back and forth from *ger* to herd with a pail, collecting and dumping milk to make *airag*. Family members took turns pumping the barrel of liquid. A young boy who had just fallen from a horse while racing and had been dragged sat bloodied and sniffling on the man side of the *ger*.

The family included the father, his elderly mother, his teenage daughter, the boy, and, somewhere in South Korea, his wife, who worked there. Relatives and helpers lived in the other two *ger*s. Bolor had seen the family a year earlier and was excited to reunite. "We are going to party!" she had said, gleefully describing (but not expecting) a pastime in which people drink *airag* until they have to press their stomach and projectile-vomit, then start drinking again.

After dark, as the main *ger* filled with people, the father took his place at the head of the room in a work *deel* and a brimmed field hat. A single bare bulb, powered by either a generator or solar panels, dimly lit the space. It didn't seem possible that so many people could fit, but fit they did, twenty men, women, and children, sitting on the beds and the ground. Everyone watched as the patriarch asked his guests questions. *Where is your husband? Do you have children? Why are there holes in your pants?* He tried to tell one visitor's fortune and gave up, saying he was no good with the palms of foreigners. Everyone passed a snuff bottle and played a hand game that resembled Rock, Paper, Scissors, and occasionally they sang. Bolor said, "Nomads are really good singers."

At one point someone asked if the family knew any dinosaur diggers. No, but one had heard that a champion wrestler once bought a complete dinosaur for 25 million *tugrik*, and that buyers came from Ulaanbaatar and paid diggers with motorcycles as well as cash. A guide named Selenge explained that a lot of Mongolians believed the law allowed them to keep fossils that weren't the subject of research or study. She knew of a guy who had a dinosaur egg at home. "Someday they can sell it," she said. "So this family keeping this egg, like treasure."

At eleven thirty, supper was ready. Bolor got back into one of the vehicles and rode across the dark pasture to the lone *ger*, where a couple lived with their small children. The wife was removing freshly roasted goat from the coals and dropping the pieces into a large aluminum bowl. Each visitor was handed a rock still slick and fragrant with goat fat and told to toss it hand to hand like a hot potato. "Good for health, especially women's health," said Selenge.

Sitting on the floor, the guests ate from the bowl. After a round of vodka sipped from a small silver cup, the drivers went outside to sleep beneath the vast and twinkling desert sky. Everyone else spread sleeping bags in a row on the floor, beneath hanging strips of drying meat. Bodies lay side by side like packaged frankfurters, with their feet to the bowl of goat bones and the moon shining through the open door.

Bolor walked out into the night to relieve herself before going to bed. Her headlamp shone a path through a patch of grass. When she flicked it off, it was too dark to see even a hand held directly in front of her face. Nearby, large, invisible creatures snorted and softly stomped in the dirt.

In the morning, the breakfast smoke puffed up and out. A young boy was already awake and training a horse. A herder drove a sheep herd by motorcycle. A man in a *deel* came riding in to capture a beautiful blond colt; he hooked her with a noose on a long pole and another guy tried to throw on a harness. They soon got the colt saddled and began the process of breaking her.

The old grandmother wore a pink *deel* and a blue headscarf and carried her orange milking stool around with her, sitting for obligatory photos in regal silence. After Bolor presented everyone with gifts, the grandmother offered curd cakes. As the party convoyed out, the captured colt could be seen clipped to the tie-line, her feet bound together, bucking.

The path from Dalanzadgad to the Flaming Cliffs runs west-northwest over pebbly plain. The drivers followed the ghost roads, which followed telephone wires, which are the best means of navigation in deep snow. They passed birds taking dust baths and the sun-bleached skulls of camels. The wind picked up, strong and cold, buffeting the sage.

The pink sky blued, signaling sunset. The caravan came to a tourist *ger* camp with a bathhouse and a dining room. The sun was dropping as if being pulled steadily by ropes from below the horizon. When a double rainbow appeared, everyone rushed onward in the vehicles to see the spectacle of Bayanzag, the Flaming Cliffs, in such remarkable last light.

The next morning, Bolor was finishing her breakfast when someone came running into the dining *ger* to say that some tourists had found dinosaur eggs. She jumped up and said, "Maybe I will make an arrest!"

At the Flaming Cliffs, she parked in a gravel lot high above the canyons, where locals set up tables and sell random stones and trinkets. The regional tourism director was already there. He followed Bolor as she followed a pair of tourists along a narrow ledge, around an outcrop, and seemingly off the edge of a cliff. The tourists—a young American couple—pointed out a place in a bluff where they had spotted unusual objects poking out of the rock. Bolor peered somberly at the site as cameras followed her every move.

"Thank you very much letting us know," she later told the tourists. "If you know, we have a lot of fossil localities. The fossils are too rich—anybody can find. But we have a poaching problem. What you found was an egg nest, *Oviraptor* nest. Are you from States?"

They said they were.

"Which city?"

Chicago.

Bolor discovered that she had just explained paleontology to a Northwestern University paleobotanist, Dr. Rosemary Bush, and her husband, a particle physicist, Dr. Steve Won. The couple was vacationing with Bush's parents and her brother and sister. Their discovery had happened the day before. Bush and Won had walked down into a wash, which Bush later described as a place "drooling with fossils." As Won, a rock climber, took a vertical approach toward a ridge he had spotted objects that looked like potatoes protruding from the rock. "Those are *eggs*," Bush had said when she saw them.

She had dug out five of the eggs, worried that if she left them behind "they would wind up in some Chinese businessman's living room." When she and Won carried them to the parking lot, their guide went wild. Dinosaur eggs were first discovered by Americans right here in this place! And now Americans had found more! This is so important for Mongolia! You come back here in five or ten years and everybody will know you! The discovery had happened at roughly the same time the double rainbow appeared, so the day was surely blessed.

Bush and Won had taken the eggs to their *ger* camp and arranged them in a row on placemats in the dining room, for photos. Bush always traveled with her Geological Society of America scale card; she laid the ruler next to each egg, noting that the largest measured roughly 8 inches long. The camp's owner hoped to keep the eggs as a tourist attraction. People were always coming to the place where so many dinosaurs had been discovered, only to see no fossils at all.

Bolor had been thinking the same thing. Maybe that's where all of this—the poaching, the politics, the *T. bataar* case, the bus, the newfound attention on Mongolia—had been leading. The Gobi Desert, scene of some of the world's greatest paleontological discoveries, had a viable opportunity for its first real dinosaur museum. Now was the time.

The next morning, as Bolor pulled out of camp, her hosts ladled horse milk onto her tires, for luck.

EPILOGUE

Roy Chapman Andrews went home to New York City and became the director of the American Museum of Natural History. Yvette had divorced him on March 31, 1931, in Paris, on grounds of desertion—exploration, it seems, had come first in his life. They had two sons, George and Kevin. In 1935, the year after taking the AMNH directorship, Andrews married Wilhelmina "Billy" Christmas, the widow of a Manhattan stockbroker. They kept a city apartment and bought an old farm in North Colebrook, Connecticut, naming it "Pondwood." By the time Andrews resigned, on November 10, 1941, he had spent over thirty years at the museum, twenty-eight of them in the field. The directorship had been an honor, but bureaucracy made him miserable. "I did not react well to confinement…," he wrote. "I was like a wild animal that had been trapped late in life and put into a comfortable cage." The *New York Times* marked his departure by praising his "unfailing gift for dramatizing his scientific adventures." Andrews "not only made the Museum a vital force in scientific education but he brought the public flocking to it, because people found its colorful halls one of the most fascinating vistas in the city scene."

In 1953, he published *All About Dinosaurs*, a book that delighted and inspired future paleontologists, including Philip Currie and Mike Novacek. The Andrewses eventually moved to Carmel Valley, California. At one point Roy was asked why the museum had never returned the fossils collected by the Central Asiatic Expeditions—Mongolia still occasionally complained that the Gobi fossils had been stolen. A "perfectly ridiculous" allegation, said Andrews, who by then was seventy-two. The Gobi dinosaur egg that he had auctioned off for $5,000 to fund future research remains at Colgate University, where it is used for scientific study. The egg is the centerpiece of the Robert M. Linsley Geology Museum, in the Robert H. N. Ho Science Center. In October 2009, it was put on public display, behind a "high-level security system," for the first time since 1957, the year that a couple of Colgate students tried to steal it.

Andrews died of a heart attack on March 11, 1960, in Carmel, and was buried in his hometown of Beloit, Wisconsin. Late in life, he had become fond of a place that most reminded him of Mongolia: Tucson, Arizona. In 2015, a small auction house in Sarasota, Florida, sold off a batch of his former possessions.

Mark Norell continues to lead the AMNH's vertebrate paleontology department, working out of the tower where Roy Chapman Andrews applied for his first job and later returned as museum director. The museum continues its annual research in the Gobi Desert in collaboration with the Mongolian Academy of Sciences. When Michael Novacek published *Dinosaurs of the Flaming Cliffs* in 1996, he noted that the opening provided by the fall of Communism wasn't "necessarily permanent," explaining, "It's a window of opportunity, one that could close without warning in this chimerical world." Thousands of Mongolian fossils remain at the American Museum of Natural History, where materials from the Central Asiatic Expeditions of the 1920s are still on exhibit, used for study, and stored in the museum's collection.

Philip Currie was feted by the Mongolian government for his part in bringing the *T. bataar* skeleton back to Mongolia. He went on working in the Gobi and in Alberta, Canada. In September 2015, the Philip J. Currie Dinosaur Museum opened, five hours northwest of Edmonton. The actor Dan Aykroyd—who had hunted fossils with Currie, as had other celebrities such as the crime novelist Patricia Cornwell and *Criminal Minds* star Matthew Gray Gubler—opened the event with a motorcycle ride. The museum "received almost every award and citation that could be thrown at it," but within a year showed signs of financial stress owing, some thought, to its remote location.

In the fall of 2015, Oyungetel "Oyuna" Tsedevdamba visited the American Museum of Natural History to demand the return of Mongolian fossils, in what appeared, to some, to be opportunistic political theater because she arrived with a film crew. She lost her position as minister of tourism when the Mongolian government changed leadership again, but continues to work as an activist and author. She is expected to one day again run for public office.

Bolor had talked about writing children's books about Mongolian dinosaurs, but it was Oyuna who got there first. As a gift to the newly created Central Museum of Mongolian Dinosaurs, she turned her impromptu dinosaur tale—the one she told at the pop-up exhibit's opening at Sukhbaatar Square—into a soft-cover book, *A Story of Tarbosaurus Bataar.* The book was printed in Mongolian and English and sold in the gift shop. As the story goes: *Once upon a time... there was one young and adventurous dinosaur called Tarbosaurus bataar. He dreamed, "One day I will be famous and make friends with children of human beings still to be born...."* Millions of years later, after the *T. bataar* became a fossil, poachers arrived, among them a tall white guy in a backpack and jean shorts. *Most of the poachers were not from the Gobi. They came from far away. So they smuggled Tarbosaurus bataar with his friends to a faraway land.* President Elbegdorj and Robert Painter saved the *T. bataar.* The unnamed smuggler wound up on his knees in a jail cell, in the presence of an American judge. A school girl named Bolor was among the children who visited the dinosaur in the museum. *Now you are a worldwide recognized hero,* children told the *T. bataar,* who answered, *I became very famous, as I wished.*

The Central Museum of Mongolian Dinosaurs opened in the summer of 2014 in the old Lenin Museum, with the *T. bataar* as the star attraction. Meanwhile, building inspectors decided the old Natural History Museum wouldn't withstand an earthquake—significant fault lines run near Ulaanbaatar—and some ten thousand natural history items were moved into storage. The spectacular "fighting pair"—one of the most important fossils ever discovered—went to an unprotected corner of the paleontological center's basement. As if to avoid being upstaged by the new Central Museum of Mongolian Dinosaurs, the Mongolian Academy of Sciences retracted an exhibit that was traveling overseas and mounted it at Hunnu Mall, a swank new shopping center on the outskirts of Ulaanbaatar with a food court, an indoor ice skating rink, and an IMAX theater.

―――――――――

Otgo went on working as the chief prepper but was looking forward to retirement. In 2015, when asked about Eric Prokopi, he said he felt bad about what had happened: "Maybe some people think I brought these people in an organized way to find fossils, but I didn't. I just showed them the sites, that's it. I fulfilled my duties and responsibilities [as a guide] and I don't feel sorry, but on the other hand, as for me, I feel sorry that the people who sold him the bones are Mongolians. If Mongolian won't sell it, he won't buy it. He paid a lot of money to get this bones to bring to the States, and he worked with this bones and prepared it for trade, and when Mongolian government made question to the Americans, he lost what he paid for. He is in some ways a victim."

Tsogtbaatar continued to run the paleontology institute and lab of the Mongolian Academy of Sciences. He called the new Central Museum his "headache," saying, "These are illegal specimens. We don't know where they came from, which layer, who found it, or when. So it's very dangerous to use it for study. It spreads bad information in the world." Tsogtbaatar was already worried that "people think Mongolians so crazy, so stupid" because of the *T. bataar* case. The new Central Museum's role in paleontology and the public's understanding of science—"it's very complicated," he said. "That museum? That's the museum of Eric Prokopi."

―――――――――

Tuvshinjargal's family kept the mint-green building on Peace Avenue. In the summer of 2015, they still had a travel agency on the fourth floor, where glass cases held huge geodes, their purple innards glittering. Bobo, Tuvshin's widow, did not want to discuss her husband or his business— she had already been detained once. International human rights watchers have urged the Mongolian government to enact statutes of limitations, making it impossible to charge a person for certain nonviolent crimes after a certain point, but so far that has not happened. The *T. bataar* case remained open. Tuvshin's eldest son, meanwhile, turned the old dinosaur basement into a hip-hop dance studio.

Fossil dealers don't like talking about Tuvshin. What's to fear? one dealer was asked. The dealer said, "Are you kidding me?" Tuvshin was the source of 90 percent of all the material coming out of Mongolia. "To me it's just like drug dealing. He knew *everything*." Jeff Falt, Oyuna's husband, hoped the investigation would continue, especially now that Tuvshin and at least two of his known associates were dead. Some suspected that Tuvshin had been working for someone richer and more powerful, "the financier on the Mongolian side," as Falt put it. "The cost of shipping, the cost of digging the dinosaurs out, and so on—someone was fronting the money for that. If you have a really conspiratorial bent of mind, you might wonder if he's getting rid of witnesses."

The commercial community's trade group, the Association of Applied Paleontological Sciences, announced plans to research and publicize international fossil-collecting laws in English, to counter rumors and misinformation that have always surrounded certain countries. "In most cases, there is no single agency for any country that has all of the correct information available," George Winters, the former AAPS president recently wrote. And information often conflicts, "with one entity claiming something is legal and another stating just the opposite." Winters wrote, "Most of the foreign agencies I have contacted cannot quote their current laws, or supply the actual documentation defining them." Winters was also working with the paleontologist Ken Carpenter, director of the Eastern Prehistoric Museum at Utah State University, to update *Collecting the Natural World*, a 1997 book by Donald Wolberg and Patsy Reinard. The book, now out of print, covered the fossil regulations of all fifty U.S. states; Winter says the next edition will include a chapter on international laws.

Not long after the *T. bataar* auction, David Herskowitz was fired from Heritage Auctions, though not, he says, for the *bataar* matter. He still works as enthusiastically as ever as a natural history broker. Before the *bataar* case, "I thought it was okay to take something from a consigner knowing that if I needed the paperwork, he'd give me the paperwork; only because I figured it was a lot of hassle to get all the paperwork together...," he said. "I didn't want to put that burden on somebody because usually they were in a hurry, on deadline and all. So I learned that you have to have the paperwork."

The buyer of the *T. bataar* skeleton was a New York City tax lawyer and developer in his seventies named Coleman Burke, who owned a former cold-storage warehouse on the Hudson River in West Chelsea that once housed a notorious 1980s nightclub called Tunnel. Burke, an avid outdoorsman, geology enthusiast, fossil hunter, and Explorer Club member, planned to mount the dinosaur on the cavernous ground floor. He called his near purchase "the ultimate antiquing."

Various auction houses killed their standalone natural history divisions after the *T. bataar* case. Heritage Auctions deleted many of its archived listings involving dinosaurs, and I.M. Chait likewise appeared to have scrubbed its website of Mongolian fossils. In June 2016, one of Izzy Chait's sons, Joseph, was sentenced to a year and a day in prison and fined $10,000 for conspiring to smuggle at least a million dollars' worth of wildlife products made with materials from protected species including elephant, rhinoceros, and coral. The case was brought by U.S. Fish and Wildlife Service Director Dan Ashe and U.S. Attorney Preet Bharara in the Southern District of New York.

Once in office, Donald Trump fired Preet Bharara, some think because Bharara was preparing to investigate him. Bharara accepted a teaching position at New York University, launched a podcast, and became an on-air political analyst for CNN. In 2015, Sharon Levin left the Department of Justice after many years to become a partner at the white-shoe New York law firm WilmerHale. On her way out the door, a former colleague called her "the Babe Ruth of forfeiture"—by the time she left the DOJ, she had overseen nearly $14 billion in asset forfeitures and returned over $9 billion to the victims of crime. Martin Bell stayed on with the

Southern District of New York as an assistant U.S. attorney, working in the unit that prosecutes public corruption.

———————

Homeland Security agents confiscated Mongolian dinosaur fossils from dealers in several states. At least two dealers were charged criminally. One of the surrendered *bataar* skulls was connected to the professional golfer Phil Mickelson—his wife reportedly had seen a specimen on display at a shop in Jackson Hole, Wyoming, and bought it for him as a birthday gift. Nicolas Cage also surrendered his *T. bataar* skull to the federal government. Prokopi heard that Leonardo DiCaprio had "traded up," swapping his *bataar* skull for *rex*. The United States altogether repatriated about three dozen Mongolian dinosaurs, many of which went on display at the Central Museum of Mongolian Dinosaurs.

———————

Hollis Butts: who knows. At last check, he was still in Japan, posting to Facebook about sea urchins, tax returns, and the vernal equinox.

———————

Chris Moore, Eric Prokopi's partner in the *T. bataar* sale, went on as a fossil hunter on the Jurassic Coast of England, never commenting publicly about the case. He declined to discuss it when approached at the big October fossil show in Munich, and again at the annual spring fossil show in Lyme Regis, and again at home in Charmouth one drizzly morning in May. The BBC recently featured him in a new documentary about fossil hunting, starring the United Kingdom's best-known naturalist, Sir David Attenborough. In 2017, Mary Anning got a new museum wing in the refurbished redbrick museum on Cockmoile Square in Lyme Regis. Many of her life mementos are on display there. Her paleontological discoveries can still be seen at the Natural History Museum in London, and her grave can be visited in the St. Michael's churchyard in Lyme, overlooking the sea.

———————

Peter Tompa, one of Prokopi's two civil attorneys, continued to work as a lawyer in Washington, DC. Michael McCullough joined with William

Pearlstein in the spring of 2014 to create the boutique New York law firm Pearlstein & McCullough, to focus on the international art market. In the summer of 2016, they took on a new partner, Prokopi's former criminal lawyer, Georges Lederman, becoming Pearlstein McCullough & Lederman.

———————

After the Tyrannosaurus Sue case, Peter Larson of the Black Hills Institute blamed everyone but himself—the government above all. Then he decided that, emotionally, he would survive a felony conviction and prison only by assuming responsibility for his part in the mess. Those who wallowed in self-pity only suffered more and caused the suffering of others, he'd noticed. Larson wanted to get on with his life with less bitterness and less of a victim mentality. He didn't know Eric Prokopi well, yet hoped that he would one day feel the same.

———————

Frank Garcia moved to South Dakota and married Deborah Vaccaro on a picturesque covered bridge. They regularly hunt fossils, and keep a karaoke machine in their basement. Recently, they opened a natural history shop in Custer, called Good Karma, where Frank, when he isn't ranting on Facebook, also sells copies of his books. To commemorate his Leisey discovery, an historical monument was erected in the spring of 2018 at the old shell pit in Ruskin, Florida. Frank once told a reporter, "When I die, don't cry for me, man, because I really enjoyed the hell out of living it. Weep for somebody that didn't get to do the things I have done."

———————

President Tsakhia Elbegdorj left office in 2017 and was succeeded by Khaltmaa Battulga, a wealthy Democratic Party cabinet minister and former pro wrestler. The Mongolian penchant for political gossip carried into the current administration. Battulga was suspected of trying to "buy the very same [voting] machines from America" in order to "count his own votes," and the University College London anthropologists Bumochir and Empson noted that Elbegdorj's supposed "'ownership' and control over the black boxes" extended to "speculation about his control over the whole

[2017] elections." They wrote that "in this murky world of political wrangling people speculate that Elbegdorj is in fact behind the appointment" of certain candidates, "having deliberately supported them at various times, and put them in place to rig the election" in order to protect himself from "future corruption charges." In March 2018, when President Trump announced a summit with North Korean leader Kim Jong Un, Elbedorj lobbied for the meeting to take place in the "neutral territory" of Mongolia. Battulga, meanwhile, had at one point dismissed the importance of diplomatic relations with the United States by saying America was "too far away." Perhaps the Trump administration felt the same way: as of May 2018, a new U.S. ambassador to Mongolia still hadn't been nominated.

Robert Painter continued to practice law in Houston and to do business in Mongolia. In early 2015, his professional relationship with the Mongolian government came into question in the Ulaanbaatar media: five members of parliament alleged that a cabinet secretary added over 17 billion *tugrik* (over $7 million today) to the federal budget in order to pay the Painter Law Firm for ongoing "legal advice" regarding the development of Oyu Tolgoi, the Gobi copper mine. A little over $260,000 had been paid so far in 2014, as Painter's annual salary, the media reported. "This is an act of embezzling public funds collected from taxpayers as well as an illegal spending on an unauthorized objective," the *UB Post* reported. Painter acknowledged that he was paid for his services, but said the $7 million figure was preposterous, calling the claims another example of political mudslinging. Regarding the *T. bataar* case, he said President Elbegdorj was chagrined to learn of the criminal charges against Eric Prokopi and wished the arrest had never happened—feelings that were never made public. "The president feels bad for him," Painter said in early 2018. "The president never wanted him to go to jail. He just wanted the dinosaur back. He respects what happens under American law, but he wasn't interested in vengeance."

Just before introducing the dinosaur bus to Mongolia, Bolor and her family took a road trip out west, stopping at Dinosaur National Monument, which straddles the Utah–Colorado border. President Woodrow Wilson created the monument in 1915 to preserve a Carnegie Museum quarry that had produced *Allosaurus, Apatosaurus, Camarasaurus, Diplodocus,* and other Jurassic dinosaurs. Bolor met a science enthusiast there named Thea Artemis Kinyon Boodhoo, who, as a volunteer under the park's longtime paleontologist, Dan Chure, was helping to digitally map the quarry, to make specimens available for public view and scientific study.

Boodhoo was the daughter of a novelist, Malcolm Brenner, who became known some years ago for his love affair with a bottlenose dolphin. It would be easy to make jokes about that or to shorthand Brenner's life—ex-Wiccan, "zoophilia advocate," etc.—but as always the truth is complicated, and, well, feel free to check all that out on your own time. Boodhoo grew up on breakfast burritos and dreamcatchers in Gallup, New Mexico, then moved to California, where the ancient forests and redwoods fascinated her. Her mother was a scholar "whose love of rare old books gave me the notion that old meant valuable," Boodhoo once wrote. "I spent days out on the cracked pavement destroying new coins with rocks and borrowed tools to make them worth more." She had hoped to become a paleontologist but had pursued advertising instead, for the financial security; lately, though, she had returned to her original love with a résumé that read "Science Advocate and Creative Professional," along with "Digital storyteller. Branding and marketing specialist."

The minute Boodhoo met Bolor in Utah, she asked permission to write about her and offered to help with the Mongolia outreach project. Before long, Boodhoo, who now lives in San Francisco, registered the Institute for the Study of Mongolian Dinosaurs in California as a 501(c)3 nonprofit and developed its public face, updating the website, logo, and social media presence. In February 2016, Boodhoo and Bolor launched an Indiegogo campaign that raised $46,000, allowing them to return to the Gobi and work on expanding the program. An engaging writer, Boodhoo also placed positive articles about Bolor in publications such as *Earth* magazine.

Bolor had backed off her crusade to bring all Mongolian fossils back to Mongolia. Now she focused on building a dinosaur museum at the Flaming Cliffs. In March 2017, she and Boodhoo traveled to Bayanzag with Dan Chure and Walt Crimm, a Philadelphia architect, to scout out siting possibilities. Boodhoo likened Bayanzag to a "sort of holy pilgrimage for dinosaur enthusiasts," though lamented the lack of information at the site—"no museum, no visitors center, no marked trails, no signs." Bolor, Boodhoo, and their colleagues envisioned a park not unlike Dinosaur National Monument or Badlands National Park, where visitors follow fossil trails and learn about prehistoric animals and plants. The museum would have to be privately funded—that part hadn't been worked out, but Bolor had already received the local government's permission to proceed.

Standing at the edge of the Flaming Cliffs, the scouting team thought about their various concerns—erosion, wind, flash floods, temperature. "All the things that make Bayanzag such an excellent place to find dinosaurs also make it an exceptionally challenging place to build a dinosaur museum," Boodhoo later wrote. The building should blend in with the environment, they all agreed. Despite the temptation to create an iconic structure "the real masterpiece at Bayanzag has already been created," she added, "by nature."

By summer 2018, Bolor and Boodhoo weren't much working together anymore, and Boodhoo had taken a full-time job as a copy writer. Bolor signed on as a National Geographic Expeditions "expert" for a twelve-day tour of Mongolia, and continued to promote the Institute for the Study of Mongolian Dinosaurs.

The court agreed to let Eric serve out the back half of his six-month sentence in Betty's garage apartment in Kingsmill, on house arrest, wearing an ankle monitor. He would be free to leave during certain daylight hours, but curfew would require him to be home by six. A wireless network would alert federal authorities if he ventured beyond Betty's property. He had decided to wear jeans until the ankle cuff was off, so the children wouldn't ask questions.

Betty was the one who picked him up at prison. Amanda was tending her booth at Bizarre Bazaar and waiting for him there, with Greyson and Rivers, on December 5, 2014, a forty-eight-degree Friday. He walked out with a cardboard box of possessions, wearing prison-issue jeans and white sneakers and a heavy beard. He set the box on the sidewalk and walked into Betty's wide-flung arms, and then she drove him to Richmond. When he rounded the corner of the Everything Earth booth and saw his children, he dropped to his knees and they jumped into his arms squealing, and stroking his funny beard. "Hey, guys!" he kept saying.

When he got to Betty's house in Williamsburg, it was as if he'd never left. Grey and Rivers hurried to show him their WELCOME HOME DADDY banner and to break out the Wii, Eric telling Grey it wasn't polite to give his avatars names like Poop. Later, he settled into the garage apartment, a nice-sized room with a kitchenette and a windowed bath. Amanda and the kids had brought in a mini Christmas tree and a poinsettia and stocked the fridge with Coke Zero, the bathroom with soap, shaving cream, and razors. Amanda came to check on him. Looking around at all the boxes he'd moved there before going to Petersburg Low, she said, "Aren't you anxious to put all this stuff away?"

"Yeah, but I haven't had a chance yet," he said. "I have to figure out what to do about a vehicle." The van was broken.

Amanda pulled Eric's diary from his prison box and glanced through it, and said, "I mean, walking the track after eating ice cream. While the rest of us are going a hundred miles an hour it's like, 'Went to lunch, had a nap.' Prison sounds *relaxing* to me at this point."

"Try taking a nap right outside the door of a public bathroom at a rest area," Eric said. "That's how relaxing it was."

Amanda forgot about the diary and went back to the topic of boxes. "I would unpack everything *right now*." Pointing, she said, "I'd be like, Goodwill, Goodwill, Goodwill."

Saturday was December drizzle but nearly 60 degrees, and Eric took the kids fossil hunting. Grey wore a T-shirt and jeans. Rivers had on a white shirt with a Peter Pan collar, a black cardigan, a plaid skirt, black tights, and boots, as if she were on her way to a party. The Colonial Parkway was browns and golds. They crossed the James River on a ferry named the *Surry* and then drove to where the road dead-ended. After passing an old red barn where someone had hung a huge Christmas wreath, they walked a rocky path through the woods, down to the river's edge. Eric and Tyler had hunted fossils on the riverbank many times on their own and with the kids. It was one of their favorite places.

The day was so overcast one of the kids asked, "Where's the moon?" Fog pressed on the river as a soft rain started to fall. An anchored skiff sat empty on the glassy water like a scene in a painting. The air smelled of pine. Eric and the kids walked the water line, stepping over slick logs and mossy rocks, pushing their way through a stand of bulrushes. Cockleburs caught on their clothes.

They came to the cut of a looming bluff, where thousands of broken fossils from the York River Formation shone in the earth. "There! No, there!" Grey yelled as they went along. "It's a tarantula egg!" (Magnolia seed.) "Mud castle!" (Cypress root.) He found a piece of a whale bone and a scrap of coral. Eric found, upturned in the mud where others had missed it, a nearly whole fossil *Chesapecten jeffersoni*.

Rivers picked up a smaller, perfectly formed shell, leaving its imprint in the sand. Eric looked at the impression and told her, "That's called a mold."

There was a lag of a couple of weeks until the ankle monitor went on and house arrest started. Eric had been calling Tyler, with little response. Normally fluent in Harlequin romance flourishes—"We declared our love for one another" and "tender moments" and "crumbled to the ground in each other's arms"—she now had just one thing to say: "Go back to your wife."

Instead, he went to a pawn shop and bought a turquoise ring for $25. He wasn't supposed to leave Virginia, but Amanda rented him a car under her name so he could drive to Gainesville and show up at Tyler's door. At Paynes Prairie, not far from Serenola, Eric pulled out the ring and asked her to marry him.

It had hurt Tyler to know that Eric had considered getting back with Amanda and lied about it, even if he'd done so to spare her feelings. By now, she had a new boyfriend and they were planning to move to North Carolina. She declined the marriage proposal and sent Eric on his way. He pawned the ring right there in Gainesville and headed straight back north.

Halfway up the Eastern Seaboard, he received a text—Tyler. She had made a mistake. Within days she had packed up her car and returned to Virginia. In Gloucester, Eric bought another pawn-shop ring. One afternoon as he and Tyler were driving down the Colonial Parkway, he stopped at the York River and proposed again, and this time she said yes.

What they wanted to do now was hang out, enjoy the kids, build up a bit of money, and one day hunt fossils again, maybe dinosaurs in Wyoming.

As the days ticked by, they thought about where to live. Typical housing wasn't an option; Eric wouldn't survive the credit check. A mortgage was also out of the question. He started watching eBay and Craigslist for a houseboat.

On January 13, 2015, a listing appeared for a big red tugboat "finished out for live aboard, huge interior, very nice and roomy." The boat was a World War II net tender, commissioned into service by the U.S. Navy on February 18, 1941. Named the USS *Noka*, it had once belonged to the class of vessels that lay anti-torpedo and anti-submarine nets, a national defense system that had become necessary after Japan's prewar aggressiveness in Manchuria, a heavily Mongolian and strategically coveted territory in northeastern China. Built in Texas, the *Noka* had motored from Port Arthur to New Orleans to Key West to Miami and had begun duty in Norfolk on the Fourth of July, five months before the attack on Pearl Harbor. The tug had served in and around Norfolk harbor throughout the war. Decommissioned on August 5, 1946, it was struck from the Naval Register and sold to successive owners, who renamed it the *Doris Loveland*, the *Russell 16*, the *Lin Clay*, and, finally, the *Bay Queen*.

Websites that kept up with these kinds of things listed the vessel's final disposition as "fate unknown." But its fate was known: the *Bay Queen* was sitting red as an unpicked cherry right there in Norfolk, 45 miles south of where Eric stared at his computer. The boat, which measured 80 feet long and 20 feet wide, had been remodeled as a floating home roughly eighteen years earlier. The engine and steering system had been removed, but you could tow it and dock it and live in it. Belowdecks it had been turned into a master bedroom and full bath. The main deck had a den, another full bath, a dining area, and a galley with a full-size electric stove and fridge. "The tiled floor follows the shear of the hull so it goes uphill as you move forward," read the ad. "After a while it seems pretty normal—you are on a boat, after all." A spiral staircase led from the den to the upper deck, where there were two more bedroom areas, another full bath, and the old wheel-house. The tug had central heating and air conditioning, hot-water heaters, and a washer/dryer hookup. Eric couldn't imagine what more they would need.

The seller wanted $29,000 and he wanted to show the boat only on weekends, but Eric couldn't wait. The name of the marina wasn't yet public, but he went on Google Earth and searched satellite images of the Norfolk shoreline until he found an unmistakable splash of red at Rebel Marina. Then he drove there and made the seller an offer. They settled on $22,000, Eric using money borrowed from his parents.

Tyler moved into the *Bay Queen*, alongside docked trawlers and sailboats, including one named *Never a Dull Moment*. Eric's probation officer inspected the tug and agreed to let Eric leave the garage apartment and move in early, partly because the boat had no motor. "Why can't you just be normal?" he asked Eric.

Eric and Tyler decorated the wood-paneled cabin with the vintage nautical pieces and magnifying glasses that Tyler liked to collect, and hung a framed Jacques Cousteau autograph over their desk. Nighttime on the water was so quiet they could hear the shrimp sizzle below them, like bacon frying.

After house arrest came a year of probation, but Eric was freer that he had been in some time. He bought a black 2005 Ford Explorer off Craigslist; it had 160,000 miles on it, three hubcaps, and a busted AC, but it was a bargain at seven hundred bucks. For work, he started a commercial fossil forum on Facebook called Natural History and went back to where he started, selling shark teeth. The day before Thanksgiving 2015, he and Tyler got married at a rented cabin in the woods of North Carolina.

The following spring, Amanda got engaged to George Bryan. They married in November—another Thanksgiving wedding. By then, the army had posted George to Savannah, and Amanda now lived at the beach, something she'd always wanted to do.

Unable to be away from the children, Eric and Tyler had made plans to follow. They liked the thought of returning to the southeast and living three hours from Gainesville, five from Land O' Lakes, where Eric's parents still lived in the house where he grow up. For a minute, they thought about selling the *Bay Queen*, then decided to keep it. Towing it to Savannah would cost a fortune, so the Prokopis planned to retrofit it and pilot it down the coast themselves, all 600 nautical miles. They emptied the wheelhouse, which Rivers had been using as a bedroom, and restored the steering system. Eric welded a rig to the back of the boat to hold a pair of Mercury outboard motors and a lift for an old Boston Whaler he'd bought for quick runs to shore. Tyler enrolled in a boating safety course to learn the meanings of buoys, the perils of shipping lanes. The trip would take weeks. Nights, they would drop anchor. On the way, they would adopt a dog and repaint the tug and have it scrubbed of barnacles.

Before leaving Virginia, they stowed most of their possessions in a storage unit and secured the onboard furniture with bungee cords. They raised an American flag on the stern. The mooring lines were released one morning in July. High in the pilothouse, Eric eased the old war boat out into the river and pointed it toward the waters of home.

ACKNOWLEDGMENTS

Fossils inspired the life's work of some of the greatest influencers of how we think about the earth's history and our place in it. As a child, Rachel Carson was said to have picked up a seashell while exploring the Allegheny River and begun wondering about the disappeared oceans and the life within them; I thought about her remarkable books often, especially *The Sea Around Us*, as I worked on this project. Elizabeth Kolbert's *The Sixth Extinction* is another gorgeously alarming meditation on the planet's past—and future. Books by the Stanford folklorist Adrienne Mayor helped me understand fossil finders' role in early American and science history. The work of Columbia University historian Morris Rossabi was crucial to the contextualization of Mongolian history and culture; I also found the anthropologist Jack Weatherford's work helpful, including his book *Genghis Khan and the Making of the Modern World*. The writings of Hugh S. Torrens, emeritus professor of the history of science and technology at Keele University in Staffordshire, England, provided important details on the life of Mary Anning, as did biographies by Shelley Emling and Sir Crispin Tickell, an Anning family descendent. Mark Jaffe's *The Gilded Dinosaur* is, to my mind, the definitive narrative text on the epic nineteenth century bone wars between Othniel C. Marsh and Edward Drinker Cope. I am also indebted to books and other published work by Edwin H. Colbert, Stephen Jay Gould, Thomas Holtz, Kirk Johnson, Mark Norell, Michael Novacek, Martin J. S. Rudwick, and Neil Shubin. Details on some of these and other resources can be found in the selected bibliography.

The AMNH archives hold a trove of stunning detail on the Central Asiatic Expeditions of the 1920s, and I'm grateful for the assistance of Gregory Raml, the museum's special collections and research librarian. At the Baylor University library, Leanna Barcelona, Ellen Filgo, and Amie

Oliver unearthed back issues of *Libertas* and other materials that without their conservatorship might have vanished. At the Academy of Natural Sciences in Philadelphia—another treasure house of historical and scientific information—Ria Capone tracked down a copy of *All in the Bones*, a wonderful biography of the original dinosaur artist, Benjamin Waterhouse Hawkins; the book's author, Robert Peck, kindly solved a minor mystery related to Hawkins. Josh Grossberg, of the USC Shoah Foundation Institute, made available an oral history of Holocaust survivors Roman and Cecilia Kriegstein, whose son asked that their name be preserved in the annals of science. Bruce Dinges of the Arizona Historical Society led me to vital information about the early days of the Tucson Gem and Mineral Show. Bob Jones, who published a history of the show, spoke with me at length about the show's background.

Eric Prokopi and his family quite simply opened their lives to me, having received no prior assurance of the result. Eric, Amanda, Tyler, and Eric's parents, Doris and Bill, spoke with me time and again—usually in person in Florida, Virginia, or Georgia—about the civil asset forfeiture case, the federal criminal charges, and the life events and choices that bracketed those legal actions. These conversations began in 2012 as I reported "Bones of Contention," the *New Yorker* story on which this book is based, and continued well into the spring of 2018. The Prokopi clan also shared scrapbooks, photos, correspondence, and other materials that shed light on the life of a fossil dealer and details surrounding the *T. bataar* case. I'm also thankful to other commercial hunters, natural history brokers, collectors, and auction-industry executives who spoke with me on the record, including Coleman Burke, Frank Garcia, David Herskowitz, Andreas Kerner, Peter Larson, Thomas Lindgren, Tony Perez, Burkhard Pohl, Greg Rohan, Kirby Siber, Mike Triebold, Shirley Ulrich, Wally Ulrich, and George Winters.

Throughout this project, I encountered scientists who were happy to share their expertise on paleontology, geology, taphonomy, stratigraphy, and a hundred other things. If I erred, it's on me, not them. The Smithsonian's Kirk Johnson, the University of Alberta's Philip Currie, and the American Museum of Natural History's Mark Norell were especially helpful, from the beginning. The vertebrate paleontologist Thomas Holtz provided timely help with the translated papers of the Russian scientist E. A. Maleev and connections to a vibrant listserv for vertebrate paleontologists. Other pale-ontologists, geologists, and fossil preparators who deserve thanks include Khishigjav Tsogtbaatar, Chultem Otgonjargal, Bruce Bailey, Barbara Beasley, Rachel Benton, J. P. Cavigelli, Clive Coy, Cari Johnson, Richard Hulbert, Eva Koppelhus, Vince Santucci, Jack Tseng, Justin Tully, and, of course, Bolortsetseg Minjin, who spoke with me time and again about her background, her role in the *T. bataar* case, and her ongoing efforts in sci-ence advocacy. Robert Painter fielded countless phone calls and questions, and I'm deeply grateful for his time. Heartfelt thanks also to Dave Rainey, Gereltuv Dashdoorov, Jeff Falt, Georges Lederman, Sharon Cohen Levin, Julie Makinen, Oyungerel Tsedevdamba, Yeweng Wong, Jeremy Xido and Amanda Burr, Selenge Yadmaa, Joe Sandler, and Betty Graham.

Journalists think deeply about how to responsibly source their work. Anonymity is not granted lightly. The overwhelming majority of sources are named in the text or in Notes but for various reasons some are not. Their help is not forgotten. Mongolia especially is a deeply fascinating place to visit—and a challenging place to report. Many thanks to those who provided information or helped parse data that informed the work.

This book exists because of my *New Yorker* editor, Daniel Zalewski, who gave the original *T. bataar* story life. DZ is that rare combination of big-picture thinker, detail hound, gorgeous writer, and bullshit detector; he's simply the best editor I've ever known, and I feel extraordinarily for-tunate to be a beneficiary of his formidable intellect and wit. I'm grateful beyond measure for David Remnick, who *writes thank-you notes!* and is never too busy to talk reporting/writing, or to provide counsel—he makes the *New Yorker* the finest home a journalist could want. My sincerest thanks also to Pam McCarthy, Bruce Diones, and fact checker extraordi-naire Elisabeth Zerofsky.

I cannot say enough about my exceptional agent, Joe Veltre, at The Gersh Agency, who is as decent a human being as he is world class at his work—I'm quite lucky to be represented by him and even luckier to know him and to call him a friend. My thanks also to his hardworking Gersh team, especially Alice Lawson, who was there at the start.

At Hachette Books, profoundest possible thanks to Mauro DiPreta for his faith in this project and to Stacy Creamer for her early support. The wonderful Michelle Howry inherited the book as its editor, and as

I continued reporting, and continued writing, and continued reporting some more, she tolerated delay after delay with much-appreciated humor and smarts—I loved working with her. The editor Paul Whitlatch in turn inherited the book as it entered the final production stages and expertly shepherded it home—I loved working with him, too. The rest of the extraordinary Hachette team included the marketing wizard Michael Barrs, the indefatigable copyeditor Ivy McFadden, and the publicity superstar Joanna Pinsker. Lauren Hummel allowed nothing to slip through the cracks and introduced me to my new favorite red pen. Alison Forner designed a stunning book jacket and Mandy Kain oversaw its creation. Michelle Aielli executed a brilliant maneuver—tote bags! VIP thanks to production editor Michael Olivo for heroically incorporating my never-ending changes into the final product. I'm so honored to have worked with all these pros.

I could not have produced this book without the support of The MacDowell Colony, a magical sanctuary that I hope to always call home. The peace and comfort I found as a MacDowell fellow allowed me to work in a way that I never knew existed—with no-strings-attached support, and without distraction. What a gift. Thank you.

My academic home, Columbia University, thrills me: I'm especially grateful to my colleagues Sheila Coronel, Laura Muha, and Steve Coll for bringing me in as the Laventhol/Newsday Visiting Associate Professor at the Graduate School of Journalism, where I'm honored to work with some of the best journalists in the business.

I completed sections of this book while an associate professor at the Missouri School of Journalism, where my excellent graduate assistants—successively Abby Johnston, Landon Woodroof, and Shawn Shinneman—were impressively helpful in organizing source materials and chasing down details and translations. I'll always be grateful to the scheduling ninjas Kim Townlain and Jennifer Rowe for accommodating my complicated schedule as a journalist, teacher, and author, and to the powerhouse Amanda Hinnant for being such a fierce role model. I still marvel at the extraordinary librarian Dorothy Carner, who tracked down one scientific paper and obscurity after another, and I remain grateful for Dean Mills, the J-school dean emeritus, who initially brought me on board. I was lucky to have overlapped with a leader who knew, respected, and championed real journalism.

My family cheerfully despised this project because it dominated my time and energies for so long. More than once, my mother, Joann, said, "Just put a period on it and say, 'The End!'" My sister, Tracey, and her husband, Terry, and their children, Avery, Anna, and Tanner, kept the sweet tea ready in Tupelo. My brother, Mike, let me talk endlessly about rabbit-hole details—without interrupting—and made the project even more enjoyable by joining me on one of my reporting trips to Europe. My nieces, Avery and Anna, jumped in for a day of freelance research, which (for me) was fun. My cousin Jill has forever been and will always be a treasured, uplifting presence. I only wish my dad—Billy to his family, Bill to everyone else—were still with us.

The great Bill Kovach, curator emeritus of the Nieman Foundation for Journalism at Harvard, provided pivotal faith and encouragement, even when he didn't realize it. Likewise, I was lucky to have known for a time in my life the love and support of Dan Chapman and his parents, Bill and Christine Chapman.

Old and dear friends, thank you. Claire Campbell, in North Carolina, and Pam Moore, in Massachusetts, took the time to read a draft and send smart, vital suggestions. Carol Leonnig was a lifeline, even while busy with her own book project and with the news she so brilliantly breaks. Pilita Clark and Peter Wilson, along with Michael Fry, drove me from London to Dorset and walked the rocky beaches with me in a spring drizzle, and made me laugh, so much. Terri Lichstein helped engineer my deadline-defying return to New York as only a treasured friend and *20/20* producer could do. Rebecca Skloot was the first person I told about this potential project, in 2009, as she was poised to release her world-changing book, *The Immortal Life of Henrietta Lacks*; her irrepressible energy and devotion continue to inspire me. Audra Melton, brilliant photographer and life guru, never flagged while traipsing to Wyoming and South Dakota with me to shoot a museum crew's *T. rex* excavation and a commercial crew's search for a juvenile *rex*. Mark Jaffe fielded what were probably too many questions about Marsh and Cope, and I'm hoping he'll take me to the ballet soon. With road trips, sushi, and music, Keri Thomas and Matt

Ward helped clear the headspace for me to summon the energy to report and write a book at all. Florence Martin-Kessler and Mark Kessler, parents of the darling Kesslerettes, provided French translations and general *esprit de corps*. Elon Green pulled off coverage of a Manhattan repatriation ceremony when I couldn't get quickly from Boston to New York. Miranda Metheny kindly and quickly translated the *GEO* article from German to English. Michael Caruso, editor of *Smithsonian* magazine, buoyed me with key assignments and fun, supportive emails. Greg Brock did me a transcendent kindness during a particularly fraught deadline period: when I needed extreme solitude, he made available his beautiful cottage in the woods of Oxford, Mississippi, my soul's home. Chin Wang, a dazzling designer and one of the best journalists I know, is, like others mentioned here, family. Sam Douglas: thank you, friend. Others who sheltered, fed, boosted, or advised me early on or along the way and who have my eternal thanks include Dannye and Lew Powell, Cindy Montgomery and Cliff Mehrtens, Melissa Hinton Benton, J. R. Moehringer, Bonnie Lafave, Jennifer McDonald, Mike Whitmer, Devin Friedman, Josh Dean, David Grann, Luke Dittrich, Ben Montgomery, Joseph Menn, Susan Orlean, and Rebecca Burns.

To all these good people I owe a debt of gratitude that I won't soon forget.

QUICK REFERENCE TO DEEP TIME

Era	Period or Epoch	When it began
	Holocene	10,000 years ago
	Pleistocene	1.8 million years ago
	Pliocene	5.3 mya
CENOZOIC	Miocene	23
	Oligocene	33.9
	Eocene	55.8
	Paleocene	65.5
		EXTINCTION
	Cretaceous	145.5
MESOZOIC	Jurassic	199.6
	Triassic	252.2
		EXTINCTION
	Permian	299
	Pennsylvanian	318
	Mississippian	359.2
PALEOZOIC	Devonian	416
	Silurian	443
	Ordovician	488.3
	Cambrian	542
PROTEROZOIC		2.5 billion
ARCHEAN		4 billion

4.6 BILLION YEARS AGO: EARTH FORMED

For the full Geologic Time Scale, see www.geosociety.org.

SELECTED BIBLIOGRAPHY

Addleton, Jonathan. *Mongolia and the United States: A Diplomatic History*. Hong Kong: Hong Kong University Press, 2013.

Alvarez, Walter. *T. Rex and the Crater of Doom*. Princeton, NJ: Princeton University Press, 1997.

Andrews, Roy Chapman. *Across Mongolian Plains: A Naturalist's Account of China's "Great Northwest."* New York: D. Appleton and Company, 1921.

———. *All About Dinosaurs*. New York: Random House, 1953.

———. *An Explorer Comes Home*. New York: Doubleday, 1947.

———. *The New Conquest of Central Asia: A Narrative of the Explorations of the Central Asiatic Expeditions in Mongolia and China, 1921–1930*. New York: American Museum of Natural History, 1932.

———. *On the Trail of Ancient Man: A Narrative of the Field Work of the Central Asiatic Expeditions*. New York: G.P. Putnam's Sons, 1926.

———. *This Business of Exploring*. New York: G.P. Putnam's Sons, 1935.

———. *Under a Lucky Star: A Lifetime of Adventure*. New York: The Viking Press, 1943.

Andrews, Roy Chapman, and William Diller Matthew. *Central Asiatic Expeditions of the American Museum of Natural History, Under the Leadership of Roy Chapman Andrews*. New York: American Museum of Natural History, 1918–1925.

Anonymous. *The Secret History of the Mongols*. Mongolia: circa 1227.

Atwood, Christopher P. *Encyclopedia of Mongolia and the Mongol Empire*. New York: Facts on File, 2004.

Bakker, Robert. *The Dinosaur Heresies: New Theories Unlocking the Mysteries of the Dinosaurs and Their Extinction*. New York: William Morrow, 1986.

Barber, Lynn. *The Heydey of Natural History 1820–1870*. New York: Doubleday & Company, Inc., 1980.

Blom, Philipp. *To Have and to Hold: An Intimate History of Collectors and Collecting*. New York: The Overlook Press, 2002.

Bramwell, Valerie, and Robert McCracken Peck. *All in the Bones: A Biography of Benjamin Waterhouse Hawkins*. Special Publication No. 23. Philadelphia: The Academy of Natural Sciences, 2006.

Carson, Rachel. *The Sea Around Us*. New York: Oxford University Press, 1951.

———. *Silent Spring*. New York: Houghton Mifflin Company, 1962.

Cleaves, Francis Woodman. *The Secret History of the Mongols for the First Time Done into English out of the Original Tongue and Provided with an Exegetical Commentary*. Cambridge, MA: Harvard University Press, 1982.

Colbert, Edwin H. *The Great Dinosaur Hunters and Their Discoveries*. New York: E.P. Dutton & Co., 1968.

Currie, Philip J., and Eva Koppelhus. *Dinosaur Provincial Park: A Spectacular Ancient Ecosystem Revealed*. Bloomington: Indiana University Press, 2005.

Cutler, Alan. *The Seashell on the Mountaintop: A Story of Science Sainthood and the Humble Genius Who Discovered a New History of the Earth*. New York: Dutton, 2003.

Cuvier, Georges. *Discourse on the Revolutionary Upheavals on the Surface of the Globe and on the Changes Which They Have Produced in the Animal Kingdom*. Paris, 1812, 1821, 1825.

Darwin, Charles. *On the Origin of Species*. London: John Murray, 1859.

Dean, Dennis R. *Gideon Mantell and the Discovery of Dinosaurs*. Cambridge, UK: Cambridge University Press, 1999.

Dingus, Lowell, and Mark Norell. *Barnum Brown: The Man Who Discovered Tyrannosaurus Rex*. Berkeley: University of California Press, 2010.

Dodson, Peter. *The Horned Dinosaurs: A Natural History*. Princeton, NJ: Princeton University Press, 1996.

Dugatkin, Lee. *Mr. Jefferson and the Giant Moose: Natural History in Early America*. Chicago: University of Chicago Press, 2009.

Emling, Shelley. *The Fossil Hunter: Dinosaurs, Evolution, and the Woman Whose Discoveries Changed the World*. New York: St. Martin's Press, 2009.

Fiffer, Steve. *Tyrannosaurus Sue: The Extraordinary Saga of the Largest, Most Fought Over T-rex Ever Found*. New York: W. H. Freeman and Company, 2000.

Fowles, John. *A Short History of Lyme Regis*. New York: Little, Brown and Company, 1983.

Gallenkamp, Charles. *Dragon Hunter: Roy Chapman Andrews and the Central Asiatic Expeditions*. New York: Viking, 2001.

Gould, Stephen Jay. *Dinosaur in a Haystack: Reflections in Natural History*. New York: Harmony Books, 1996.

Gould, Stephen Jay, and Rosamond Wolff Purcell. *Finders, Keepers: Eight Collectors*. New York: W. W. Norton & Co., 1992.

Harris, Bev. *Black Box Voting: Ballot-Tampering in the 21st Century*. Seattle: Self-published online, 2004. blackboxvoting.org.

Hutton, James. *Theory of the Earth*. Edinburgh, UK: Royal Society of Edinburgh, 1796.

Jaffe, Mark. *The Gilded Dinosaur: The Fossil War between E. D. Cope and O. C. Marsh and the Rise of American Science*. New York: Crown, 2000.

Johnson, Kirk, and Ray Troll. *Cruisin' the Fossil Freeway: An Epoch Tale of a Scientist and an Artist on the Ultimate 5,000-Mile Paleo Road Trip*. Golden, CO: Fulcrum, 2007.

Jones, Bob. *A Fifty-Year History of the Tucson Show*. Tucson, AZ: The Mineralogical Record Inc., 2004.

Katusa, Marin. *The Colder War: How the Global Energy Trade Slipped from America's Grasp*. New York: Wiley, 2014.

Kielan-Jaworowska, Zofia. *Hunting for Dinosaurs*. Cambridge, MA: The MIT Press, 1970.

Klein, Naomi. *The Shock Doctrine: The Rise of Disaster Capitalism*. New York: Picador, 2007.

Kolbert, Elizabeth. *The Sixth Extinction: An Unnatural History*. New York: Henry Holt and Company, 2014.

Kaplonski, Christopher. *Truth, History and Politics in Mongolia: Memory of Heroes.* London: Routledge, 2004.

Lawson, Russell M. *Science in the Ancient World: An Encyclopedia.* Santa Barbara, CA: ABC-CLIO, 2004.

Long, John, and Peter Schouten. *Feathered Dinosaurs: The Origins of Birds.* New York: Oxford University Press, 2008.

Lyell, Charles. *The Principles of Geology.* New York: D. Appleton & Co., 1857.

Mayor, Adrienne. *The First Fossil Hunters: Paleontology in Greek and Roman Times.* Princeton, NJ: Princeton University Press, 2000.

———. *Fossil Legends of the First Americans.* Princeton, NJ: Princeton University Press, 2005.

McPhee, John. *Annals of the Former World.* New York: Farrar, Straus and Giroux, 1981.

Mooallem, Jon. *Wild Ones: A Sometimes Dismaying, Weirdly Reassuring Story About Looking at People Looking at Animals in America.* New York: Penguin Press, 2013.

Morgan, David. *The Mongols.* Malden, MA: Blackwell Publishers, 1986.

Murray, David. *Museums: Their History and Use.* Glasgow, UK: James MacLehose and Sons, 1904.

Norell, Mark. *Unearthing the Dragon: The Great Feathered Dinosaur Discovery*, with photographs and drawings by Mick Ellison. New York: Pi Press, 2005.

Novacek, Michael. *Dinosaurs of the Flaming Cliffs.* New York: Anchor Books, 1996.

Ostrom, John. *Marsh's Dinosaurs: The Collections from Como Bluff.* New Haven, CT: Yale University Press, 1999.

Paul, Gregory S. *Princeton Field Guide to Dinosaurs.* Princeton, NJ: Princeton University Press, latest edition 2010.

Pearce, Susan, Rosemary Flanders, and Fiona Morton. *The Collector's Voice: Critical Readings in the Practice of Collecting.* London: Routledge, 2016.

Pick, Nancy, and Frank Ward. *Curious Footprints: Professor Hitchcock's Dinosaur Tracks & Other Natural History Treasures of Amherst College.* Amherst, MA: Amherst College Press, 2006.

Preston, Douglas. *Dinosaurs in the Attic.* New York: St. Martin's Press, 1986.

Prothero, Donald R. *The Story of Life in Twenty-Five Fossils: Tales of Intrepid Fossil Hunters and the Wonders of Evolution.* New York: Columbia University Press, 2015.

Rainger, Ronald. *An Agenda for Antiquity: Henry Fairfield Osborn & Vertebrate Paleontology at the American Museum of Natural History, 1890–1935.* Tuscaloosa: University of Alabama Press, 1991.

Randall, Lisa. *Dark Matter and the Dinosaurs: The Astounding Interconnectedness of the Universe.* New York: Ecco, 2015.

Rossabi, Morris. *The Mongols: A Very Short Introduction.* New York: Oxford University Press, 2012.

———. *The Mongols and Global History.* New York: W. W. Norton & Company, 2010.

Rudwick, Martin J. S. *The Meaning of Fossils: Episodes in the History of Paleontology.* Chicago: University of Chicago Press, 1985.

Scott, Thomas M., et al. "Geologic Map of the State of Florida." Tallahassee: Florida Department of Environmental Protection, Florida Geological Survey, 2001.

Sellers, Charles Coleman. *Mr. Peale's Museum.* New York: W. W. Norton & Co., 1980.

Semonin, Paul. *American Monster: How the Nation's First Prehistoric Creature Became a Symbol of National Identity.* New York: New York University Press, 2000.

Shubin, Neil. *Your Inner Fish: A Journey into the 3.5-Billion-Year-History of the Human Body*. New York: Pantheon, 2008.

Simmons, John E. *Museums: A History*. Lanham, MD: Rowman & Littlefield, 2016.

Simpson, George Gaylord. *Fossils and the History of Life*. New York: Scientific American Books, 1983.

Sonoda, Naoko. *New Horizons for Asian Museums and Museology*. Singapore: Springer, 2016.

Stegner, Wallace Earle. *Beyond the Hundredth Meridian: John Wesley Powell and the Second Opening of the West*. New York: Houghton Mifflin, 1954.

Sternberg, Charles Hazelius. *The Life of a Fossil Hunter*. New York: Henry Holt and Company, 1909.

Tickell, Crispin. *Mary Anning of Lyme Regis*. Lyme Regis, UK: Lyme Regis Philpot Museum, 1996.

Warren, Leonard. *Joseph Leidy: The Last Man Who Knew Everything*. New Haven, CT: Yale University Press, 1998.

Weatherford, Jack. *Genghis Khan and the Making of the Modern World*. New York: Crown Publishers, 2004.

Weishampel, David B., Peter Dodson, and Halszka Osmólska. *The Dinosauria*. Berkeley: University of California Press, 2004.

Wilford, John Noble. *The Riddle of the Dinosaur*. New York: Alfred A. Knopf, 1985.

Winchester, Simon. *The Map That Changed the World: William Smith and the Birth of Modern Geology*. New York: HarperCollins, 2001.

NOTES

EPIGRAPHS

Rachel Carson, whose *Silent Spring* and *The Sea Around Us* are classics on nature and environmentalism, was said to have been inspired by fossils as a child in Pennsylvania: "Springdale residents who remember Rachel as a young girl tell the story, perhaps true, perhaps apocryphal, that her romance with the ocean began one day when she found a large fossilized shell in the rocky outcroppings on the family's hillside property. It provoked questions that Rachel wanted answers to. She wondered where it had come from, what animal had made it and lived within it, where it had gone, and what happened to the sea that had nurtured it so long ago." See Linda Lear, *Rachel Carson: Witness for Nature* (New York: Henry Holt & Co., 1997).

INTRODUCTION: ORIGINS

xv   Nate Murphy: For more about Murphy see the *Billings Gazette* series by Ed Kimmick, which starts with "Discovery & Deception: Spectacular Finds, Criminal Charges," published May 3, 2009. For sentencing information in the Bureau of Land Management case see "U.S. Judge Gives Dino Collector Nate Murphy 120 Days in Pre-Release Center," also by Kemmick, *Billings Gazette*; June 24, 2009. David Trexler, a paleontologist who had worked with Murphy, told the *Gazette* that when accusations against Murphy surfaced he ignored them, "figuring Murphy 'was being persecuted because he didn't have a degree, and it was a personality thing.'" After Murphy pleaded guilty, Sue Frary, director of the Great Plains Dinosaur Museum in Murphy's home town of Malta, told the newspaper that Murphy "did this to himself, and that is what is so tragic."

xvi   Less than one percent: "Some fossils have been turned to stone, or petrified; many others are preserved, without any change other than the loss of soft tissues," S. J. Olsen once wrote for the Florida Geological Survey. My brief account of fossilization is reductionism in the extreme. The process owes to complex factors, which you can read about in a variety of scientific texts, such as in chapter 7 of the paleontologist David D. Gillette's book *Seismosaurus: The Earth Shaker*, published in 1994 by Columbia University Press, New York. (For an interesting update on that dinosaur see "Whatever Happened to *Seismosaurus*?" by Brian Switek, smithsonian.com; August 17, 2010, in which Switek overviews the subsequent conclusion that *Seismosaurus* was "really an especially large *Diplodocus*." He writes, "Misidentifications are sometimes made—'pygmy' species have turned out to be juveniles of known species and partial skeletons of giants have been discovered to be difficult-to-interpret parts of more modestly sized animals—but science self-corrects as it goes along.") See also "Most Species That Disappear Today Will Leave No Trace in the Fossil Record," by Patrick Monahan, *Science*; March 15, 2016.

xv   Leonardo: The dinosaur skeleton was discovered on July 27, 2000, and was so named because graffiti on a nearby rock read "Leonard Webb and Geneva Jordan, 1917." The discovery was later presented at the annual Society of Vertebrate Paleontology meeting, showing that at the time of his premature death Leonardo stood twenty-two feet long, weighed up to two tons, and was covered in small polygonal scales. For a brief overview, see " 'Mummified' Dinosaur Discovered in Montana," by Hillary Mayell, National Geographic News; October 11, 2002. See also "Dinosaur Spills His Guts," by Robin Lloyd, Live Science; September 25, 2008.

xv   "skin, scales": " 'Mummified' Dinosaur Discovered in Montana," by Hillary Mayell, National Geographic News; October 11, 2002.

xv   One percent, continued: Yet while fossils are a nonrenewable resource, there are untold millions of fossils out there. We've documented more than 1.2 million species in the 250 years that we've been assigning taxonomic classifications. An estimated 86 percent of the planet's existing species and 91 percent of ocean life "still await description." See "How Many Species Are There on Earth and in the Ocean?" by Camilo Mora, Derek P. Tittensor, Sina Adl, Alastair G. B. Simpson, and Boris Worm, *PLOS*; August 23, 2011; journals.plos.org.

xvi   For an overview of *Glossopteris* and continental drift, see "Alfred Wegener: Building a Case for Continental Drift," published online in December 2014 by the University of Illinois, http://publish.illinois.edu/alfredwegener/.

xvi   Western Interior Seaway: A hundred million years ago, during the Cretaceous period of the Mesozoic era—the time of the terrestrial dinosaurs—the Farallon Plate of the Pacific Ocean pushed beneath the North American continent, raising mountains and volcanoes. As the middle of the continent sank, water from the Gulf of Mexico and the Arctic Ocean flowed into the depression, creating a vast, shallow sea. The sea covered Florida, Texas, Colorado, Wyoming, North Dakota, Mississippi, Louisiana, and parts of Montana, Utah, New Mexico, Oklahoma, Arkansas, Tennessee, Alabama, Georgia, South Carolina, Nebraska, Kansas, Minnesota, and most of South Dakota. (Picture boating from Idaho to Iowa.) Extraordinary creatures lived in and around this sea for tens of millions of years. The land animals included dinosaurs. The marine animals included enormous plesiosaurs and mosasaurs. The flying reptiles included pterosaurs. "...You had mountains popping up, and as they popped up they got eroded on top, and the sediments came down onto a flat plain," I once watched P. J. Cavigelli, collections manager of the Tate Geological Museum, in Casper, Wyoming, tell a small group of tourists visiting the Lance Formation, near Lusk, where he and his team were digging out an adult *T. rex*. "These rushing rivers came down and lost their velocity and started dropping sediment, and deposited fossils—I compare this to what's going on in Bangladesh right now, in the Himalayas. The ocean was just a hundred miles thataway." He pointed, talking about the Western Interior Seaway. "Every now and then you'd get catastrophic floods that buried things. That's where we get our good skeletons."

xvii   "I've been in people's houses": "A Tyrannosaur of One's Own," by Laurie Gwen Shapiro, *Aeon*; January 28, 2016.

xviii   Microscopic fossils: The oldest known fossils are cyanobacteria and date to 3.5 billion years ago. For an overview, see "Introduction to the Cyanobacteria: Architects of earth's atmosphere," by the University of California Museum of Paleontology, http://www.ucmp.berkeley.edu/bacteria/cyanointro.html.

xix   Fossil vertebrates and invertebrates: See "What You Should Know about Vertebrate Fossils," Society of Vertebrate Paleontology, vertpaleo.org.

xix   Overseas museums...have no problem buying commercially: In June 1984, the Academy created the Committee on Guidelines for Paleontological Collecting, under the auspices of the Board on Earth Sciences of the Commission on Physical Sciences, Mathematics, and Resources. The committee, whose thirteen members included paleontologists, the commercial fossil hunter Peter Larson, and representatives from the surface mining industry, state/federal government, and natural history museums, was appointed to serve through June 30, 1987. The committee was created "because of increasing concern by scientists over conflicts between collectors of fossils and land managers, developers, and other constituencies. For many years, the major area of conflict involved vertebrate fossils and various, mostly federal, land-managing agencies. However, the conflict has now expanded to include all of paleontology, and it is having an impact on all sciences that use fossils. Even hobby collectors have been involved in the conflict." See "Paleontological Collecting," by the Committee on Guidelines for Paleontological Collecting, National Academy Press, Washington, DC, 1987.

xix   Shell casings: Not to get too far afield here, but for information on forensic evidence see "Types of Forensic Evidence," by the National Institute of Justice, nij.gov.

xix   Big Bone Lick: In July 1739, thirty-seven years before the birth of the United States, an expedition left Montreal for New Orleans in a fleet of war canoes. The party consisted of more than a hundred French soldiers and more than three hundred natives, commanded by a Canadian military officer, full name Charles Le Moyne, Baron de Longueuil. The group intended to defeat the Chickasaw Indians, allies of the English, and control the Mississippi River. The expedition paddled south to the Allegheny River. If the Indians were Algonquin Abenaki, as it's believed that they were, the party would have traveled in large birch-bark canoes and kept completely silent while in transit, the better to slip up on game and avoid detection by enemies or "water monsters." (See *Fossil Legends of the First Americans*, by Adrienne Mayor, published in 2005 by Princeton University Press.) Upon reaching the lower Ohio River, the party made camp east of what is now Louisville, Kentucky. An Indian hunting party went out to find game. Following buffalo tracks to a swamp that reeked of sulfur, they came upon massive bones, some as tall as a man, protruding from the earth like an untended graveyard. Native Americans already had legends about big bones often found in the earth. The legends referenced giant animals that resembled a magnified version of bison. The French often get credit for discovering America's first significant fossils but the Iroquois alone had been talking about fossils for at least two hundred years. As their oral histories showed, Native Americans were not dumbfounded by very old bones, as white people later claimed, but rather intuited their connection to prehistoric life. American Indian explanations for fossils varied according to where they lived and what they saw, and to the stories their ancestors had told. The significance of "magic horn," on the other hand, was universal. Magic horn meant ivory, which was useful in the making of amulets and tools. Ivory meant money. Standing at the edge of that reeking swamp, Algonquians would have recognized tusks as a commodity treasured by European traders, who were eager to compete against Siberian ivory. At the bog that day, the hunters recovered as much as they could manage: three massive teeth, along with a tusk and femur. They hauled the fossils back to camp, where the French found them so curious they packed the materials up and shipped them to Paris. In 1740, the fossils were delivered to the Jardin des Plantes, the royal botanical gardens, and stored in Louis XV's royal Cabinet du Roi, his cabinet of curiosities. In his report, Longueuil acknowledged "les Sauvages" for finding the bones, yet never mentioned the name of the tribe or credited the discoverers by name. Eventually, the discovery site was mined for salt, and became known as Big Bone Lick. Today it's a state park in the town of Union, Kentucky, about twenty-five miles southwest of Cincinnati. (See *The Sixth Extinction: An Unnatural History*, by Elizabeth Kolbert, published in 2014 by Henry Holt and Company.) The welcome sign reads "Birthplace of American Vertebrate Paleontology," though that is a tricky claim to make. The Big Bone Lick fossils weren't the first discovered on the continent, but it was true that they were the ones that had made the biggest international impression. Nearly a hundred years after the hunting party entered the bog, the Canadian paleontologist Edward M. Kindle suggested that Native Americans be credited with the first significant fossil discoveries in America. Scientists weren't so well inclined. Among them was George Gaylord Simpson, an American Museum of Natural History curator and celebrated paleontologist who taught zoology at Columbia, then at Harvard. Simpson considered all Indian fossil discoveries as "casual finds without scientific equal." He wrote, "Indians certainly found and occasionally collected bones... but these discoveries are no real part of paleontological history." His point: Just as "Columbus discovered America," Longueuil discovered America's first fossils. The favored narrative held. The park signs and literature, and most historical accounts, still credit Longueuil, underscoring Simpson's idea that scientific discovery is more about results than about getting there first.

xix   Large three-toed footprints: The longest river in New England, the Connecticut, runs through the wild green western wing of Massachusetts. In 1800, a nine-year-old boy named Pliny Moody was plowing his father's land there, in the town of South Hadley, when his blade struck stone. Unearthing the object, he found a set of strange impressions in rock. Three-toed, like a turkey, the prints measured a foot long, as if a large bird had walked across wet cement. Pliny hauled the slab home to his parents, Ebenezer and Lois, and the Moody family installed it as a decorative doorstep. Townspeople debated what creature had made the tracks, most attributing them to "Noah's raven," the Biblical bird sent forth from the ark during the Flood to find land. Eventually the tracks became known to Edward Hitchcock, the son of a poor local hatter who turned to the ministry, "having been led by my trials to feel the infinite importance of eternal things..." (Edward Hitchcock's papers are archived at Amherst College. An excellent resource is *Curious Footprints: Professor Hitchcock's Dinosaur Tracks &*

Other Natural History Treasures at Amherst College, by Nancy Pick and Frank Ward, published by Amherst College, 2006.) Hitchcock became Amherst College's professor of chemistry and natural history in 1825, hoping to impart to his students the principles of "natural theology," or how geology intersected with religion, though at the time the school owned "not even a skeleton" through which students could learn human anatomy. Hitchcock's wife, Orra, an artist, illustrated his lectures. In 1830, when the state appointed him to make a geological survey of Massachusetts, something that only one other state, North Carolina, had done, Hitchcock relished the opportunity to walk and improve his fragile health. Traveling around Massachusetts, he documented the landscape and stones, using Charles Lyell's recently published *Principles of Geology* as a reference.

The State of Massachusetts printed twelve hundred copies of Hitchcock's survey, one for every town, plus two for Amherst and two each for the state's colleges, Williams and Harvard. Within the next few years, twenty states would make geological maps of their own, producing a tool crucial to locating and interpreting rock formations, sediments, and soils. Geological maps allowed for the planning of communities and roads, the identification of mineral and ore deposits, and the finding of water, and helped civic leaders understand how to prepare for natural disasters such as landslides and earthquakes. Along the way, Hitchcock didn't just come to depend on fossils; he became transfixed by them.

In the nearby town of Greenfield a quarrier in his late twenties was busy working odd jobs. Dexter Marsh had dark, curly hair and a long nose and a beard that cupped his chin without covering it. He came up rough and had not even an elementary school education, but he could read and write and townspeople knew him as "a man of great force and originality, one of the strongest thinkers and closest reasoners with whom I ever became conversant," wrote Lorenzo Langstroth, a teacher and pastor. Marsh's daybooks, though marked by "Spartan brevity," were filled with fragmented descriptions of his activities and expenses. *May 24: "boards for fence .46." August 11: "tickets to see the model of ancient Jerusalem .40."* Marsh didn't mention his first wife's death in his daybooks, or his second marriage, or the birth of any of his five children, but he noted the "first plum tree blown" and a "bonnet for Arabella," along with the letters he sent, the snuff boxes he purchased, and the antislavery efforts he supported in Greenfield, a stop on the Underground Railroad. All his life, Marsh had been a hard and clever worker. When he yoked his first oxen he was so small "he had to stand upon a block to do it." He saved money on hand-made rabbit traps by building them with wooden pegs instead of nails. He was considered a "mechanic"—a handyman. In his daybooks he often labeled the wood-chopping or the ditch-digging or the stone-laying or the apple-picking or the snow-shoveling simply "work." At Town Hall, he was the janitor. At the Second Congregational Church, he rang the bells for weddings and funerals. He also quarried sandstone. In 1835, he was installing a sidewalk on a hill near the courthouse when he noticed birdlike tracks in one of the slabs. "This was an hour of perplexity," wrote Oliver Marcy, a natural sciences professor who went on to found Northwestern University's Museum of Natural History. "To that time he was wholly ignorant of geology, and possessed only the common notion of the formation of the earth; but being a man of accurate observation and logical order of thought, he was convinced that the print before him was the print of a bird's foot. But the print was in solid rock, quarried from several feet beneath the surface. How it came there he could not decide." Marsh began collecting the trackways, including those that had lain in plain sight for years. The stones that everyone assumed were "imperfect flags"—irregulars—now "were taken up as valuable." The slabs were the color of rust; they were rough, with jagged borders. Some were small enough to hold, and others were as heavy and broad as a tabletop.

As Marsh quarried the trackways he'd prop the slabs against his fence and ask people what animal they thought had made them. Some of the prints lay in the flagstone walkway of a man named William Wilson, who showed them to a local doctor, James Deane, who became fascinated with them and described them to Edward Hitchcock in a letter. Hitchcock traveled the twenty miles north, from Amherst, to see the "footmarks" for himself, finding what would become his life's work. "In the six following years I brought out five papers in the journals, containing over a hundred pages and 26 plates, describing 32 species, including my first paper, before any one else had described one species, and before they had scarcely been noticed by any other writer," he wrote. At first, Hitchcock doubted the tracks were footprints—the world's only confirmed fossil tracks were in Scotland, possibly left by an ancient tortoise. But the more slabs he saw the more he decided that Marsh might be right: the tracks belonged to birds, some of them small and others "almost incredibly large."

But why were no bones ever found along with the tracks? These creatures seemed to have strolled through muck and then vanished, their path frozen for all time. At first, no one agreed with Hitchcock. What bird on earth grew to such sizes? "The whole length of the foot…is sixteen or seventeen inches!" Hitchcock wrote at one point, saying the bird's stride measured "between four and six feet!" Yet as Hitchcock went on amassing a collection and publishing what he found, geologists found the evidence to be in his favor, and other scientists began to come around, at a price. "…The disclosures made by my writings attracted others into the field who became uncompromising competitors in the way of collecting, and with some it became a matter of trade," he wrote. "The consequence was that the value of specimens rose to almost fabulous prices."

The quarryman Dexter Marsh was the largest supplier. He told Hitchcock it was his ambition to assemble the largest collection of fossil footprints in the world. Marsh could look at a trackway and "not only tell the direction of a bird, but its comparative speed, the condition of the mud, whether the weather was rainy or not, whether the bird making the track was walking on shore or in the water, and when the bird passed from shore into the water," Marcy wrote. "He came at very definite conclusions concerning the weight and height of the birds."

When Hitchcock heard about the Pliny Moody tracks, in South Hadley, he went to see those, too. By now Elihu Dwight, a Dartmouth graduate and one of the first doctors in town, had acquired the slab from the Moodys; now Hitchcock acquired it from Dwight for the Amherst collection, where it remains on display today. Hitchcock's analysis of the trackways was published in the July 1836 issue of the *American Journal of Science*, establishing the scientific disciplines of ichnology (the study of trace fossils and imprints) and ornithichnology (the study of stony bird tracks). Dexter Marsh wasn't mentioned in the paper, and neither was Pliny Moody.

Word spread quickly from America to Europe that giant fossil bird prints had been found in New England. "This animal turns out to have been one of the most common of all that trod upon the muddy shores...," Hitchcock wrote. "I regarded it as the giant ruler of the valley." Eventually becoming president of Amherst College, he received a visit from Charles Lyell and at least one letter from Charles Darwin. Dexter Marsh, meanwhile, built a "cabinet" at his home and, starting in January 1846, opened it to the public as a free museum. People came from Germany, Turkey, Baltimore, Natchez, and New Orleans to see the world's finest collection of fossil trackways. The poet Oliver Wendell Holmes visited, as did the Scottish chemist James F. W. Johnston, a cofounder of the British Association for the Advancement of Science. After viewing the collection Johnston wrote that Marsh was "only a common mason and gardener, but he has, nonetheless, spent more time and money in searching for and digging up the bird-tracks of this region, and possesses a larger and finer collection of them, than any other person or institution in the United States." He added, "In looking at this collection made by a working man, dug up either with his own hands, or by men working along with him—at his expense, under his direction, and in spots which his own sagacity indicated as likely to reward research—I could not refrain from admiring the enthusiasm and perseverance of their owner, and regretting that, even in this intellectual State, science was too poor, not only to engage such a man wholly in its service, and to add to its treasures by employing him unremittingly in his favorite pursuit, but that it was unable even to purchase the fruits of his past labours, and add them to the public collections..." Johnson wrote, "I must add...what all collectors will well understand, that Mr. Marsh looks upon these slabs of stone as so many children...Mr. Marsh has living feet gathering now in plenty around his daily table...for their sake, these great stones should be converted into bread." Marsh is buried beneath a simple stone at the Federal Street Cemetery in Greenfield. His collection was auctioned, some to Harvard, some to Yale, some to the Boston Society of Natural History, and some to Hitchcock, for Amherst, with funds contributed by John Tappan, an early settler of Brookline; the philanthropist and developer David Sears; and Gerard Hallock, an owner of the New York *Observer*. The Marsh fossils, along with Hitchcock's trackways, are now beautifully displayed at the Beneski Museum at Amherst, along with the Pliny Moody slab.

A quarrier named Roswell Field, who lived in Gill, picked up where Dexter Marsh left off, extracting the most trackways on record. "His prices have indeed been generally high, but when the specimen was unique, I must give him what he asked, or leave it for some one else," Hitchcock wrote, adding, "To persons not familiar with the value of natural history specimens, the idea of giving $150 for a broken slab of stone a few feet square...seems extravagance and folly." The Marsh auction had had an unintended side effect, producing "an impression of the great value of these relics throughout the Valley." Now, "exorbitant prices were attached to them wherever found." To save money, Hitchcock sometimes dug fossils himself and transported them on his personal wagon, arriving in Amherst in "evening, because, especially of late, such manual labor is regarded

by many as not comporting with the dignity of a professor." But it was Roswell
Field who called the trackways what they were. In 1859 he traveled to Spring-
field, to present a paper to the American Association for the Advancement of
Science, saying, "If I can rightly decipher these fossil inscriptions, impressed on
the tombstones of a race of animals that have long since ceased to exist, they
should all of them be classed as Reptilia." He was correct. They were dinosaur.
Hitchcock bought Field's collection, too. The person who finally confirmed
that that dinosaurs made the "curious footprints" of the Connecticut Valley was
a Philadelphia Quaker named Edward Drinker Cope, whose thirst for fossils
contributed to the "bone wars" of the late 1800s, the greatest dinosaur-hunting
showdown in history.

xx   "among the greatest fossil collectors": "Harley Garbani dies at 88," by Dennis
McLellan, *Los Angeles Times*; April 24, 2011.

xx   Stan Sacrison: Buffalo, the county seat, is a blip on the two-lane highway, with
a great bar, Saloon 3. The skull of one of Stan's finds, Tyrannosaurus Stan, has
been called the finest *rex* head on record. It's on display at the Black Hills Institute of Geological Research, a commercial hunting, prepping, and casting company in Black Hills, South Dakota. BHI excavated and prepped the skeleton and
now sells casts of it for a hundred thousand dollars apiece. See "Boneheads: A
Tale of Big Money, Prison, Disney World, and the World's Foremost Dinosaur-
Hunting Twins," by John Tayman, *Outside*; May 2001.

xx   Contributions of commercial fossil hunters: For one perspective, see Neal L.
Larson, Walter Stein, Michael Triebold, and George Winters, "What Commercial Fossil Hunters Contribute to Paleontology," published in the *Journal
of Paleontological Sciences*, a publication of the Association of Applied Paleontological Sciences, the trade group for commercial fossil dealers. AAPS is the
commercial counterpoint to the Society of Vertebrate Paleontology. The international organization, whose membership consists primarily of scientists, pub-
lishes the peer-reviewed *Journal of Vertebrate Paleontology*. Larson et al. wrote,
"Like our academic colleagues, we love the field of paleontology and the pursuit
of the unknown. Most of us are driven by our passion and not by profits." And
most, they added, "would be pleased to have closer relationships with academic
paleontologists."

xxi *Vulcanodon*: The early Jurassic sauropod, one of the long-necked plant eaters,
lived in what is now Zimbabwe. The Natural History Museum, in London, has
a great interactive dinosaur timeline at http://www.nhm.ac.uk.

xxi "The Nation's *T. rex*": The skeleton will be on loan to the Smithsonian for fifty
years. See "'Nation's T. rex' Stands Upright for the First Time in 65 Million
Year—and He's Scary," by John Woodrow Cox, *Washington Post*; October 1,
2015. There's great video with that one. See also "Track the Nation's T-Rex as It
Arrives at the Smithsonian," by Kirstin Fawcett, smithsonianmag.com; April 14,
2014.

xxi "Whether or not it's okay to sell": "The Greatest Challenge to 21st Century
Paleontology: When Commercialization of Fossils Threatens the Science," by
Kenshu Shimada, Philip J. Currie, Eric Scott, and Stuart S. Sumida, *Palaeonto-
logia Electronica*, March 2014.

xxii "rocks that can talk to you": Interviews with Kirk Johnson, Sant Director,
National Museum of Natural History, Smithsonian Institution.

CHAPTER 1: "SUPERB TYRANNOSAURUS SKELETON"

3   "Foul was the evil...": Saint Augustine, *The Confessions of St. Augustine*, trans. John K. Ryan (New York: Image Classics, 1960).

4   His right eye: Eric Prokopi's eye problem was later diagnosed as herpetic keratitis, a painful condition that, if left untreated, can lead to blindness. Amanda blamed "dinosaur dust" from the many hours of prep work Eric did, but the condition is more like a cold sore of the eye.

4   Giant ground sloth: The animal evolved around 35 million years ago in South America and went extinct during the last ice age. (The six current species are tree sloths.) *Megatherium* was the largest of its kind—over twelve feet tall when standing on its hind legs. *Megalonyx jeffersonii*, one species of extinct sloth, was named for President Thomas Jefferson, an avid fossil collector who displayed his collection at the White House, after someone sent him a set of bones found in a West Virginia cave. Based upon the cave bones, Jefferson prepared what's been called America's first scientific paper and delivered it in 1797 at the American Philosophical Society.

4   World's largest natural history shows: The Tucson show takes place in late January and early February. Denver is in September. Munich is in October. Tokyo, December. These shows attract many thousands of natural history vendors and buyers; many of the buyers are private or overseas museums, scientists, teachers, and private collectors. Public museums in the United States generally don't buy commercial anymore.

4   "Before a big piece sells...": Interviews with Amanda Prokopi and her mother, Betty Graham. I spent time with Amanda Prokopi in Gainesville, Florida, and Williamsburg, Virginia, and kept up by phone, between 2012 and 2018. I spent time with Betty Graham in Virginia in late 2014.

5   The one species whose name everyone gets right: Holtz said this in a lecture, "The Life and Times of *Tyrannosaurus rex*," at the Burke Museum in Seattle, a video of which was uploaded to YouTube on March 19, 2013. See https://www.youtube.com/watch?v=sqkqkxYGNZc.

5   Dominated Earth for 166 million years: Dinosaurs lived in the Mesozoic era, which lasted from 251 million years ago to roughly 66 million years ago. Its three periods, oldest to most recent, were the Triassic, Jurassic, and Cretaceous. See "International Chronostratigraphic Chart," International Commission on Stratigraphy, stratigraphy.org.

5   Birds: From March 2016 to January 2017, the AMNH showed an exhibit, curated by AMNH vertebrate paleontologist Mark Norell, called *Dinosaurs Among Us*. The exhibit and corresponding web presence explained that the fossil record of the link between dinosaurs and birds "grows richer by the day. So rich, in fact, that the boundary between the animals we call birds and the animals we traditionally called dinosaurs is now practically obsolete." Extensive research exists on the bird-dinosaur connection. See Mark Norell, *Unearthing the Dragon: The Great Feathered Dinosaur Discovery*, with photographs and drawings by Mick Ellison; and John Long and Peter Schouten, *Feathered Dinosaurs: The Origins of Birds*. Two indispensable books on dinosaur species in general are the *Princeton Field Guide to Dinosaurs*, by Gregory S. Paul, and the massive *The Dinosauria*, edited by David B. Weishampel, Peter Dodson, and Halszka Osmólska.

5   "Dinosaurs are the gateway to science": Interviews with paleobotanist Kirk Johnson, Sant Director, Smithsonian Institution National Museum of Natural History.

6   *T. rex* bite force: "The Biomechanics behind Extreme Osteophagy in Tyrannosaurus rex," *Scientific Reports* 7 (May 17, 2017). See also Nicholas St. Fleur, "Between a T. rex's Powerful Jaws, Bones of Its Prey Exploded," *New York Times*, May 18, 2017.

6   "likely exploded": Ibid.

6   "SUPERB TYRANNOSAURUS SKELETON" and description: Heritage Auctioneers & Galleries Inc., catalog for Heritage Signature Auction No. 6068, May 20, 2012, New York.

6   Hunting: "Almost any paleontologist will attest that, although of course he collects fossils to advance the science of paleontology, his personal drive also arises at least as much from the joy of the hunt and of outdoor life, often in remote camps and with some apparent hardships that are actually part of the pleasure," the late AMNH paleontologist George Gaylord Simpson wrote in *Fossils and the History of Life*. Simpson, who became an AMNH paleontologist in the late 1920s, worked on Mesozoic mammals, including those found in Mongolia. He spent his academic career at Columbia, Harvard, and, after moving to Tucson, the University of Arizona. In *Fossils and the History of Life*, he wrote, "As a monitor of geological resources and an arbiter of evolutionary history, paleontology has played an essential and sometimes a crucial part in varied pursuits of great economic importance. It has even greater values in a more human and less strictly commercial sense. It helps us to understand the world in which we live and to understand ourselves, our origins, our relationships, and our natures."

6   His mounts: Prokopi's major sales included a handful of giant ground sloths, one of which sold via Sotheby's in Paris, where the skeleton loomed in the front window on rue Saint-Honoré.

7   Rainbow: Bolor told me Mongolians name their children for natural wonders. Her name, for instance, means "quartz." *Chuluu* means "stone." *Chuluutse* means "stone flower." *Oyuna* means "copper."

7   Eternal blue sky: A guide named Selenge Yadmaa told me Mongolians have ninety-nine skies. "We're very close to nature because we're nomadic," she said. One day in the Gobi she and I sat for a while in the Land Cruiser of her driver, Bat-Ezdene, whom everyone called "Eeigii." Selenge, who had been a miner and a TV journalist before becoming a guide, told me there were different types of prayer cairns, some made by shamans in reverence to earth and sky. As people pile stones on the cairns they also leave silk sashes. Sashes can be given as gifts, in symbolic colors, as signs of respect. Give a teacher yellow. Give white to dignitaries. Shamans used black against curses. Blue was all-purpose—"You can't go wrong with blue." The cairns often stood many feet high. Every stone in the bed of the hill wants to be at the top of the hill, Selenge said. The higher the stone the happier the person placing it. Eeigii sat listening to all of this. I thought he didn't speak English until he answered, in English, a question meant for Selenge. "You understood everything we were just saying?" I asked, learning that Mongolians say it's important to be lower than grass. "Listen but not talk," Eeigii said. "Like spy."

8   Ulaanbaatar: Say it *OOH-lan-BAT-are*. *Tarbosaurus bataar*, on the other hand, is generally Anglicized to *bat-TAR*. The capital has also been spelled Ulan Bator. Until 1924, the city was known as Urga. In Mongolian: Улаанбаатар хот. A good general resource is *Encyclopedia of Mongolia and the Mongol Empire*, by Christopher P. Atwood.

8   "the largest dinosaur fossil reservoir in the world": "Cretaceous Dinosaur Fossil Sites in the Mongolian Gobi," UNESCO, December 19, 2014.

8   "no place for kids": Many of Bolor Minjin's quotes and personal details are from interviews or observation. To hear her talk about her projects, see "Science Stories: Bolor Minjin on Making a Museum" (https://www.fi.edu/file/science-stories -bolor-minjin-making-museum) and "Bolor Minjin Visits the Gobi Desert" (https://www.fi.edu/file/science-stories-bolor-minjin-visits-gobi-desert), via the Franklin Institute.

9   "Oh it's so hard": Ibid.

9   "I'm a paleontologist": Ibid.

10 Philip Currie's poach-pit count: For more information, see Philip J. Currie, "Fossil Bounty Hunters' Days May Be Numbered," *New Scientist*, June 16, 2012; and Currie, "Dinosaurs of the Gobi: Following in the Footsteps of the Polish–Mongolian Expeditions," *Palaeontologia Polonica 67* (2016).

11 Fender Stratocaster: You'd be surprised how many news-writing pun lovers refer to Mark Norell as a "rock star." Norell is also a prolific art collector. "I like the research aspect," he once told *Aeon*. See Laurie Gwen Shapiro, "A Tyrannosaur of One's Own," *Aeon*, January 28, 2016; and Annie Correal, "How Mark Norell, a Paleontologist, Spends His Sundays," *New York Times*, February 26, 2016.

11 "lightweight": I interviewed Mark Norell at length, on multiple occasions, for this project. This interview took place on July 10, 2012, over lunch near the AMNH.

11 "how we come up with ideas": Interview with Norell.

11 "go feral": Interviews with Kirk Johnson, happily confirmed by Norell. I interviewed Johnson numerous times between 2012 and 2018, by phone or in person in Denver and Washington, DC. For information on Mongolian fossils's place within the AMNH see John Noble Wilford, "Presenting a Fight to the Death from 80,000,000 B.C.," *New York Times*, May 19, 2000.

12 "touring with the Stones": Interviews with Norell. The AMNH scientists who work in the Gobi no longer allow journalists on their expeditions. A few reporters managed to get in before the banishment, including Donovan Webster, who wrote about the experience in an article in the July 1996 issue of *National Geographic*. See "Dinosaurs of the Gobi" for references to standing naked in a sandstorm, drinking the blood of horses, *Lawrence of Arabia*, and stunning illustrations and photos, including one of Norell teasing a fossil from the earth with "surgical tenderness."

12 Gobi: For centuries, the Chinese had one word for the vast desert (*shamo*) and Mongolians another (*gobi*). The Mongolian term prevailed, and as of 2015 the term "Gobi desert" was "not recognised in China's geographical lexicon," Troy Sternberg wrote in "Desert Boundaries: The Once and Future Gobi," a paper for the March 2015 issue of *The Geographical Journal*.

12 "clobbering": Interviews with Norell.

12 straddles the Mongolia–China border: The Gobi is primarily associated with Mongolia. Unless otherwise noted, I refer to the Mongolian Gobi when I reference the desert.

12 "I'm not one of the people": Interviews with Norell.

12 "almost a Silk Road": Shapiro, "A Tyrannosaur of One's Own."

13 Skull of *Saichania*: This was being offered as "FANTASTICAL ANKYLOSAU-RID SKULL," and was expected to sell for up to $80,000. Not long after the auction, Heritage scrubbed its website of the major dinosaurs listed in this auction, and the pages weren't available via internet archives. Other items remained in Heritage's online archive, showing the final sales prices. They included "Fine Bird-Dinosaur Skeleton," *Jinfengopteryx elegans*, from "Central Asia," which sold for $32,000, and "A Superb Tyrannosaurus Tooth with an Erupting Crown," from the Nemegt Formation of the Mongolian Gobi, which sold for $37,500. The items can also still be found in the hard-copy version of the auction catalog.

13 Central Asia: The borders of Central Asia have long been under debate. The region is often defined as the five "stans"—Kazakhstan, Uzbekistan, Turkmenistan, Kyrgyzstan, and Tajikistan, and sometimes Afghanistan. While "East Asia" is a more apt descriptor, I reference Central Asia here partly because the United Nations Educational, Scientific, and Cultural Organization (UNESCO) defines the term as "Afghanistan, northeastern Iran, Pakistan, northern India, western China, Mongolia, and the former Soviet Central Asian republics."

13 "no legal mechanism": The letter from Mark Norell to Heritage Auctions, dated May 17, 2012, can be found in *United States of America v. One Tyrannosaurus Bataar Skeleton* a/k/a LOT 49315 LISTED ON PAGE 92 OF THE HERITAGE AUCTIONS MAY 20, 2012 NATURAL HISTORY AUCTION CATALOG, 12-CIV-4760, U.S. District Court, Southern District of New York.

13 "The auctioning of such specimens": Ibid.

13 "might not otherwise be realized": Dia Art Foundation, https://diaart.org.

14 "boulder" of gold: Catalog for Heritage Signature Auction No. 6068.

14 David Herskowitz: "Museums, especially in this country, really don't have the money to spend on specimens, so they rely on philanthropists," he said in one interview. "So we are hoping that a philanthropist or a museum trustee would be able to put up the money and purchase this, and then donate it to a museum." See Wynne Parry, "For Sale: Tyrannosaurus skeleton at NYC Auction," Live Science, May 17, 2012. Watch "Preview of Auction of Dinosaur Remains Found in Gobi Desert," a three-minute video of the skeleton and Herskowitz talking about it, at https://www.youtube.com/watch?v=Oy6czapJKd8.

14 "crown jewel": From video shot by Painter Law Firm staff member Andrew King.

14 "Mongolian fossils are spectacular": The online petition was started on change.org by Neil Kelley, a California graduate student in paleontology.

14 "no impropriety exists": The letter from Heritage attorney Carl R. Soller, of Cowan, Liebowitz & Latman, P.C., can be found in 12-CIV-4760.

15 "We need a lawyer": Interviews with Bolor Minjin and Oyuna Tsedevdamba.

15 Forty-eight seconds: Auction video.

CHAPTER 2: LAND O' LAKES

16   Flat Florida: See Thomas Scott et al., "Geologic Map of the State of Florida."
      Also see the Florida Geological Survey Open File Report No. 80, 2001, in coop-
      eration with the Florida Department of Environmental Protection. Another
      good source is *The Geology of Florida*, edited by Anthony F. Randazzo and
      Douglas S. Jones (Gainesville: University Press of Florida, 1997).

16   Liquid Florida: Assessing the water resources of the United States is no joke. The
      U.S. Department of the Interior published a valuable bibliography of studies,
      *Bibliography of U.S. Geological Survey Studies of Lakes and Reservoirs—the First
      100 Years*, by Thomas C. Winter, in 1982.

16   Dorothea "Doris" Trappe: Interviews with Doris Prokopi in Land O' Lakes.

16   Cologne: On May 30, 1942, bombers "passed over the city at the rate of one
      every six seconds, dropping a total of 1,500 metric tons of high-explosive and
      incendiary bombs.... When at length the all-clear sounded, about 600 acres
      (240 hectares) of Cologne had been flattened, including 90 percent of the
      central city, 5,000 fires had been ignited (the glare of the flames was visible to
      returning RAF aircrews up to 150 miles away), 3,300 homes had been destroyed
      and 45,000 people left homeless. The casualty toll reached 474 killed and 5,000
      wounded..." If not for the "air-raid shelters and the deep cellars under so many
      homes in old Cologne," the death toll would have been much higher. For a mere
      glimpse of what Cologne experienced during World War II, see Max G. Trethe-
      way, "1,046 Bombers but Cologne Lived," *New York Times*, June 2, 1992.

17   *Freikörperkultur*: Also known as FKK and Free Body Culture. One of many,
      many resources on this is "Nudity in Germany: Here's the Naked Truth," by
      Marcel Krueger, CNN.com, October 10, 2017.

17   Lake Como: "Welcome to Lake Como" flyer ("We are a family nudist club.
      Please bare with us..."), received in person but with no disrobing, Lutz, Florida,
      April 2015. Activities include tennis and volleyball tournaments, hiking, a 5K
      run, concerts, pub crawls, and a classic car show. See lakecomonaturally.com.

17   "extra thirty minutes for small talk": See Diana Everett, "Tender Loving Care,"
      Tampa Tribune, March 30, 1981.

17   "I'm going to Canada": Interviews with Doris and Bill Prokopi, in Land O' Lakes.

18   "like gypsies": Interviews with Doris and Bill Prokopi.

18   Once an orange grove: In *The Orchid Thief*, Susan Orlean described Florida as
      "infinitely transformable," writing, "It is as suggestible as someone under hypno-
      sis. Its essential character can be repeatedly reimagined. The Everglades soil that
      is contaminated by intractable Brazilian pepper trees is now being scraped up
      in order to kill the invader trees, and then the sterilized soil is going to be piled
      high, covered with plastic snow, and turned into a ski resort.... The flat plain-
      ness of Florida doesn't impose itself on you, so you can impose upon it your own
      kind of dream." See Bibliography.

18   "That's for the husband": Interviews with Doris Prokopi.

18   Eric wasn't averse: When asked what his son was like as a child Bill tended to
      repeat two stories, both about mischief. In one, Eric slipped beneath the church
      pews and tickled ladies' feet. In the other, he sat on the roof at Halloween, dan-
      gling a fake spider in front of arriving trick-or-treaters.

18   Del Borgo's "Canterbury Overture": Land O' Lakes High School Spring Concert program, Tuesday, May 23, 1989. Papers of Doris Prokopi.

18   "I don't like it": Interviews with the Prokopi family.

19   "You choose": Interviews with Doris Prokopi.

19   "Prokopi, the 'unknown swimmer'": *The Laker* community newspaper, October 16, 1991.

19   "Swimming is in his blood": Ibid.

19   "trouble with relays": Mick Elliott, "Gators literally a 1-man team in swimming," *Tampa Tribune*, November 9, 1989.

19   "Defiance of authority": District School Board of Pasco County Notice of Suspension, May 9, 1989. Papers of Doris Prokopi.

20   "Hardly a roadcut or realignment": S. J. Olsen, *Fossil Mammals of Florida*, Florida Geological Survey, Special Publication No. 6, 1959.

21   Sharks and three thousand teeth: One source is *Florida's Fossils*, by Robin C. Brown (Sarasota, FL: Pineapple Press, February 2008).

21   "With patience": Margaret C. Thomas, *Let's Find Fossils on the Beach* (Sunshine Press, 1962). Another book, *Let's Go Fossil Shark Tooth Hunting*, by B. Clay Cartmell (Natural Science Research, June 1978), reminded hunters that shark teeth could be found in a number of states, even the landlocked ones, but that in Venice someone could "walk less than 100 feet from his parked car and begin picking up fossil teeth."

21   "The important thing to remember": Olsen, *Fossil Mammals of Florida*.

CHAPTER 3: GARCIA, KING OF THE ICE AGE

22   "You'll never be a brain surgeon": Frank Garcia told me this in interviews but he's also written about it, variously in autobiographies and on Facebook. He has self-published four books, at last count. The first, in 1974, was *An Illustrated Guide to Fossil Vertebrates*, which he illustrated himself. The others are memoirs. This chapter and part of the next were informed by interviews with Garcia in the Florida towns of Ruskin and Venice, and by two of his self-published memoirs, *Sunrise at Bone Valley* (1988) and *I Don't Have Time to Be Sane* (2007).

22   "That's Amore": On YouTube you can see or hear Garcia's velvety stylings of various songs, including "Besame Mucho" (https://www.youtube.com/watch?v=mvvnbn4uv0U). In one 2009 video (https://www.youtube.com/watch?v=vsQqqm3YdkY), he takes the stage at the Tampa mayor's Hispanic Heritage Celebration, in sunglasses, and performs his original tune "Corazon de Tampa." At one point during the reporting of this book, I fleetingly imagined Garcia in concert with other singers mentioned here, such as the auctioneer and jazz singer Izzy Chait and the herder family I met one night in the Gobi.

22   "most interesting man": Jeff Klinkenberg, "Give a Dog a Bone," *Tampa Bay Times*, June 19, 2011.

23   Dinosaurs had never been found in Florida: Olsen, *Fossil Mammals of Florida*.

23   "We had *camels* in Florida?": Klinkenberg, "Give a Dog a Bone"; and interviews with Garcia interviews.

23   Joe Larned: My favorite Larned story: In a 1977 newspaper essay he predicted a string of brutally cold Florida winters that would "bottom out" in January 1985. When he turned out to be right the *Lakeland Ledger* ran a front-page story on

Larned, headlined, "He Told You So." One guy told Larned he'd lost his groves two years running and had to quit citrus. "Why didn't you tell me two years ago?" the guy asked Larned. Larned replied, "Jesus, why didn't you read the newspaper eight years ago?" See Mike Capuzzo, "He Feels the Future in His Bones," *Miami Herald,* February 24, 1985. For more on Larned, see Martha F. Sawyer, "A Touch of the Prehistoric Lives in Polk," *Lakeland Ledger,* March 24, 1982; and Barbara Donaghey, "Fossil Museum a Fine Place to Bone up on the Past," *Lakeland Ledger,* May 14, 1979. These papers are publicly archived via Google Newspapers.

23   "man's slash and burn tactics": Capuzzo, "He Feels the Future in His Bones."

24   "You realize you're the first human being to ever see that": Sabrina Porter, "The Fossil Hunt Is On and the Larneds Lead the Posse," *Lakeland Ledger,* December 7, 1980. For more information on Bone Valley, check out the Mulberry Phosphate Museum, in Mulberry, Florida, "the phosphate capital of the world," www.mulber ryphosphatemuseum.org. The museum bought Joe Larned's collection.

24   "I think what impressed me most": *Sunrise at Bone Valley,* Florida West Coast Public Broadcasting Inc., 1990.

24   "the world's largest dolphin skull": Frank Garcia's Facebook page.

25   Lifted as a whole: For more on jacketing fossils, see "Techniques in the Field," AMNH, http://preparation.paleo.amnh.org.

25   "Did you learn": Interviews with Garcia, and Garcia's memoirs.

26   "field associate": This is per Daryl Domning, a longtime Howard University professor and Smithsonian affiliate. Domning said Garcia worked with him "extensively" in the 1980s, "collecting in Bone Valley and nearby areas of Florida, with his field expenses funded by my NSF grants." Garcia also had a relationship with Clayton Ray, the longtime curator of late Cenozoic mammals and fossil marine mammals in the NMNH's paleobiology department.

26   "heroic": Domning said Garcia "provided the overwhelming bulk of NMNH's collection of Bone Valley sirenian fossils. These included the first in-situ specimens of a new species that I named after him in 2008 (*Nanosiren garciae*)."

26   *Illustrated Guide to Fossil Vertebrates*: Garcia published the book in 1974 and at last check it was available through Amazon for $9.99.

26   Leisey shell pit: Florida West Coast Public Broadcasting Inc. made a one-hour documentary about Garcia's Leisey discovery in 1990, called *Sunrise at Bone Valley*, which is also the title of one of Garcia's memoirs. Garcia posted the documentary to YouTube at https://www.youtube.com/watch?v=FWJyE2Wq96Q. You can see the vastness of the Bone Valley terrain and the drag lines at work, along with Garcia teaching and talking about fossil hunting.

27   "extremely significant": Gil Klein, "Amateur Digger May Have Made the Biggest US Fossil Find Ever," *Christian Science Monitor,* January 16, 1984.

27   "new chapter": Ibid.

27   "You've been offered a great deal of money": "Amateur Fossil Hunter Frank Garcia Interviewed," *Today,* NBC, April 17, 1985. NBCUniversal Media LLC. Provided by NBCUniversal Archive.

28   "an amateur in the best sense of the word": Klein, "Amateur Digger."

28   "In twenty years": *Sunrise at Bone Valley* documentary, Florida West Coast Public Broadcasting Inc., via YouTube.

28 For more information on Florida's fossil-collecting law, see floridamuseum.ufl.edu.
28 Association of Applied Paleontological Sciences: AAPS was first created in 1977–78 as the American Association of Paleontological Suppliers, to serve as a "united voice" for the fossil trade and to promote "ethical collecting practices and cooperative liaisons with researchers, instructors, curators, and exhibit managers in the academic and museum paleontological community." In the early 1990s, a group called the International Association of Paleontological Suppliers was created "to help foreign businesses organize and become aware of legislation in various countries regarding the import and export of fossils." The IAPS joined the AAPS in 2002. See aaps.net.
28 "unaware that the commercialization": Shimada, Currie, et al.
28 "misguided perceptions": Ibid.
30 "Mystical, magical": Frank Garcia, *Sunrise at Bone Valley* (self-pub., 1988). As he wrote the book, Frank was excavating "the world's largest llama site" at the Leisey pit, a find that had been indirectly prognosticated by a seer he met via an acquaintance called "Mama Fish."
30 "Imagine a character": Frank Garcia, *I Don't Have Time to Be Sane: The Life Story of One of the Most Notorious Fossil Hunters in America* (self-pub, Fossil Finder Books, 2007). This line was written by Don Miller, president of the Delaware Paleontological Society, who authored the preface.
30 . "His lectures are always educational": Garcia, *Sunrise at Bone Valley*.
30 "You are never too old": Ibid.

CHAPTER 4: DIVE

31 Leisey shell pit descriptions: interviews with Frank Garcia, Doris Prokopi, and Eric Prokopi; Garcia memoirs; photos, videos, and news archives. The story was widely covered in the 1980s by Florida newspapers and magazines including *Time* and *Newsweek*. A good overall source of Leisey information is the Florida Museum of Natural History, whose website, floridamuseum.ufl.edu, notes: "Prior to the discovery of the Leisey Shell Pit 1A locality, the commercial shell pits of South Florida had not been considered a significant source for sizable concentrations of vertebrate fossils. Leisey changed that view…"
31 Technological advances: For information on tech in paleontology, see John A. Cunningham, Imran A. Rahman, Stephan Lautenschlager, et al. "A Virtual World of Paleontology," *Trends in Ecology and Evolution* 29, 6 (June 2014).
32 "He *smells* them": Interviews with Doris Prokopi.
32 "just pick out whatever we wanted": Porter, "The Fossil Hunt Is On."
32 "We wouldn't even stop": Ibid. One of the books in Eric Prokopi's eventual personal library was *Ocean Realm Diving Guide to Underwater Florida*, by Ned DeLoach, (Miami: Ocean Realm Publishing Corp., 1983). His copy was so worn it was held together with Scotch tape.
32 Diving: I must have read five hundred newsletter articles and book excerpts on the experience of river diving. I liked one *Tampa Bay Fossil Chronicles* piece from September 1992, by Leslie Newberry, a certified divemaster, describing the Withlacoochee River, which is considered dark water because of all the tannic acid: it leaches from cypress roots, "turning the water the color of tea or coffee."

Newberry wrote about the morning mist curling off the water, the sputtering boat motor, the pungent damp-wood smell, and the overall feeling of "entering a primitive place during an era long past. Out in the distance, you almost expect to see a mammoth or mastodon elephant on the banks of the river, taking an early morning drink."

32 "sinker wood": One source is "1800s-Era Sunken Logs Are Now Treasure; Here Are the Men Who Find Them," by David Zucchino, *Los Angeles Times*, July 13, 2014.

34 "priests, government officials, kings, emperors, slaves": Russell M. Lawson, *Science in the Ancient World: An Encyclopedia.* An excellent source of information is Adrienne Mayor's *The First Fossil Hunters: Paleontology in Greek and Roman Times.* Mayor, a Stanford folklorist who has called herself a "historian of 'science before Science,'" spent two decades on the work, which blends classical studies and paleontological science, connecting Greek myths and monsters like griffins to, for example, illiterate nomads' ancient interpretations of the dinosaur bones they found in the Gobi Desert. "The tasks of paleontologists and classical historians and archaeologists are remarkably similar—to excavate, decipher, and bring to life the tantalizing remains of a time we will never see," Mayor wrote. In a new introduction for the book, which has become an interdisciplinary staple in university curricula, she wrote, "The sensation of losing track of time and place while submerged in libraries and museums often put me in mind of Jacques Cousteau's phrase *l'ivresse des grandes profondeurs*, 'rapture of the deep,' to describe the giddy intoxication experienced by divers exploring hitherto inaccessible undersea realms."

34 Anaximander lived from 611 to 546 BCE.

34 Xenophanes may have been the first: G. S. Kirk and J. E. Raven, *The Presocratic Philosophers: A Critical History with a Selection of Texts* (Cambridge University Press, 1984).

34 Aristotle: Ibid. Mayor wrote, "It is often suggested that Aristotle's 'fixity of species' idea was a deathblow to rational speculation about evolution and extinction in classical antiquity and the Middle Ages. This misleading view unfairly conflates two very different cultures and eras. The notion of immutable species created in one fell swoop was not a monolithic principle in classic antiquity—it only became so in the Middle Ages when Aristotelian thought was merged with biblical dogma in Europe."

34 Pliny the Elder: This guy was a "night-worker," an instant napper, a rabid bather, and a vital link in the history of science. The ten volumes of *Natural History* covered astronomy, meteorology, geography, ethnography, anthropology, human physiology, zoology, botany, agriculture, horticulture, pharmacology, magic, water, mining, and mineralogy. Pliny understood the scope and potential legacy of his attempt, writing, "There is not one of us who has made the same venture, nor yet one Roman who has tackled single-handed all departments of the subject." No one, it was true, had written a book with a chapter on strange rain and on "Milk, Blood, Flesh, Iron, Wool, Tiles, and Bricks." No one, at least not quite in Pliny's way, had dismissed rainbows, which "fortel not so much." *Natural History* influenced scientists for centuries to come, and two thousand years later is still engaging reading, especially the bits infused with superstitious customs

"established by those of old, who believed that gods are present on all occasions and at all times." To wit: before eating at table, it is customary to remove one's ring. To calm anxiety, dab a bit of saliva behind the ear. To signal approval, give a thumbs-down. To worship lightning, cluck the tongue. If, during a banquet, you happen to stupidly mention fire, reverse the omen by pouring water beneath the table. Women, never twirl your spindles while walking down a road or else you'll invite a blighted harvest. Never sit with your fingers interlaced while visiting a pregnant woman, lest you be accused of sorcery. To cure epilepsy, feed the patient the flesh of a wild animal slain with the same iron weapon that killed a human being. Also good for epilepsy: sex. Should you happen to swallow some quicksilver (and who doesn't, from time to time?) chase it with lard. And etc., involving goat dung, ivory shavings, ant eggs, sea frogs boiled in vinegar, "complaints of the anus," and knots tied while saying the name of a widow. Love, by the way, "is killed by a bramble toad worn as an amulet in a fresh piece of sheep's skin." In case you wanted to know.

Two years after finishing *Natural History*, Pliny was fifty-six, overweight, and suffering from asthma. It was the year 79. Now the commander of a Roman fleet, he lived in Misenum, on the western side of the Bay of Naples. Like a lot of Romans, he enjoyed coating his skin in oil and bathing in the sun. This is what he did on the 24th of August, after which he fixed himself a "light luncheon" and went "back to his books," his nephew, Pliny the Younger, later wrote in a letter to the historian Tacitus. Around one o'clock, Pliny's sister asked him to come have a look at a strange dark cloud. Pliny went outside and gazed across the bay, where, less than six miles away loomed Mount Vesuvius. The cloud of superheated ash and stone that had formed above it was moving south, toward Pompeii, as was fast-flowing lava, incinerating everything in its path. It would take centuries for scientists to fully understand that the eruption consisted of molten rock, gases, and pumice ash traveling at a force of one and a half tons per second. Some 1,866 years later, the United States would drop an atomic bomb on Hiroshima, Japan, unleashing a blast whose pressure wave, fire, and radiation killed over sixty-six thousand people—the Vesuvius blast was a hundred thousand times stronger than that. Anyone who didn't evacuate Pompeii during the first few hours stood no chance of surviving incineration or suffocation or roof collapse or flying rock.

"This phenomenon seemed to a man of such learning and research as my uncle extraordinary and worth further looking into," wrote Pliny the Younger, an eyewitness who authored the only surviving account. Pliny the Elder summoned a small naval fleet. As he prepared to leave aboard a fast cutter, he received urgent word from Rectina, a friend living near the volcano, who needed rescuing, and who cautioned that the only way out was by sea. His nephew wrote, "He steered his course direct to the point of danger."

One of the towns along that beautiful coast was Herculaneum, which lay just south of Mount Vesuvius. Herculaneum was closer to the volcano than was Pompeii, but had avoided damage, so far. Then, the winds changed. The searing cloud barreled toward Herculaneum at a hundred miles an hour, burying it in sixty feet of volcanic ruin. Pliny believed he would find his friends farther down the coast, at Stabiae, so on he sailed. The closer he got to shore, the hotter it got. As flaming debris rained down on his boat and "vast fragments" tumbled down the mountain, threatening to block the shore, a crew member told him to turn back.

"Fortune favors the brave," Pliny answered, and ordered the crew on.

He went ashore at Stabiae and found one of his friends, Pomponianus, but not Rectina. Pomponianus was panicked, so Pliny hugged him and soothed him. Then he had a bath and they all sat down to dinner.

That night, Vesuvius glowed red in the dark, and the cinder and rock continued to fall. Most of the people in Pliny's party were too nervous to sleep, but Pliny snored so loudly the servants could hear him from outside.

He woke in the night as the courtyard filled with so much ash and stone that it threatened to block any escape. Houses were being "rocked from side to side with frequent and violent concussions as though shaken from their very foundations." The roof groaned with the weight of a growing load of stone. Deciding that they would be safer outdoors, Pliny's party fled for the open fields with pillows tied to their heads. That didn't save Pomponianus, who was struck by a flying rock and died.

By now it was day but the ash cloud made it dark as night. Torches moved through the gloom. Pliny ran to the beach to see if he could sail. The winds were still too strong and the waters "extremely high, and boisterous." When Pliny said he felt weak, his men spread out sailcloth for him to sit on and brought him two cups of cold water. Then "the flames, preceded by a strong whiff of sulphur, dispersed the rest of the party, and obliged him to rise," his nephew wrote. "He raised himself up with the assistance of two of his servants, and instantly fell down dead."

Three days later, after the air had cleared, Pliny's men returned to where they had last seen him, and found his corpse buried beneath pumice. "His body was found entire," his nephew told Tacitus, "and without any marks of violence upon it, in the dress in which he fell, and looking more like a man asleep than dead."

That was the end for Pliny but not for his work. *Natural History* would canonize the idea that nature wasn't just a part of life; it *was* life.

35   The vipers of Malta: Stephen Jay Gould and Rosamond Wolff Purcell, *Finders, Keepers: Eight Collectors.* For a treat, see Agostino Scilla's writings on Malta, published in 1670 in his treatise *Vain Speculation Undeceived by Sense.*

35   "dug up": In 1565, a Swiss naturalist, Conrad Gessner, published *On Fossil Objects (De Rerum Fossilium)*, which, to the British geologist and science historian Martin J. S. Rudwick, signaled the beginning of the discipline to be called paleontology. The term "initially included minerals, interesting concretions, and curiosities, as well as organic remains," the University of Pennsylvania paleontologist Peter Dodson wrote in the introduction to Adrienne Mayor's *The First Fossil Hunters.* He continued, "Rudwick emphasized the difficulty of the task of interpreting ancient remains. Nothing in nature comes with a label attached..."

35   "For using dragon's bones": Ernest Ingersoll, *Dragons and Dragon Lore* (New York: Payson & Clarke Ltd., 1928), a signed copy of which sells, used, for nearly four hundred dollars.

35   "How can so many Americans still disbelieve in evolution?": Kevin Holden Platt, "Dinosaur Fossils Part of Longtime Chinese Tonic," *National Geographic News,* July 13, 2007.

36   For more on Steno, see Troels Kardel and Paul Maquet, eds., *Nicolaus Steno: Biography and Original Papers of a 17th Century Scientist* (New York: Springer, 2013); and Alan Cutler, *The Seashell on the Mountaintop*. In a synopsis of Steno's significance, UC Berkeley's Museum of Paleontology notes that at first a mission to understand Earth's layers "may not seem like important work, but consider this: if you wanted to know about the evolution of life on Earth, you would need a fairly accurate timeline. Questions such as: 'How long did something stay the same?' or 'How fast did it change?' can only be assessed in the context of time." See http://www.ucmp.berkeley.edu.

36   "pass over without regard": Robert Hooke, *The Posthumous Works of Robert Hooke.* (London: Sam. Smith and Benj. Walford, 1705).

36   "a solid naturally enclosed in a solid": Nicolaus Steno, *De Solido intra Solidum Naturaliter Contento Dissertationis* Prodromus [Forerunner to a Dissertation on a Solid Naturally Contained within a Solid] (1669). This quote is found in a number of texts on Steno, including Martin J. S. Rudwick's *The Meaning of Fossils: Episodes in the History of Paleontology,* and Donald Prothero's *Bringing Fossils to Life: An Introduction to Paleobiology,* 2nd ed. (New York: McGraw-Hill, 2004).

36   "inclined to the horizon" and "continuous over the surface of the Earth": Ibid.

37   "pointing to a hillside": Gould and Purcell, *Finders, Keepers.* Gould was a Harvard evolutionary theorist known for both his brilliance and his arrogance, the *New York Times* reported when he died in 2002, of cancer. "An entertaining writer credited with saving the dying art form of the scientific essay, Dr. Gould often pulled together unrelated ideas or things," the *Times* noted. His "research, lectures, and prolific output of essays helped to reinvigorate the field of paleontology." See Carol Kaesuk Yoon, "Stephen Jay Gould, 60, Is Dead," *New York Times,* May 21, 2002.

37   Earth's age: A 2014 Gallup poll found that four in ten Americans believed God created the earth between six thousand and ten thousand years ago. "Religious, less educated, and older respondents were likelier to espouse a young Earth creationist view," Live Science reported, adding, "While most Americans have a healthy respect for science, many could use a refresher course in the basics." See Tia Ghose, "4 in 10 Americans Believe God Created Earth 10,000 Years ago," by Live Science, June 5, 2014. By May 2017, Gallup was reporting that the percentage of U.S. adults who believed "God created humans in their present form at some time within the last 10,000 years or so—the strict creationist view" had dropped to 38 percent. Most now believed in "some form of evolution." See Art Swift, "In U.S., Belief in Creationist View of Humans at New Low," Gallup News, May 22, 2017. The AMNH paleontologist George Gaylord Simpson wrote in *Fossils and the History of Life,* "There is no necessary conflict between religion and science. Among religions, only the bigoted fundamentalist sects have a dogma condemning evolution. Many religious teachers and laymen accept the fact of evolution. Many evolutionists are religious. Evolutionists may also be creationists, but in a very different and truer sense than that of those who call themselves creationists. . . . The concept of evolution suggests, and the fossil record confirms, that all organisms, past and present, are parts of one extremely long and extremely branching family tree. Life on earth is a single phenomenon with many millions of manifestations. From this widened point of view, *Homo sapiens* is just one small twig on the tree of life."

37 "observation over speculation": John Woodward, "An Essay Towards a Natural History of the Earth and Terrestrial Bodies, Especially Minerals, and Also of the Sea, Rivers, and Springs" (London: A. Bettesworth, W. Taylor, 1723).

37 "I do not know how the sea was able to reach so far inland": Agostino Scilla, *La Vana Speculazione Disingannata dal Senso* [Vain Speculation Undeceived by Sense] (Naples, Italy, 1670). At the end of his treatise, Scilla added thirty beautiful, engraved plates bearing the images of fossils, "figures that enhance the intrinsic beauty of fossils, with an artist's sense of balance and placement," Stephen Jay Gould wrote. The Sedgwick Museum at Cambridge University houses these, along with Scilla's surviving specimens and plates. They represent some of the earliest intersections of art and science.

38 "Here's what makes him a hero": David Quammen, "A Passion for Order," *National Geographic*, June 2007. For more on Linnaeus as our holotype, also see David Notton and Chris Stringer, "Who Is the Type of *Homo sapiens*?," International Commission on Zoological Nomenclature.

38 "so full of fossils and chemical apparatus": Edmond A. Mathez, *Earth: Inside and Out* (New York: The New Press, 2001).

38 "The mind seemed to grow giddy": "Biographical Account of the Late Dr James Hutton, F.R.S. Edin.," read by John Playfair, January 10, 1803, Transactions of the Royal Society of Edinburgh. In *Fossils and the History of Life*, George Gaylord Simpson introduced me to a new word: chronophobiac, coined by the historian and philosopher of geology Claude Albritton in *The Abyss of Time* (San Francisco: Freeman, Cooper and Company, 1980). Albritton was referring to those who "cringe in fear when pondering geological time," Simpson wrote.

38 "no vestige of a beginning": *James Hutton, Theory of the Earth* (Edinburgh: 1795). A full reading can be found at gutenberg.org. The last sentence of volume I reads, "...we shall thus be led to admire the wisdom of nature, providing for the continuation of this living world, and employing those very means by which, in a more partial view of things, this beautiful structure of an inhabited earth seems to be necessarily going into destruction."

39 "Consider the Earth's history": John McPhee, *Basin and Range* (New York: Farrar, Straus and Giroux, 1981). McPhee's masterwork on geology started in the pages of *The New Yorker* and is collected in his book *Annals of the Former World*. See Bibliography.

39 geologic chart: See "International Chronostratigraphic Chart," International Commission on Stratigraphy, stratigraphy.org. There, you can also see a slideshow of gorgeous, extreme stratigraphy like the Grand Canyon.

39 "Digging into Florida's Past": Bone Valley Fossil Society newsletters, personal papers of Eric Prokopi, provided to author. The contents of a collection of these newsletters informs other passages in this chapter. Other information was found in the *Tampa Bay Fossil Club Chronicles* newsletter, January 2001. Papers of Eric Prokopi. Frank Garcia was one of the club's founders. Its sponsors included the University of South Florida Department of Geology, the Leisey Shell Corp., and Tampa's Museum of Science and Industry.

39 "riding the couch": *Tampa Bay Fossil Club Chronicles*, April 2001. Papers of Eric Prokopi.

40 "Why Janey and Johnny": In November 1991, Dr. Warren Allmon, then a University of South Florida geologist, complained in the Tampa Bay newsletter

that students and Americans in general knew appallingly little about evolution, "the most basic and important theory in all of biology," because they had been exposed to the concept "only briefly and inaccurately, if at all." Allmon was an entertaining regular in the newsletter. In one issue he critiqued Baltimore's aquarium and other national aquaria as misleadingly incomplete and inaccurate. "If we communicate that a rainforest can be 'built' with some plants and mist and a sloth, and a coral reef with some fish and fiberglass, then we run the risk of communicating the idea that these systems are not really distinct, complicated, or special," he wrote, adding that "if we destroy them we can easily fix them if only Disney imagineers could have a go at it." And when critiquing the World of Energy exhibit at Disney's Epcot Center, he ridiculed the public presentation of dinosaurs and their world as "so outdated, so cliche, so outstandingly awful, that it cancels out any possible positive effect." Why were all the sauropods still standing in swamps? Where were the herbivores? "If, when we think of the past, the first image that comes to mind is of a hellish place of dragons and giants, then we will find difficult to accept the idea that we can study it scientifically," he wrote. "The Mesozoic was neither a time of smiling Flintstones-type dinosaurs nor of aliens from another world. It was neither extreme of the pendulum's swing, the assertions of Hollywood (and its surrogates in Orlando) notwithstanding. It was our world, the one we now inhabit."

40   "The word dinosaur": Roy Chapman Andrews, *All About Dinosaurs*. Andrews's other popular children's books included *In the Days of the Dinosaurs* (New York: Scholastic, 1975) and *All About Strange Beasts of the Past* (New York: Random House, 1956).

CHAPTER 5: DEAL

41   *Dell'historia naturale*, (Naples, Italy: Nella stamparia à Porta Reale per Costantino Vitale, 1599). The fold-out engraving is something to see, and the cabinet itself must have been even more so, anchored by that ceiling croc. "Crocodiles were popular in the cabinets of curiosities because of their extravagant size and monstrous appearance and because they were enigmatic dwellers of both land and water," John E. Simmons wrote in *Museums: A History* (New York: Rowman & Littlefield Publishers, 2016).

41   "antlers, horns, claws": Gabriel Kaltemarckt, "Bedenken wie eine Kunst-cammer Aufzurichten seyn Möchte" [Thoughts on how a kunstkammer should be formed], 1587, reprinted in Susan Pearce, Rosemary Flanders, and Fiona Morton, *The Collector's Voice: Critical Readings in the Practice of Collecting*, vol. 2, wherein Kaltemarckt's work was called a "classic text in the museological tradition."

41   "fast-walking messenger" and other Peter the Great matter: Gould and Purcell, *Finders, Keepers*.

42   "Order Proboscidea": *Bone Valley Fossil News*, the newsletter of the Bone Valley Fossil Society, October 1991. Papers of Eric Prokopi.

44   *You shoult go get toes shocks*: Interviews with Doris Prokopi.

44   "Eric is a very conscientious guy": Rod Gipson, "Making Waves: LOL's Prokopi Plans to Become Gator Swimmer," *Tampa Tribune*, June 3, 1992.

44   *Les mauvaises terres*: If you ever have some time to kill, see the twelve-volume *Reports of Explorations and Surveys, to ascertain the most practicable and economical*

route for a railroad from the Mississippi River to the Pacific Ocean, made under the direction of the U.S. Secretary of War in 1853–4, for the U.S. Congress. Beverley Tucker, Printer, Washington, DC, 1855. Published by the Government Printing Office, 1855–61. The University of Michigan calls the reports a "cornerstone piece of Americana." The details are wonderful. As one artist sketched California pectins he was stung by a mother scorpion with a load of babies on her back. Parties passed through places called Bed Dog, Gouge Eye, and Dutch Flat. Everyone was urged to carry a geological hammer; no. 9 birdshot; 5 gallons of alcohol and blotting papers, for pressing plants; arsenic in 2-pound tea canisters, for treating the moist flesh of fresh kills; and a butcher knife, needle, and thread, for "skinning and sewing up animals." The correspondents often mentioned fossils but without much detail. ("Fossils of the cretaceous or jurassic formation were found in the creeks crossed to-day.") On a brackish stretch of the Colorado River, one party found seashells on hilltops. Near Sacramento, gypsum caves of "dazzling whiteness." In the Rio Grande Valley, horizontal strata of sandstones and limestones. Fossil ferns near a coal seam not far from Fort Belknap, Montana. Fossil oysters near the Llano River. Concretions "as big as oranges" on a tributary of the Missouri. On the Posuncula River they found shark teeth, bone fragments, and fossil wood 500 feet above sea level, seeing for themselves that strata may appear as horizontal as cake layers or as pitched as a funhouse floor. See http://www.cprr.org/Museum/Pacific_RR_Surveys/.

44 "very skeleton": Ibid. By then, *paleontology* had been a word since 1822, having been coined as *palaeontologie* in *Journal de Phisique*, in France.

44 "Women of great age": *Reports of Exploration and Surveys*.

44 "From the uniform": Ibid.

45 Larson brothers: Peter was inspired by a local fossil collector named June Culp Zeitner, who started a dozen gem and mineral clubs and encouraged all the U.S. states to each adopt an official gem, mineral, fossil, and/or rock. At the International Gem Show of 1976 she was named the "First Lady of Gems" at age ninety, and received a prestigious award from the Carnegie Museum of Natural History in Pittsburgh. For more information, see Steve Fiffer's *Tyrannosaurus Sue* and Peter Larson's memoir *Rex Appeal: The Amazing Story of Sue, the Dinosaur That Changed Science, the Law, and My Life* (Montpelier, VT: Invisible Cities Press, 2012), coauthored by Kristin Donnan, with a foreword by the paleontologist Robert Bakker.

45 Black Hills Institute: Pete Larson's business partners included his brother, Neal, and an old college friend, Bob Farrar.

45 Rock shops and backyard museums in general: In his book *Cruisin' the Fossil Freeway*, the Smithsonian's Kirk Johnson tells a story about a commercial hunter named Lee Campbell, who had "sold a nearly complete ankylosaur to the Hayashibara Museum in Japan." After Johnson saw the skeleton "laid out in the backyard of a Tucson hotel in 1991" Campbell asked him and the illustrator Ray Troll, "Do you want to go to church?" Johnson wrote, "Lee led us to a mildly dilapidated white wooden church. The pews had been removed and replaced with large shelves that sagged beneath the weight of dinosaur bones. The choir area was full of plaster jackets. Workbenches laden with partially cleaned bones lined the walls. Dinosaur skeletons were the only active members of this

congregation.... We asked Lee how he'd ended up in this particular situation, and he told us that he was being bankrolled by a Kentucky dentist who loved digging fossils more than cleaning teeth."

46   "I don't know if it's a sixth sense": Sue Hendrickson with Kimberly Weinberger, *Hunt for the Past: My Life as an Explorer* (New York: Scholastic, 2001). Hendrickson grew up in Munster, Indiana. As a child she was so shy her mother would drop her off at birthday parties and return to find her still sitting alone on the doorstep, having never gone inside. She earned all A's in school, and in Girl Scouts she learned to love nature. "People often ask me whether my interest in searching for buried items started in childhood. The answer is yes," she once wrote. "My earliest memory of finding 'treasure' is when I was around four years old": in a pile of alley garbage she found a brass perfume bottle embellished with a tiny white heart. Eventually Hendrickson dropped out of school and hit the road with her boyfriend and wound up in the Florida Keys, diving for tropical fish to sell to aquariums and pet shops. Then a friend asked her to help raise a sunken freighter, and she started working as a shipwreck salvager. In 1974, after earning her GED in Seattle, she traveled to the Dominican Republic to work a shipwreck and discovered amber mining, marveling that a golden tomb could fit in the palm of a hand. Seeing an amber-trapped creature was "almost like looking at a photograph taken 23 million years ago," she later wrote in a memoir. Hendrickson taught herself entomology, which helped her identify which amber specimens would most interest scientists. Returning to the Dominican Republic again and again, she searched miners' slag piles, bought the best pieces, and sold them to universities and museums, later expanding her search to Chiapas, Mexico. It's said that only six butterflies have ever been found in amber, and that Hendrickson found three of them. On the day she checked out the bluff near Faith she was walking with her golden retriever, Gypsy. Only twelve *T. rex* skeletons existed at the time and Pete Larson had never seen a *T. rex* in the ground, but he recognized the honeycomb texture of the materials as camellate, a distinctive structure found in the bones of theropods. Dinosaur bone is porous; if you touch it to your tongue, it sticks. This stuck.

46   "swallowed my kids": Larson and Donnan, *Rex Appeal*.

46   FBI raid: Descriptions derive from Peter Larson's memoir, *Rex Appeal*; the extensive news coverage of the seizure of Tyrannosaurus Sue; and my interviews with Vince Santucci, a U.S. National Park Service paleontologist and senior geologist who was once called the nation's only "pistol-packing paleontologist." Investigators were getting tougher because "they believed that an arrest of a collector with a high profile might deter others from stealing fossils from public lands," Steve Fiffer wrote in *Tyrannosaurus Sue*. At one point Santucci said, "In a way, the dealers *are* protecting the fossils, but they're destroying their research value by not letting scientists do it."

47   "divided on this issue": Fiffer, *Tyrannosaurus Sue*.

47   "unanimous in condemnation": Ibid.

47   "extremists": Ibid. Bob Bakker's groundbreaking book *The Dinosaur Heresies*, which portrayed some dinosaurs as agile and fleet, was a bestseller in the 1970s, ushering in a "renewed fascination in science—and a higher demand for fossils," as Larson and Donnan put it in *Rex Appeal*. Like his Yale teacher John Ostrom,

who, in the 1960s revolutionized dinosaur science by picking up the century-old argument of Thomas Henry Huxley that dinosaurs had more in common with birds than reptiles, Bakker believed it scientifically dangerous, and wrong, to exclude amateur hunters from paleontology.

47   "fell from his head": Larson and Donnan, *Rex Appeal*.

47   "'the *real* sciences'": Ibid.

48   Guilty: The Paleontological Resources Preservation Act passed in 2009, providing the first unified measures and criminal penalties against illicit fossil collecting; but when the Sue case came along there was no relevant statute on fossils. The Larson brothers and other BHI associates were charged with violating the Antiquities Act of 1906, which protects "any historic or prehistoric ruin or monument, or any object of antiquity, situated on lands owned or controlled by the Government of the United States." In *Cruisin' the Fossil Freeway*, Kirk Johnson wrote that "a swarm of lawyers started getting paid to think about the legalities of dinosaur ownership." In *Tyrannosaurus Sue*, Steve Fiffer noted that seven of the twelve jurors who convicted Peter Larson later held a news conference in which they said that if they "had it to do all over again, they would now acquit the defendants of everything." One juror publicly criticized federal prosecutors for "spending millions of dollars on a glorified trespassing case." In the media, the pervasive story was binary, Larson later wrote: "We were scientists; we were shifty fossil brokers with no scientific sensibilities. We were law-abiding, simple ranch boys making a living; we were sophisticated thieves and liars. We saved precious bits of history; we took advantage of history and people to make a buck." Everywhere he went, some well-meaning person tried to offer legal advice. With more than a little hyperbole, he noted that the case "eventually would be described in the same breath in our part of the country as Waco or Ruby Ridge." Johnson, the Smithsonian chief, once speculated that Larson legally collected enough bones at one site, on the South Dakota property of an old rancher and lifelong fossil collector named Ruth Mason, to make "fifty or so" *Edmontosaurus* skeletons alone. He wrote, "Regardless of how history treats him, Pete Larson will go down in the books as one of the most prolific dinosaur hunters of all time."

48   "unspeakably fresh": Rita Reif, "Declaration of Independence Found in a $4 Picture Frame," *New York Times*, April 3, 1991.

48   "The Declaration of Independence is just a piece of paper": James Barron, "Public Lives; He's Auctioned the 1776 Declaration, Twice," *New York Times*, July 4, 2000. Redden also said, "If it's the right property, the right circumstance, people are desperate to buy." See Lynn Douglass, "Legendary Sotheby's Auctioneer Talks Selling Duchess of Windsor's Jewels, Magna Carta," *Forbes*, July 3, 2012.

48   "world treasure": See "A Dinosaur in Manhattan? T. Rex Fossil Goes on Block at Sotheby's," *Associated Press*, September 28, 1997. For the sales catalog, the paleontologist Philip Currie, by this time curator of dinosaurs at the Royal Tyrrell Museum of Paleontology in Alberta, described Sue as the "standard against which other dinosaurs are measured" (Fiffer, *Tyrannosaurus Sue*).

48   McDonald's and Disney: The two corporations had recently launched a long-term marketing partnership—the restaurant chain was to sponsor the Dino-Land USA exhibit at Disney World's new Animal Kingdom park in Orlando. A Disney communications executive thought a cast of Sue would make a great

centerpiece for Year 2000 promotions, saying, "I had been struck by something Bill Clinton had said about the millennium: 'If you are interested in celebrating the future, try to honor the past'" (Fiffer, *Tyrannosaurus Sue*).

48   "'Da Bears'": Larry McShane, "Museum Pays $8.4 Million for T-Rex," *Washington Post*, October 5, 1997. See also J. Freedom du Lac, "The T. rex That Got Away: Smithsonian's Quest for Sue Ends with a Different Dinosaur," *Washington Post*, April 5, 2014.

48   "The day they sold Sue": Interviews with Kirk Johnson.

49   "I don't want to offend": Ben Marks, "Skeletons in Our Closets: Will the Private Market for Dinosaur Bones Destroy Us All?," *Collectors Weekly*, April 24, 2014.

49   "the smoking gun": Interviews with Frank Garcia.

49   "an ever larger": Letter from Clayton Ray to Frank Garcia, dated July 8, 1992.

49   "extremists": Ibid. For more information on the tension between paleontologists and commercial fossil hunters, see "Statement of Principle, Committee on the Guidelines for Paleontological Collecting," Board of Earth Sciences, National Research Council, 19 National Academy of Sciences, National Academy Press, 1987. See also "Management of Archeological and Paleontological Resources on Federal Lands: Hearing Before the Subcommittee on Public Lands, Reserved Water, and Resource Conservation of the Committee on Energy and Natural Resources, United States Senate, Ninety-ninth Congress, First Session, on how Effectively the Land Managing Agencies are Carrying Out Their Responsibilities to Manage, Protect, and Preserve Archeological and Paleontological Sites and Objects," October 14, 1985. The U.S. Government Printing Office published the report in 1986. All these materials are available online.

49   "Okay, if you not doing the swimming anymore": Interviews with Doris Prokopi.

50   "find anything": Interviews with Eric Prokopi.

50   "Once you're dealing": Interview with Richard Hulbert, FMNH vertebrate collections manager. He said museum staff remembered Prokopi and his hunting buddy as "the two Erics" because they shared a name. The two Erics fell out over fossils at one point and stopped hunting together. "These collectors, they'll be feuding and fighting and having vendettas against each other for perceived slights—'You raided my fossil site, blah, blah, blah'—and five years later they're buddies again," Hulbert told me. "It's like bad marriages or something."

50   Volunteering at the Florida Museum of Natural History: Eric examined tooth after tooth, labeling each describing each taxon on a paper collection label. "Description of *Paleocarcharadon orientalis* tooth," he subject-lined one museum memo, about a pygmy white shark. "The coarsely serrated teeth of *P. orientalis* are designed for efficient cutting of large prey animals…" "Description of *Paleocarcharadon orientalis* tooth…" Eric Prokopi memo to Dr. Robert Purdy, October 5, 1994. One dugong bone that he donated was still on display at the FMNH when I visited in the late summer of 2012.

50   "just basically rocks": Eric Prokopi said this in our first interview, in 2012, at Serenola, in Gainesville.

51   He stuck to the rivers: Florida has a long history of divers who search for relics. Scouring the waters of rivers like the Santa Fe, the Aucilla, and the Suwannee, they found thousands of items fashioned out of flint, coral, bone. Some of the finds went to museums, others to private collections, and occasionally the worlds

of the hobbyist and the archaeologist overlapped. In one joint project between scientists and sport divers, amateurs "learned the importance of quantitative data and careful record keeping for documenting their archaeological finds," one study found, in 1996. "This project has had a lasting effect on the good relationship between many of the State's professionals and amateurs with common archaeological interests." By early 2014 hunters were swept up in a Florida Fish & Wildlife sting called Operation Timucua and basically charged with collecting out of bounds. For information, see Ben Montgomery, "North Florida Arrowhead Sting: What's the Point?," *Tampa Bay Times*, January 2, 2014. Also Daniel Ruth, "Ridiculous 'raiders of the lost artifacts,'" *Tampa Bay Times*, January 10, 2014. Ruth noted that "in one case, a defendant allegedly sold a box of about 90 assorted artifacts to an undercover agent for a grand total of $100— not quite Pablo Escobar territory." Ruth argued that the state was never able to "produce evidence the defendants ever dug into a state archeological site to obtain any ill-gotten artifacts. Yet several defendants are looking at substantial prison time..." One suspect, William Barton, of Leon County, committed suicide. Ruth wrote, "Breaking a law is not necessarily the same thing as criminality. It's a concept embodied by Themis, the Greek goddess of holding the scales of justice—an image of compassion and fairness as old as antiquity."

51 "The process of building": *The Pony Express*, Florida Museum of Natural History newsletter, 1995. To bring fossil skeletons back to "life" is by any measure an act of patience and devotion. Anatomy tells the prepper how the creature looked in life; physics suggests a pose. If the hunter must know how to see, the prepper must learn to envision. I once met a woodcarver on the coast of North Carolina who kept in his yard a pile of white-pine logs. When I asked how he made his beautiful duck decoys he said, "Just cut off the part that ain't duck."

CHAPTER 6: TUCSON

52 The definitive history of the Tucson gem, mineral, and fossil show is *A Fifty-Year History of the Tucson Show*, by Bob Jones, a lifelong mineral collector who taught eighth-grade science in Scottsdale. He started collecting as a boy in Connecticut, after visiting the Yale Peabody Museum. His class field trip went to see dinosaurs but it was the minerals that knocked Jones out. One looked like a sword, another like a hedgehog. "I says, 'My God, look at these things!' After that, I went home and got my father's claw hammer and screwdriver—I was gonna set the world on fire, finding minerals," he told me. "It's the eye appeal and the thrill of the hunt that makes it such a popular hobby. And it costs you nothing." By the time I spoke to Jones, in November of 2015, he hadn't missed a Tucson show in over forty years. For more, see *A Fifty-Year History*. Details in the Bibliography.

53 "One does not call Escoffier a chowhound": Wayne King, "Polished or Not, Rocks Draw Fans," *New York Times*, February 17, 1984.

53 "Museum curators": Daniel E. Appleman, "Paul E. Desautels (1920–1991)," *Rocks & Minerals* (January/February 1992).

53 "curtained off portion": Jones, *A Fifty-Year History*.

54 "quiet closed-door," "American entrepreneurial spirit," "sleeping room," etc.: This comes from both Bob Jones's book and from my interview with Jones.

54 "It was pretty much agreed that any mafioso": Interview with Jones.
54 "good *rocks*": Ibid.
54 "lap-carried on the plane": Jones, *A Fifty-Year History.*
55 "Museum curators would spread the word": Interview with Jones.
55 "New York Stock Exchange": Sources include "A Wonder-Filled World of Minerals," by Dan Pavillard, *Tucson Daily Citizen*; February 7, 1970.
55 "Some of the things I saw": See Malcolm W. Browne, "Clash on Fossil Sales Shadows a Trade Fair," *New York Times*, February 15, 1994.
55 "This whole campaign": Ibid.
55 "sort of a self-policing thing": Interview with Jones.
56 "Okay, what do you got in the bathroom?": Ibid.
56 "These are the minerals that are quietly sold": Ibid.
56 Economic impact: For more, see "Characteristics and Economic Impact of the Tucson Gem, Mineral & Fossil Showcase Tracking Study," by FMR Associates Inc., Tucson; Steven Spooner, "Gem Show Travelers Bring Economic Boom to Tucson," *Daily Wildcat*, February 6, 2017; and "World's Biggest Gem Show—$120 Million Economic Impact," *BizDESIGN*, Winter 2015.
57 Big Trade events: Prokopi kept track of the shows on a paper calendar that spanned three years, marking little more in the squares than "Tampa Bay Show," "Gulfport Show," "Aurora Show," "Fossil Mania," "Paleofest," "Munich," "Tucson." Personal papers of Eric Prokopi.
57 "I just dove" and "Today, in spite of his young age": This comes from an article that either Doris or Eric Prokopi clipped and saved. The newspaper and date are unknown, but it's possibly the *Lakeland Ledger*, published between December 1996 and August 13, 1997.
57 "I've seen it all": Email from Eric Prokopi to Amanda Graham, September 5, 1999. Papers of Amanda Prokopi.
57 "Wow!": Ibid.
58 "If you want lunch," etc.: Email from Eric Prokopi to Amanda Graham, September 6, 1999. Papers of Amanda Prokopi.
58 "spike dogs": David Zucchino, "1800s-Era Sunken Logs Are Now Treasure; Here Are the Men Who Find Them," *Los Angeles Times*, July 13, 2014.
58 "I don't quite know" and deadheading passage: *How to Do Florida*, Episode 501, Deadhead Logging, produced by How to Do Florida Inc., Crawford Entertainment. Posted to YouTube on July 17, 2014.
60 "He drives a *truck*?": Interviews with Amanda Prokopi and Betty Graham.
60 "Your constant smile": This and other details come from *Never a Dull Moment*, a craft photo book compiled by Jill Hennessy Shea, a Prokopi family friend in Gainesville.
60 Go big or go home, "good Virginia silver," "treat your friends," etc.: Interviews with Amanda Prokopi and Betty Graham.
61 "Mom, he has Dad's work ethic!": Ibid.
62 "You've been hiding *that*?": Interviews with Amanda Prokopi.
62 "I miss you": Email from Eric Prokopi to Amanda Graham. Papers of Amanda Prokopi.
62 "Before I met you": Ibid.

CHAPTER 7: BIG GAME

63   *Pfff, this is easy*: Interviews with Amanda Prokopi.

63   dump a body: In *The Orchid Thief,* Susan Orlean wrote, "Florida was a different kind of wild than Western wild. The pioneers out west were crossing wide plains and mountain ranges that were too open and endless for one set of eyes to take in. Traveling west across those vacant and monumental spaces made human beings look lonely and puny, like doodles on a blank page. The pioneer-adventurers in south Florida were traveling inward, into a place as dark and dense as steel wool, a place that already held an overabundance of living things. The Florida pioneers had to confront what a dark, dense, overabundant place might have hidden in it. To explore such a place you had to vanish into it."

63   "Eric is perfectly proportioned," etc.: Interviews with Amanda Prokopi.

63   "Let's go Friday": Ibid.

63   "I don't love you because you are beautiful": Note from Eric Prokopi to Amanda Graham. Papers of Amanda Prokopi.

64   "did you want to get engaged": Interviews with Amanda and Eric Prokopi.

64   "commercial paleontologist": Prokopi-Graham wedding announcements. Papers of Amanda Prokopi.

64   "Now I'm sure everyone knows": Amanda Graham's written wedding toast. Papers of Amanda Prokopi.

65   "Especially in what I call the good old days": Interview with Andreas Kerner, September 25, 2014, Branchville, New Jersey.

65   Sheikh Saud bin Mohammed al-Thani: *ARTnews* named the sheikh, the cousin of the ruling emir of Qatar, the world's top art collector in 2011. As a member of Qatar's royal family, al-Thani was in charge of developing his country's museums, including the Natural History Museum. He "wasn't a big deal in art buying circles—he was massive," reported the BBC. One UK collectibles dealer, Paul Fraser, wrote, "He always got what he wanted." The sheikh's interest in collecting started in boyhood, with stamps, and eventually extended to "Western art, furniture, classic cars, bicycles, meteorites ('space sculpture' to collectors), three fossilized dinosaurs acquired in Wyoming, a complete edition of Audubon's 'Birds of America,' and the Graves watch, a handmade gem known among collectors as the holy grail of timepieces...," the *New York Times* reported. In 2005 the sheikh was placed on house arrest and stripped of his position on the national culture council after allegedly misusing public funds to buy art and collectibles, charges that eventually were dropped. He died unexpectedly at his home in London on November 9, 2014, cause of death unknown. See Paul Vitello, "Saud bin Mohammed al-Thani, Big-Spending Art Collector, Is Dead," *New York Times*, November 17, 2014; Sara Hamdan, "An Emirate Filling up with Artwork," *New York Times*, February 29, 2012; Will Gompertz, "Qatari Art Collector Sheikh Saud bin Mohammed Al-Thani Dies," BBC, November 11, 2014; and Paul Fraser, "The Vast and Spectacular Collections of Sheikh Saud al Thani of Qatar," paulfrasercollectibles .com, June 8, 2011. David Herskowitz told me al-Thani was a "very good client" and that he brought the sheikh "into the industry." He said, "When he heard about my auctions he contacted me and started to talk about things that were on the market. He goes, 'Oh, you can buy this stuff? Can you get me this? Can

you get me that?' So then when my auctions came he was on the phone with me, bidding. He would've bought at least seventy-five percent of the auction if I didn't tell him to stop bidding. That's the kind of guy I am, by the way. When he's bidding for an ammonite that I know is a three-hundred-dollar piece and he's already up to four thousand five hundred dollars, I'm gonna tell him, 'Okay, Your Highness, please don't go any further, I'll get you another one.'"

66 "Building natural history museums": Interviews with Kirk Johnson.

66 "*That's* art": Interviews with Amanda Prokopi.

66 "*National Geographic* for your house": Ibid.

66 "Eric lives in organized chaos": Ibid.

66 Bizarre Bazaar: This show is so popular that at Christmas people make a day of it, renting a limo and wearing matching sweatshirts ("Naughty or Nice?") and funny, homemade holiday hats. Shoppers are so obsessed, they wait in line for the doors to open as if for concert tickets or Black Friday at Best Buy. Amanda especially liked selling the hand-made jewelry she had taught herself how to make.

66 Guernsey's auction: "Dinosaurs & Other Prehistoric Creatures" catalog, June 24, 2004. The auction alarmed paleontologists, and it attracted buyers, from a six-year-old kid who successfully bid on the partial brow bone of *Triceratops*, to star medical examiner Michael Baden, who bought three *Edmontosaurus* bones, telling a reporter, "What's very interesting is how they are similar to our bones today. We probably have eighty percent to eighty-five percent of the same DNA today as these dinosaurs that are extinct." See John J. Goldman, *Los Angeles Times*, "An Auction of Prehistoric Proportions," June 25, 2004.

67 Sold diamonds in Korea: Interviews with David Herskowitz.

67 *Jurassic Park*: The years 1990 to 1993 were a sort of launching point for dinosaurs. First *Jurassic Park* premiered, following the book by the same name by Michael Crichton. Then Tyrannosaurus Sue sold for over $8 million, launching the first bone rush since the nineteenth-century feud between Marsh and Cope. Also, the Soviet Union was breaking up, which, as you'll read in later chapters, had an unexpected impact on paleontology.

68 How could a *fly* get into a *gemstone*?: That's what Herskowitz wondered, but while amber is often considered a gem, it's actually fossil tree sap. More specifically it's an example, like ivory, of what the International Gem Society calls organic amorphous materials. See gemsociety.org for more information.

68 The aunt: Interviews with David Herskowitz.

68 "People fight over stuff": Interviews with Herskowitz.

68 "I loaded up the truck and moved to Bever-ly": "Hills, that is; swimming pools, movie stars." See "The Ballad of Jed Clampett," the theme song for the long-running television sitcom *The Beverly Hillbillies*, performed by the bluegrass legends Lester Flatt and Earl Scruggs.

68 "I must have been a good bullshitter": Interviews with David Herskowitz. For more on the first history auction, see Brian Jerkins, "Natural history auction likely to be a bone-a-fide success," CNN.com; December 2, 1995.

69 "Henry was more of a free spirit": "Henry Galiano's Got Hundreds of Skeletons in His Closet," *People*, January 20, 1986.

69 "niche within the public's innate interest": maxillaandmandible.com.

69 "If you've given someone": *People*, "Henry Galiano's Got Hundreds of Skeletons…"

69 "Everybody lives in apartments": William R. Greer, "Beneath Columbus Avenue, Bones Become Art," *New York Times*, December 20, 1986.

69 "slaughterhouses": Ibid.

70 "buffalo heads": Ibid.

70 "There's no life down here": Ibid.

70 "paleontologists, entomologists": maxillaandmandible.com

70 GONE DIGGING: "Maxilla & Mandible Closes Its Doors for good," *West Side Rag*, August 30, 2011.

70 "wasn't just a store," etc.: Ibid.

70 Galiano: For Herskowitz, Galiano "cleared the items for auction": Jerkins, "Natural History Auction…" He once said, "I'm not concerned about the pricing of these things, simply because if you don't put a value on it, it's worth nothing."

71 Chinese dealers' advertisements: Eric Prokopi kept most if not all of the Tucson show guides, a review of which provided the detail for this passage.

71 "You talk to a paleontologist": Interviews with Kirk Johnson.

72 "Fellow Fossil Hunters": According to the letter from Friends of New Jersey State Museum, dated March 2, 2001, the trip cost $4,500 per person and included airfare, lodging, meals, museum admissions, and ground transportation. The group would travel from July 23 to August 3. The letter was signed by W. B. (William) Gallagher, New Jersey State Museum curator and Rider University adjunct professor. Papers of Eric Prokopi. The Liaoning region was hemorrhaging fossils at that point. This wasn't long after the smallest known theropod dinosaur, *Microraptor*, had been found in the area, and around the time that a team of Chinese and American Museum of Natural History scientists, led by Ji Qiang and Mark Norell, had announced the discovery, by farmers, of a 130-million-year-old fossil dinosaur that had been covered entirely in primitive feathers and downy fluff. Nothing like it had ever been seen, and Norell later said the small, fleet dromeosaur offered "the best evidence yet that animals developed feathers for warmth before they could fly." See Cliff Tarpy, "Keeping an ear cocked for voices in the dark," *National Geographic*, August 2005.

72 "ANOTHER AWESOME": Eric Prokopi eBay archives, which I pulled in 2012.

72 "We had a ball at Lowes": Annual holiday newsletter, papers of Amanda Prokopi.

73 "This is not a tire store": Interviews with Amanda Prokopi.

73 "dive into adulthood": Holiday newsletter.

73 Giant ground sloths: Prokopi sold one sloth to Mace Brown, an investor and financial planner from the Charleston area. It is a mark of the interconnectedness of the scientific and commercial fossil worlds that Brown is friends with Bolor Minjin's husband, Jonathan Geisler, a South Carolina native and a paleontologist who went on to teach on Long Island, New York. Brown started collecting natural history in middle school: rocks first, and later fossils. After amassing more than three thousand objects he decided he needed "an endgame," and created the Mace Brown Museum of Natural History at the College of Charleston. It's located on the second floor of the School of Sciences and Mathematics Building. At least one of the Prokopi's finds can be seen there.

73 "unusual" stuff: Various interviews, mostly with dealers.

73   Micanopy: The Timucua tribe inhabited the area before Hernando de Soto showed up in 1539. Michael J. Fox's 1991 movie *Doc Hollywood* was shot there.

76   See Barbara A. Beasley, "Appendix B: Paleontological Damage Assessment and Commercial Value Determination for Commercially Valuable Fossil Resources Seized in the Paleontological Violation Case No. 03-02-7453733," USDA Forest Service, November 2003.

77   "biggest amethyst druse": See Andreas Guhr's biography at http://www.earth dancer.co.uk/authors/andreas-guhr/.

77   "This extremely valuable log": redgallery.com

77   "led expeditions": Ibid.

77   "What a magical word!": Uwe George, "Das Grab der Drachen (The Grave of the Dragons)," *GEO*, July 1993.

77   "A new kind of dragon hunter," etc.: Ibid.

78   "QVC's Indiana Jones": Interview with Tom Lindgren.

78   "art of presentation" and "People come here every year": Ibid.

78   "their own private time": Ibid.

78   "You'd shake his hand": Ibid. Lindgren had moved to Tucson by the time I talked to him at the booth of his company, GeoDecor. "I don't think Eric meant to be bad. I don't think Eric meant to be the black side," he told me. "I think in his heart everything was good intentions. He wanted to have the right to live the American dream, which is: You work hard and you make a discovery, or whatever. But he also was naive enough to believe what he'd been told: It's legal and free to have this." I asked why anyone would risk it. "Because the potential payback was huge," Lindgren said. It has to be about more than money, I said. Lindgren said, "There's some prestige that comes along with selling a million-dollar fossil: 'Here's the guy that can sell the big items. This is the guy that when we have something major, let's go to him and get him to sell it for us.' It's the approval and admiration of your peers."

CHAPTER 8: MIDDLEMAN IN JAPAN

79   Dinosaur: The documented history of dinosaur discoveries began in the 1600s in the quarries of Stonesfield, in the county of Oxfordshire, England. The aptly named village produced extraordinary slate. Mining usually began around the holiday of Michaelmas, at the end of September, and continued until Christmas: quarriers hauled large blocks, or pendles, to the surface and covered them with dirt or water, to keep them moist. At the onset of a hard frost the mine operators sounded the church bells, summoning men to come spread out the rock, exposing it to the elements. The absorbed water froze and thawed, froze and thawed, a cycle that split the stone thinly into slabs that proved useful, and beautiful, as roof tiles. "One week of hard frost in January ensured well-split slates and plenty of work to keep the 'slatters'—the men who shaped and holed the thin layers to produce the batchelors, whippets, muffities, short, middle, and large cocks used for roofing tiles—in employment until the following Michaelmas," Nina Morgan, a geologist, wrote for the Geological Society of London. Stonesfield slate could be seen in roofscapes throughout the Cotswolds and, nearer by, at Oxford University. Stonesfield sat at the crest of an escarpment on the River Evenlode. The quarried slate was actually a middle Jurassic limestone from the

Taynton formation, which lay packed with a wealth of "marine exuviae" that included fossil shark teeth, nautili, scales, spines, and ammonites, but also trees and ferns and seeds and reeds and branches and leaves. In 1676, someone found a fossil fragment measuring 2 feet long and 15 inches in diameter, and weighing almost 20 pounds. Doubly bulbous at one end, the fragment would someday be temporarily (and quite erroneously, if colorfully) called "Scrotum humanum," but for now no one could even guess what the thing was, which was *Iguanodon*, one of the three first dinosaur discoveries, which eventually led the British zoologist Richard Owen to a pivotal conclusion. Comparing the bulbous wonder with two other specimens found in England, *Megalosaurus* and *Hylaeosaurus*, he decided they were all similar enough to each other—and different enough from everything else—to deserve their own grouping. Drawing on the Greek terms for "terrible" (*deinos*) and "lizard" (*sauros*), he created the umbrella term *Dinosauria*, thus giving humankind a new word, and a new world: dinosaur. The other key players here were the country doctor Gideon Mantell and his wife, Mary Ann, both of whom are credited with discovering *Iguanodon* and Robert Plot, a chemistry professor at Oxford and the first curator at the university's Ashmolean Museum of natural history. Plot had been working on a systematic study of local fossils, rocks, and minerals for his book, *Natural History of Oxfordshire*. (His sketches—of "Moon-stone," "Thunder-bolts," "Cockle-stones," and "Stones resembling parts of Men, or things of Art"—were some of the earliest and most beautiful examples of illustrated fossils.) After receiving the bulbous object he sketched it and included it in his book, in what may have been the first published illustration of a dinosaur fossil. I would be remiss if I didn't tell you at least a little about the influential William Buckland, who grew up fossil hunting on the Dorset coast, at the coastal bluffs of exposed Triassic and Jurassic limestone and shale at Lyme and Charmouth. The cliffs were his "geological school," Buckland wrote, saying, "They stared me in the face, they wooed me and caressed me, saying at every turn, pray, pray be a geologist." In 1801, he started at Oxford. Graduating in 1808, he became both an ordained minister and a geologist. Full-lipped and cleft-chinned, Buckland hunted fossils on a black mare, keeping his specimens and rock hammers in a large blue saddlebag. He became Oxford's first professor of geology. His lectures were popular with students and faculty for their weirdness and passion. "He paced like a Franciscan preacher up and down behind a long showcase...He had in his hand a huge hyaena's skull," one former student wrote. "He suddenly dashed down the steps—rushed skull in hand at the first undergraduate on the front bench and shouted 'What rules the world?' The youth, terrified, threw himself against the next back seat, and answered not a word. He rushed then on to me, pointing the hyaena full in my face—'What rules the world?' 'Haven't an idea,' I said. 'The stomach, sir,' he cried (again mounting the rostrum) 'rules the world. The great ones eat the less, the less the lesser still.'" Buckland became known for "zoophagy," a nineteenth century fad wherein people tried unusual foods. Buckland quite simply would eat anything, including, supposedly, a puppy. He dined on the flesh of a porpoise head (it tasted like "broiled lamp wick") and earwigs ("horribly bitter"), and was said to have licked strange drippings off a church floor and declared the "martyr's blood" nothing but bat urine. It stretches the imagination to think that the following really occurred, as recounted by the author Augustus Hare: "Talk of strange relics led to mention of the heart of a French King preserved at Nuneham in a silver casket. Dr. Buckland, whilst looking at it, exclaimed, 'I have eaten many strange things, but have never eaten the heart of a king before,' and before anyone could hinder him, he had gobbled it up, and the precious relic was lost for ever." Charles Darwin wasn't impressed. "Though very good-humoured and good-natured [he] seemed to me a vulgar and almost coarse man," Darwin wrote of Buckland. "He was incited more by a craving for notoriety, which sometimes made him act like a buffoon, than by a love of science." It was Buckland who identified the bulbous Stonesfield bone as *Megalosaurus*. At this point an international network of geologists and paleontologists was taking shape, with many of the scientists attached to museums and universities, their ideas circulating with increasing speed thanks to rapid improvements in the printing and distribution of newspapers and magazines. Georges Cuvier, celebrated and sought after as the father of paleontology, began to mentor a young English geologist named Charles Lyell. Noting that England was more "parson-ridden" than any of Europe's countries except Spain, Lyell declared his intention to "free the sciences from Moses." Building upon James Hutton's concept of deep time, Lyell described a planet that underwent slow changes over long periods time and published it in *The Principles of Geology*. The geology book became a bestseller. Public demand took Lyell from England to America, where thousands of people sought tickets to one Boston appearance alone.

79   "happy tourist": Hollis Butts's public Facebook page.

79   "bone wars": E. D. Cope—slender and handsome, with a wide English mustache—was a hotheaded Quaker and the brusque young son of a Philly shipping magnate. He had taught himself paleontology by studying with the esteemed University of Pennsylvania anatomist Joseph Leidy at Philadelphia's Academy of Natural Sciences, the country's first natural history organization. Marsh—beefy and balding, with a brushy beard—was nine years Cope's elder. Growing up poor on a farm in Lockport, New York, he had become fascinated by fossils and collected them. Marsh was thinking about becoming a carpenter or a teacher when his millionaire uncle, George Peabody, a London financier, plucked him off the farm and sponsored his education at Phillips Exeter Academy and then at Yale, where he was able to pursue his true interests and kept his "large mineral and fossil collection under lock and key." Marsh and Cope became friends after meeting in Berlin. Marsh was the shy and methodical one, Cope the brash, fast one. Marsh noticed geological formations; Cope noticed beauty. Marsh's men carried guns; Cope refused offers of firearms despite working in remote, dangerous areas. Marsh was funded; Cope came from a patrician family but had nothing. Both men paid skilled fossil hunters like Charles Sternberg to prospect and excavate on their behalf, but both also went into the field. Marsh took an unpaid faculty position at Yale as the nation's first chair of paleontology, securing his future by persuading his generous uncle to make an endowment founding what is now the Yale Peabody Museum of Natural History. Cope, devoted to the Academy of Natural Sciences, taught at Princeton. They died within two years of each other in the late 1890s; by that time their rivalry very nearly embodied Tennyson's poetic notion of "Nature, red in tooth

and claw." But the competitiveness produced amazing new specimens, including Marsh's important discovery of a primitive bird with teeth, a find that supported the ideas of Charles Darwin and "Darwin's bulldog," the English biologist Thomas Henry Huxley, who suspected birds evolved from creatures not unlike dinosaurs. Cope ultimately discovered fifty-six species of dinosaur, and Marsh discovered eighty-six, including *Apatosaurus, Stegosaurus, Triceratops,* and *Camarasaurus.* "Americans have always felt inferior to Europeans—culturally, politically, economically," Steven Conn, a historian, once said. Paleontology finally seemed to be the place where Americans could "be better than the Europeans." During the bone wars there was an obvious commercial market for fossil vertebrates, with "Professor Marsh himself as the most active figure in that market," Charles Schuchert and Clara Mae LeVene wrote in *O.C. Marsh, Pioneer in Paleontology* (New Haven, CT: Yale University Press, 1940). The best source on the Marsh-Cope bone wars is *The Gilded Dinosaur,* by Mark Jaffe. Disclosure: Jaffe and I are friends; but no matter what, I would feel the same about his deeply reported, beautifully written book. This chapter owes a debt to the detailed research Jaffe produced in telling his important story about two of the most pioneering, influential (and altogether colorful) figures of the nineteenth century. You can see him talk about Marsh and Cope in *Dinosaur Wars,* an American Experience documentary by PBS (WGBH; January 17, 2011.)

79 First major dinosaur site: Two Union Pacific Railroad workers tipped Marsh off about fossils that "extend for seven miles & are by the ton at Como Bluff." In a letter one told him, "We are working men and not able to present them as a gift..." See John Ostrom, *Marsh's Dinosaurs: The Collections from Como Bluff.*

79 "It was *Jurassic Park* finally": PBS, *Dinosaur Wars.*

80 Bone Cabin Quarry: Up the road is Sinclair—the town and the oil company. The company still uses the green *Brontosaurus*-looking creature as a mascot. The connection goes way back. Sinclair had papier mâché dinosaurs at the 1933–34 Chicago World's Fair. Spinoff projects included dinosaur stamp books, "Brontosaurus" soap, and partnerships wherein the company donated geological materials to libraries and schools. Another, larger, exhibit followed, at the 1964–65 New York World's Fair; the dinosaurs were based on the research of AMNH's Barnum Brown and Yale's John Ostrom. One autumn day in 1963, crowds of New Yorkers gathered at the Battery to watch nine fiberglass dinosaurs pass the Statue of Liberty on a barge marked "Sinclair Dinosaurs on Way to N.Y. World's Fair." After that, the models went on tour, then to permanent homes at national monuments, state parks, and other sites in six states, although the *Ornitholestes* was stolen and never recovered.

80 "erecting such a skeleton": See Phil Roberts, "The Builder of the 'World's Oldest Cabin,'" University of Wyoming Department of History, uwyo.edu.

80 "the world's oldest building," etc.: Ibid.

80 Bone Cabin closed: At one point, Eric Prokopi went to see Bone Cabin because he heard it was for sale. The owner wouldn't sell to a commercial hunter, but Prokopi enjoyed seeing the attraction, Como Bluff, and Medicine Bow. The town is known as the setting for Owen Wister's 1902 novel *The Virginian,* considered the first significant Western of the genre. Wister dedicated the book to Theodore Roosevelt and began with a note to the reader, saying that "Wyoming

between 1874 and 1890 was a colony as wild as Virginia had been a hundred years earlier, with the same primitive joys and dangers. There were, to be sure, not so many Chippendale settees." On the day I stopped by Bone Cabin, in 2016, everything was shut down, including the house where Boylan once lived. The stables were empty, the weeds high. Inside Bone Cabin I could see trash strewn about; the glass display case was broken. Security cameras, if they worked, watched whomever rolled up to have a look.

80 "With nothing much to do": Hollis Butts's public Facebook page.

80 "true America": Ibid.

81 "Breithaupt is active": Ibid.

81 "Well, I was the scumbag": Ibid.

81 "The seas came in": Willow Belden, "A Conversation with BLM Paleontologist Brent Breithaupt," Wyoming Public Media, March 29, 2013. This seven-minute radio interview is worth your time. Hear Breithaupt, a Bureau of Land Management paleontologist based in Cheyenne, talk about his mandate to protect fossils found on federal lands, and about Wyoming's wealth of prehistory, including trackways laid down in the middle Jurassic, 165 million years ago, when, as he puts it, hundreds of small to medium-size meat-eating dinosaurs walked across an ancient tidal flat, "leaving their footprints."

82 "I had a map": Hollis Butts's public Facebook page. Other details about Butts come from the Pacifica High School yearbook and U.S. military and census records.

82 "really old" Japanese furniture: Hollis Butts told me this in an email by cutting and pasting one of his Facebook posts, which noted that he also sold "minerals (to a lesser extent) and a lot of 'nature goods' (ostrich eggs, for example)." He told me he moved to Japan after being "kicked out of Iran (where I had been working) after the revolution." He traveled Asia until he had "almost no money left and then decided that Japan might be a good place to work." He hunted fossils as a child when his father was stationed at Fort Hood, Texas (nearby there are "outcroppings of chalky soft Cretaceous rock full of shells"), and later hunted in Germany ("black Jurassic shale"). "Years later, while living in Japan, I was on a vacation in the US with my family, and at Ouray, Colorado, I chanced into the Columbine Rock Shop, a place full of interesting fossils. I thought I should be selling fossils in Japan."

82 "Is life now so regulated": Hollis Butts's public Facebook page.

83 "for being able to solve": John Colapinto, "Brain Games: The Marco Polo of Neuroscience," *The New Yorker*, May 11, 2009.

83 "You buy it": Rex Dalton, "Paper Sparks Fossil Fury: Paleontologists Criticize Publication of Specimen with Questionable Origin," *Nature* news, February 2, 2009.

83 "bull-like appearance": See Clifford A. Miles and Clark J. Miles, "Skull of *Minotaurasaurus ramachandrani*, a new Cretaceous ankylosaur from the Gobi Desert," *Current Science* 96, 1 (January 10, 2009).

83 "laws were indeed broken": Dalton, "Paper Sparks Fossil Fury." *Nature* noted that there was "no clear paper trail that guarantees the fossil was acquired through legal channels" and quoted AMNH's Mark Norell, who complained that it was "totally inappropriate to publish on this specimen; it is stolen patrimony."

Another story about Hollis Butts involved a tyrannosaurid skeleton sold to an eye surgeon and paleontology enthusiast in Hingham, Massachusetts, Dr. Henry Kriegstein. Kriegstein envisioned the skeleton in the living room of his house on Martha's Vineyard, alongside a 7-foot-long *Triceratops* skull he and his daughter had found in Montana. When Kriegstein learned that the dinosaur might have come from the important feathered-dinosaur beds of Liaoning, China, which were being poached to death, he sent photos to the paleontologist Paul Sereno at the University of Chicago. Despite the fossil's commercial origins Sereno agreed to study it on the condition that Kriegstein surrender the skeleton to science. He decided the skeleton represented a "punk size" progenitor of *T. rex*—suggesting that tyrannosaurs developed as far back as 125 million years ago, much earlier than the fossil record indicated. A team of researchers, including the Black Hills Institute's Pete Larson, whom some paleontologists consider a tyrannosaur expert despite his lack of a doctoral degree, reanalyzed the fossil and disputed Sereno's conclusions, arguing that the dinosaur was in fact a juvenile or subadult *Tarbosaurus*. Hollis Butts could've told them that: it was just how he had presented the fossil at Tucson. Paleontologists were furious that the specimen was taken seriously at all, considering its commercial history. Despite the debate on whether the skeleton represented a new species, Sereno named the specimen *Raptorex kriegsteini,* after Kriegstein's parents, Roman and Cecilia Kriegstein, Holocaust survivors. "In the normal course of things this fossil could have ended up on someone's mantelpiece or been forgotten in an attic somewhere, and lost to science," Sereno said. Now "Dr. Kriegstein has found immortality for his family."

CHAPTER 9: HOLLYWOOD HEADHUNTERS

85   Paynes Prairie is a glorious place. In 1774, the naturalist William Bartram described the Alachua savannah as a "level green plain...fifty miles in circumference, and scarcely a tree or bush of any kind to be seen on it," but today it is lush and full of life. The Prokopis enjoyed going there as a family. I went to Paynes Prairie with them in the summer of 2012. We walked the long boardwalk, out into the marsh. I hoped to see an alligator. Amanda said that if we got chased we should run "in a zigzag pattern" because "gators can't cut." The Prokopis pointed out a semidistant lump in the unmoving water, just beyond a stand of reeds. We waited for the creature to show itself, and when it didn't, Eric walked off the marked path and down to the water's edge. He began pulling up reeds and throwing them at the lump like spears. This made me various kinds of uneasy. First, I did not want him to be chased and eaten in front of his family; also, it seemed unwise to be uprooting the reeds in hopes of uprooting the wildlife. Later, I looked up the park rules, which read, "All plants, animals, and park property are protected. The collection, destruction, or disturbance of plants, animals, or park property is prohibited." The lump in the water submerged with barely a ripple, then resurfaced, just as quietly, in basically the same spot. We moved on. For more information on Paynes Prairie see Elizabeth A. Bohls and Ian Duncan, *Travel Writing 1700–1830: An Anthology*, (Oxford, UK: Oxford University Press, 2006); and floridastateparks.org.

86   "Yeah, we have to live here": Interviews with Amanda and Eric Prokopi.

86 Serenola renovation: You can see a slideshow tour of the property on YouTube, via Coldwell Banker M.M. Parrish Realtors, posted on December 12, 2012. https://www.youtube.com/watch?v=DCOzwOq5ij4.

87 David Herskowitz wasn't sure what to think of Eric Prokopi, and vice versa. "I don't like people that I don't know, and it was hard to know Eric because he was so quiet and to himself," Herskowitz told me. He said he decided to work with Prokopi only after a trusted dealer described him as "very moral, honest." Herskowitz didn't like it that Prokopi once raised his fee at a point of delivery but otherwise said, "He was never shady in the fossil business." This comes from interviews with Herskowitz in January 2016, in Tucson.

87 I.M. Chait: At his gallery, Izzy Chait kept a piano. In the 1990s, after having set aside jazz and blues for twenty-five years, he decided to pursue a "musical legacy." He formed an Izzy Chait Quartet and started singing again in public, covering George Gershwin medleys and legends like Sinatra. He recorded eight albums. His wife, Mary Ann, who had been his high school sweetheart, played bass.

87 "So I did gun shows, swap meets, the Rose Bowl": David Rosenfeld, "Izzy Chait: A Life of Extraordinary Things," *Westside People*, May 15, 2014.

87 "As a committed supporter of Asian art": See "A Personal Message from I.M. Chait," chait.com.

87 "Americans went crazy": Rosenfeld, "Izzy Chait."

87 "Egyptian mummy's hand; lion, hyena": See Roja Heydarpour, "And to the Winners Go the Dinosaur Skull and the Mummified Hand," *New York Times*, March 26, 2007.

87 "perfect for a New York City apartment": Ibid.

88 Skull buyer: For coverage of the Cage-DiCaprio bidding war see Julie Miller, "Nicolas Cage Outbid Leonardo DiCaprio for a Dinosaur Skull That May Have Been Stolen," vanityfair.com, October 29, 2013; Ben Mirin, "Nicolas Cage's Dinosaur Skull May Be Hot Property but He Shouldn't Have Bought It in the First Place," Slate, October 29, 2013; and David A. Keeps, "A Cooling-Off," *Los Angeles Times*, October 11, 2008. Cage's collecting habits have been widely covered. A favorite category: comic books. See Andy Lewis, "Nicolas Cage's Superman Comic Nets Record $2.1 Million at Auction," *Hollywood Reporter*, November 30, 2011.

88 "stunning views and extreme privacy": This according to Los Angeles real estate agent Wilson Chueire, chueiregroup.com. Property records and media reports showed that DiCaprio bought the home in 1999.

88 DiCaprio environmental causes: In 1998, DiCaprio set up the Leonardo DiCaprio Foundation to protect endangered wildlife from extinction. In 2017, the foundation announced plans to give $20 million in grants to some hundred environmental causes. See Rebecca Rubin, "Leonardo DiCaprio Foundation Awards $20 Million in Environmental Grants," *Variety*, September 19, 2017. The *Los Angeles Times* has published numerous articles related to DiCaprio and his environmental causes, as has the *New York Times*. Also see Suzanne Goldenberg, "How Leonardo DiCaprio Became One of the World's Top Climate Change Champions," *The Guardian,* February 29, 2016; and Stephen Rodrick, "Inside Leonardo DiCaprio's Crusade to Save the World," *Rolling Stone*, February 28, 2016. DiCaprio once told *Wired*, "I've been interested in science and

biodiversity ever since I was very young, probably from watching films about the rain forest at the Natural History Museum." See Robert Capps, "The Nine Lives of Leo DiCaprio," *Wired*, January 2016. In 2015, artnet.com wrote that DiCaprio kept an "important collection of fossils in his home, consisting mainly of predatory dinosaurs." See Daria Daniel, "Take a Look Inside Leonardo DiCaprio's Growing Art Collection," artnet.com, March 11, 2015.

88   "up from none five years ago": See Kelly Crow, "The Oldest Crop," *Wall Street Journal*, June 8, 2007.

89   He got word from Butts: In an October 2012 email, just after Eric Prokopi's arrest, Butts told me, "In fairness, I should point out that we do not like each other but we do business and are civil. And he does have a wife and kid to support so I wish the government would instead go and arrest some bankers, universally detested at the moment."

CHAPTER 10: THE WARRIOR AND THE EXPLORER

93   "There was no reason": Among the excellent scholarly resources on Mongolian history and culture see David Morgan, *The Mongols*; and Jack Weatherford, *Genghis Khan and the Making of the Modern World*. The Columbia University scholar Morris Rossabi has published numerous invaluable books on Mongolia, including *The Mongols and Global History* and *The Mongols: A Very Short Introduction*, altogether the product of over forty years of scholarship. See a short video of Rossabi talking about Mongolia's global, historical significance here: https://vimeo.com/65237984. Jonathan Addleton's *Mongolia and the United States: A Diplomatic History* provides crucial historical context.

93   Mongol Empire ruled: The Mongol Empire invaded northern China and, in the 1260s, Genghis's grandson, Kublai Khan, became emperor. He focused on agriculture, new technology, and aid to orphans, widows, and the elderly, Rossabi wrote in *The Mongols*. He protected artisans, built a shrine to Confucius, and elevated the status of merchants and science (especially prizing astronomy), and encouraged "unprecedented contact" between East and West. He established good relations with Korea—which continue today, even with North Korea—by sending one of his daughters to marry the king. Kublai built relationships with Muslims because he needed their "invaluable skills as tax collectors and financial administrators." In China, the Ming dynasty replaced Mongolian rule in 1368, remaining in power until the early 1640s.

93   "empire was so huge": Morgan, *The Mongols*. In his book *The Mongols: A Very Short Introduction*, Morris Rossabi wrote, "What actually provoked the Mongols to initiate what turned out to be the greatest conquests in world history? One explanation is the precariousness of their economy. Droughts, cold winters, or diseases among their animals threatened their survival. Under these circumstances, they either had to trade or raid for essential goods."

93   The empire eventually fractured: The Mongol empire became too vast—and the communications and transportation networks were too rudimentary—to govern from a central administration. The khanates struggled for power, some arguing that Mongolia should remain pastoral and others arguing for sedentary settlements that allowed for commerce and craft. See Rossabi, *The Mongols*.

93 Waited for its chance: This is necessary reductionism. The Mongolia-Russia-China story is a fascinatingly long and complex one. See the Bibliography for books on the subject.

94 "adventurers, missionaries, or merchants": Jonathan Addleton, *Mongolia and the United States.*

94 "the rolling prairies of Kansas and Nebraska": Addleton, *Mongolia and the United States*. In his book *The Mongols: A Very Short Introduction*, Morris Rossabi wrote that Mongolians' devotion to their horses and their aversion to washing—water being scarce—"were perceived as evidence of their barbarism."

94 A crucial step toward survival: Christopher Kaplonski, *Truth, History and Politics in Mongolia: Memory of Heroes.*

94 "the world's most perilous": Rossabi, *The Mongols: A Very Short Introduction.*

94 "friendly cooperation": Addleton, *Mongolia and the United States.*

94 "I don't think this country can be compared": See Fred C. Shapiro, "Starting from Scratch," *The New Yorker*, January 20, 1992.

94 *The Secret History of the Mongols:* The text, written in the original Mongolian, is widely considered the most important book ever published in Mongolia and the country's oldest surviving literary work. In one English translation, Igor De Rachewiltz noted that "no other nomadic or semi-nomadic people has ever created a literary masterpiece like it, in which epic poetry and narrative are so skillfully and indeed artistically blended with fictional and historical accounts." See Igor de Rachewiltz, "The Secret History of the Mongols: A Mongolian Epic Chronicle of the Thirteenth Century," December 2015; shorter version edited by John C. Street, University of Wisconsin–Madison. Books and Monographs. Book 4. h9p://cedar.wwu.edu/cedarbooks/4. Gereltuv Dashdoorov, a translator of Roy Chapman Andrews books and an executive and guide at Mongolia Quest, a tourism company in Ulaanbaatar, once told me, "It's our *bible*."

94 "semi-mythical": Rossabi, *The Mongols: A Very Short Introduction.*

94 "one of the great literary monuments": *The Secret History of the Mongols, for the First Time Done into English out of the Original Tongue and Provided with an Exegetical Commentary*, by Francis Woodman Cleaves, Harvard University Press, 1982. Cleaves was the first to fully translate *The Secret History* into English, and did so in collaboration with Fr. Antoine Mostaert. The text was published in 1982 by Harvard University Press, decades after the translation work occurred.

94 democratic hero or genocidal terror: In *The Mongols*, Rossabi writes, "His attacks resulted in the indiscriminate and brutal killing of at least tens of thousands of people and in the maiming of hundreds of thousands, and the recent depiction of him as a great heroic figure and as a believer in democracy and in international law clashes with historical reality." He added that the Golden Horde, as the Mongol army was known, "introduced a level of violence that had scarcely been seen." During his political career, President Tsakhia Elbegdorj would continually cite Genghis's reputation as a "rule of law" leader even though that legacy is somewhat unclear.

95 poisoned by Tatars: this according to a number of historical accounts, including Rossabi's *The Mongols.*

95 "fantastic monsters": Kaplonski, *Truth, History and Politics in Mongolia.*

95 death of Genghis: Rossabi wrote that the leader's death and burial, in August 1227, became both mystery and legend. Maybe he died of an arrow wound,

maybe of injuries suffered during a hunt. Rossabi wrote, "A more bizarre account states that the captured wife of an enemy leader hid a knife in her vagina and stabbed Chinggis to death during sexual intercourse." See Rossabi's *The Mongols: A Very Short Introduction.*

95   Russian Revolution: At least one Westerner saw an opening in Mongolia around this time. An American attaché to the State Department urged Washington to install a U.S. consulate there, saying Mongolia held valuable commercial opportunities. This was "the psychological moment for the inauguration of American activity in Mongolia," the attaché urged. Washington again declined. See Addleton, *Mongolia and the United States.*

95   Sükhbaatar: Mongolian names often have varying spellings. The spellings in this book are consistent with those found in the *Encyclopedia Britannica.*

95   "flamboyantly crazy": See the foreword that AMNH paleontologist Michael Novacek wrote for Charles Gallenkamp's *Dragon Hunter: Roy Chapman Andrews and the Central Asiatic Expeditions.*

96   "all the adjuncts that contribute to happiness": This quote appears, unattributed, in *The WPA Guide to Wisconsin*, originally published as part of the Works Progress Administration Federal Writers' Project series, which began in 1937. Trinity University Press reprint, October 2013.

96   "He told me it was men of desperate fortunes": Daniel Defoe, *Robinson Crusoe* (London: William Taylor, 1719).

96   "bible": Roy Chapman Andrews, *Under a Lucky Star: A Lifetime of Adventure.*

96   "minerals, fossils, stuffed animals": Gallenkamp, *Dragon Hunter.*

96   "I was born to be an explorer": Roy Chapman Andrews, *This Business of Exploring.*

96   Beloit College: In *Under a Lucky Star*, Andrews wrote that his alma mater was considered "the Yale of the West."

97   "repurposed pipes and plumbing fixtures": "A Dinosaur by Any Other Name," American Museum of Natural History website, August 13, 2012, amnh.org.

97   "It didn't feed on flesh": "Big Thunder Saurian Viewed and Approved," *New York Times*, February 17, 1905.

97   AMNH visitors: Some five million people visit the American Museum of Natural History annually. See the AMNH Annual Report, 2016, available at AMNH.org.

97   "The magic city": Andrews, *Under a Lucky Star.*

97   "baronial splendor": Robert A. M. Stern, Gregory Gilmartin, and John Massengale, *New York 1900: Metropolitan Architecture and Urbanism 1890–1915* (New York: Rizzoli International Publications Inc., 1983).

97   "birds, mammals, reptiles": Gallenkamp, *Dragon Hunter.*

98   "penetrating some of the earth's remotest areas": Ibid.

98   "in which men worked who to me were as gods": "Unusual Men in the Public Eye," *Popular Science Monthly*, July 1929. At first, the AMNH was housed in the Arsenal Building on Central Park South. The building at Seventy-Seventh Street and Central Park West opened on December 22, 1877. One can imagine Andrews roaming the museum, marveling at the creatures that gazed back at him across time. The romance of discovery held such a grip on the public imagination that the Explorers Club had recently been founded, in Manhattan, to "preserve the instinct to explore."

98   Charles H. Sternberg: In the 1800s, Sternberg dropped out of college in order to hunt full time for Philadelphia's Edward Drinker Cope, the self-taught hunter

and prolific, scholarly author whose mad rivalry with the Yale paleontologist Othniel C. Marsh led to the epic "bone wars" of the late 1800s. Sternberg and his family collected some of the finest skeletons ever unearthed in the U.S. and Alberta, Canada, many of which can be seen in the world's top museums. "The object of my life has been to advance human knowledge, and that could not be accomplished if I kept my best specimens to gratify myself," Sternberg wrote in his 1909 memoir, *The Life of a Fossil Hunter*. "They had to go, and they went, often for less than they cost me in labor and expense, into the hands of those who could give authoritative knowledge of them to the world and preserve them in great museums for the benefit of all." Sternberg also wrote, "I demand that my name appear as collector on all the material which I have gathered from the rocks of the earth," a sentiment some modern hunters knew well.

98   "But one had to take him in context": Edwin H. Colbert, *A Fossil-Hunter's Notebook: My Life with Dinosaurs and Other Friends* (New York: Dutton, 1980).

98   "pseudo-scientific work of white supremacism": See Jedediah Purdy, "Environmentalism's Racist History," newyorker.com, August 13, 2015.

99   "the evolutionary 'staging ground'": Douglas Preston, *Dinosaurs in the Attic*.

99   "solve the geographical": See "Young Americans Seek to Erase 'Black Spots' off Map," *New York Times*, June 1, 1912.

99   "the unknown section of North Korea": Ibid.

99   "the most vivid example": Ibid.

99   "Andrews's career was a straight line from Beloit": Barry Gewen, "Protoceratops Lays an Egg," *New York Times*, June 3, 2001.

100   Yvette Borup Andrews: The American Museum of Natural History had recently hired her brother, George, as a curator of geology. But while George had survived the "severe test" of polar exploration, he soon drowned, with a friend, in a Connecticut canoeing accident. See "Death of George Borup, Revised Plans of the Crocker Land Expedition," *Bulletin of the American Geographical Society* 44, 6 (1912). "It was his ambition to devote his life to science and scientific exploration," the bulletin reported.

100   "Darkest China": See Martin Thomas and Amanda Harris, eds., *Expeditionary Anthropology: Teamwork, Travel, and the "Science of Man"* (New York: Berghahn Books, 2018).

100   "supreme trophy": Roy Chapman Andrews, "Hunting the Great Ram of Mongolia," *Harper's*, February 1921.

100   "almost as smooth as a tennis court": Roy Chapman Andrews, *The New Conquest of Central Asia*.

100   "The world has other sacred cities": Andrews, *The New Conquest of Central Asia*. The women's headdresses were so colorful and elaborate they reminded him of a flock of butterflies. He wrote, "The wife of one of the great khans in particular was the most magnificently adorned creature I have ever seen. According to the custom of the northern Mongol women, she had her hair plaited over a frame into two enormous flat braids, curved like the horns of a mountain sheep and reinforced with bars of gold. Each horn ended in a gold plaque, studded with precious stones, and supporting a pendant braid like a riding-quirt. This was enclosed in a long cylinder of gold, heavily jeweled. On her head, between the 'horns,' the lady wore a gold filigree cap studded with rubies, emeralds,

and turquoises, and surmounting this, a 'saucer' hat of black and yellow, richly trimmed with sable. Just above her ears great ropes of pearls hung from her gold cap halfway to her waist..."

100 "Every house and shop": Ibid.

101 "indescribable mixture of Russia, Mongolia, and China": Andrews, *Across Mongolian Plains: A Naturalist's Account of China's "Great Northwest."*

101 "There is no similar area": Roy Chapman Andrews and Yvette Borup Andrews, *Camps and Trails in China: A Narrative of Exploration, Adventure, and Sport in Little-Known China* (New York: D. Appleton, 1918).

101 Gobi: There are two ways to look at the Gobi's ranking as the third-largest desert. At 500,000 square miles, the Gobi is actually the fifth largest if you count the polar deserts Antarctic (5.5 million square miles) and Arctic (5.4 million square miles); it's third behind the non-polar deserts Sahara in northern Africa (3.5 million square miles) and the Arabian, on the Arabian Peninsula (1 million square miles). Marco Polo called the Gobi "the abode of many evil spirits which amuse travellers to their destruction with most extraordinary illusions." See Troy Sternberg, "Desert Boundaries: The Once and Future Gobi," *The Geographical Journal* (March 2015). See also Edward Wong, "How China's Politics of Control Shape the Debate on Deserts," *New York Times*, October 27, 2016.

101 Knowledge of Eastern Asia's fossils: Roy Chapman Andrews, "New Expedition to Central Asia," *Natural History* 20 (September–October 1920).

101 "crossed and recrossed" and "studied by the exact methods": Andrews, *The New Conquest of Central Asia.*

101 "unusual obstacles": Andrews and Andrews, *Camps and Trails.*

101 "roaring train": Andrews, *Across Mongolian Plains.*

102 "relic of the Pleistocene": Andrews, *The New Conquest.*

102 "great flat feet" and "natural road-makers": Andrews, *On the Trail of Ancient Man.*

102 Central Asiatic Expeditions: The American Museum of Natural History published a nine-volume work titled *Central Asiatic Expeditions of the American Museum of Natural History, Under the Leadership of Roy Chapman Andrews.* Volume 1, published in 1918, was called *Preliminary Contributions in Geology, Palaeontology and Zoology, 1918–1925.* On Amazon, I recently saw a hardcover set titled *Collected Works of the Central Asiatic Expeditions to Mongolia and China*; bound in bright yellow, the collection was on offer from a rare-books dealer for $6,033.

102 "We should try to reconstruct": Andrews, *Under a Lucky Star.*

102 "As we sat in the mess tent at night": Andrews, *The New Conquest.*

102 photos and film: Some footage was lost, but a remarkable amount of archival materials remain in the American Museum of Natural History's excellent library. To see some of J. B. Shackelford's stunning expedition photos, go to http://lbry-web-007 .amnh.org/digital/index.php/collections/show/10.

103 "dry achievements of science": Helena Huntington Smith, "Hunter of the Snark," *The New Yorker*, June 29, 1929. Smith called Andrews an "excellent amateur," saying his chief contribution to the Central Asiatic Expeditions was as organizer and public enthusiast.

103 "the covered wagon of the Gobi Desert": Dodge Brothers Inc. brochure, 1925. The Mongolian terrain was exceedingly difficult but Andrews liked Dodge cars for the flexible chassis and the engine's more than 28 horsepower. "A Dodge had

climbed the Twin Peaks of San Francisco higher than any other car, and was the first automobile to reach the floor of the Grand Canyon and climb back out under its own power," wrote Clive Coy, the University of Alberta's chief vertebrate paleontology technician, and a brilliant, thorough collector of Roy Chapman Andrews memorabilia. On his archived blog, *Whales, Camps & Trails,* you can see historic photos and much more; see http://whalescampsandtrails.blogspot.com.

103   "Most people derive a thrill": Smith, "Hunter of the Snark."
103   "contribution of large value": Roy Chapman Andrews, "A New Search for the Oldest Man," *Asia* ("The American Magazine on the Orient"), November 1920.
103   "friendly relations" and "destined to increase the prestige": Ibid.
104   "Soon it became a small city": Andrews, *The New Conquest of Central Asia.*
104   "There is almost unending bargaining": Roy Chapman Andrews, "Scientific Work in Unsettled China," *Natural History* 22 (May–June 1922).
104   "It is a delightful Aladdin's Lamp": As quoted in Preston, *Dinosaurs in the Attic.*
104   "the roof of the world": Andrews, *The New Conquest.* Kalgan, by the way, is today called Zhangjiakou. The city of over four million is a scheduled skiing, snowboarding, and biathlon venue for the 2022 Winter Olympics, to be headquartered in Beijing.
104   "Red hills and buttes": Ibid.
104   "I knew that something unusual had happened": Ibid.

CHAPTER 11: THE FLAMING CLIFFS

105   "one of the most picturesque places": Andrews, *The New Conquest of Central Asia.*
105   "vast pink basin," "for when seen," and "there appeared": Ibid.
105   "almost as though led by an invisible hand": Ibid.
105   "obviously reptilian": Ibid.
106   "Water that was up to our ankles": See Douglas J. Preston, "A Daring Gamble in the Gobi Desert Took the Jackpot," *Smithsonian,* December 1987.
106   "the clear impression": Preston, "A Daring Gamble."
106   "In his heart": Ibid.
106   "Mr. Andrews is at bottom that ancient type": Smith, "Hunter of the Snark."
106   "I hope you know": See Vincent L. Morgan and Spencer G. Lucas, *Walter Granger, 1872–1941, Paleontologist,* bulletin 19 (Albuquerque: New Mexico Museum of Natural History and Science, 2002). This is a fascinating account of Granger's importance to paleontology and his relationship to Roy Chapman Andrews. Andrews had a serious scholarly mission, as Mike Novacek once put it, but by all accounts Granger was the real (and relatively unheralded) scientist. In 1941, the American Museum sold the *T. rex* holotype found by Barnum Brown to the Carnegie Museum of Natural Museum in Pittsburgh. See Christopher Joyce, "Bone to Pick: First T. Rex Skeleton, Complete at Last," *Morning Edition,* NPR, September 14, 2011. See also the Carnegie Museum's website, https://carnegiemnh.org/tyrannosaurs-rex/.
107   "universally held in esteem": Ibid.
107   Coaxing bones from dust: If a poorly collected or damaged fossil reached the laboratory in New York "it was said to have been 'RCA'd.'" See Gallenkamp, *Dragon Hunter.*

107 "At Granger's core lay the master craftsman": Morgan and Lucas, *Walter Granger*.
 They wrote that Granger would be only "barely mentioned in most accounts of
 paleontological exploration and even in American Museum retrospectives..."

107 "Do your utmost": Andrews, *The New Conquest of Central Asia*.

107 "lapdog-sized predator covered in feathers": See Stephen Brusatte, "New Fossil
 Reveals Velociraptor Sported Feathers," *Scientific American*, July 17, 2015.

108 "With mounting excitement": Preston, *Dinosaurs in the Attic*.

108 Dinosaur eggs: My favorite line about dinosaur eggs may be the one John Updike
 wrote in the spring of 1958. After a trade with a museum in Aix-en-Provence,
 France, the American Museum of Natural History now had on display a Jurassic
 egg, probably sauropod. "The egg sat alone in a glass case, thousands of miles
 and millions of years from Mother," Updike wrote. See John Updike, "Dinosaur
 Egg," *The New Yorker*, April 19, 1958.

108 "egg thief": For the original paper see Henry Fairfield Osborn, Peter C. Kaisen,
 and George Olsen, "Three new Theropoda, Protoceratops zone, central Mongolia," *American Museum Novitates* 144 (1924).

109 "What are the darned things worth?" and related: "The Eggs and I," by Roy
 Chapman Andrews, ran in newspapers across the country, including in the *Los
 Angeles Times* on August 31, 1952.

109 "spread the gospel": Andrews, *Under a Lucky Star*.

109 "a sucked orange": Ibid.

110 "That's you": Ibid.

110 "all-American expedition": Gallenkamp, *Dragon Hunter*.

110 "Imperial objectives": Ronald Rainger, *An Agenda for Antiquity: Henry Fairfield
 Osborn & Vertebrate Paleontology at the American Museum of Natural History,
 1890–1935* (Tuscaloosa, AL: University of Alabama Press, 1991).

110 "ranger hat," "especially white," and "embodied": Ibid.

110 "paleontological Garden of Eden": See Colbert, *The Great Dinosaur Hunters and
 Their Discoveries*, and also Evans Clark, "Tracing the First Man in Asia," *New
 York Times*, September 13, 1925.

111 John Barrymore: Andrews declined to give him one but did give Barrymore
 "a dozen bits of shell about the size of my thumb nail." See Andrews, *Under
 a Lucky Star*. For more on Barrymore's collection obsession, see my *New
 Yorker* story "The Tallest Trophy," about the time he took a Tlingit totem pole
 from Alaska and erected it in his backyard in Hollywood. It ran in the April
 20, 2015, issue and can be found at https://www.newyorker.com/magazine/
 2015/04/20/the-tallest-trophy.

111 "more than anything else": Andrews, "The Eggs and I."

111 "grand publicity stunt" and related: Andrews, *Under a Lucky Star*.

111 "We have got a perfectly good 'corner'" and related: Ibid.

111 "Dinosaur Egg 100,000,000 Years Old for Sale": *New York Times*, January 8,
 1924.

111 "none had ever before been bought," "greatly exceeded": Ibid.

112 "Illustrated London": Andrews, *Under a Lucky Star*.

112 Austen B. Colgate: A young Colgate, Bayard, worked on the Central Asiatic
 Expeditions as a field mechanic. Colgate University uses the egg, laid by an
 oviraptorid, in research to this day. "Its importance cannot be overemphasized

because the egg (and its sibling eggs) provided the first definitive evidence of how some dinosaurs reproduced, opening up a whole new area of research on dinosaurs," Colgate geology professor Connie Soja wrote on the school's Dinosaur Egg Research page. In 2004, she visited Bayanzag, aka the Flaming Cliffs, to see the egg's discovery site. After hiking to the top of the formation at dawn, she wrote, "What an adventure to be there as the sun was coming up . . . Knowing our egg's historical significance, not just to Colgate but to paleontology, added an emotional element to walking in Roy Chapman Andrews's footsteps . . ." For more information, go to colgate.edu.

112   "The original Easter egg!": Andrews, *Under a Lucky Star*.

112   Everyone would assume: Ibid.

112   "all lands and resources": Mongolian Constitution, 1924. See *The Constitutions of Mongolia: 1924, 1940, 1960, 1992*, edited by B. Chimid and Ts. Sarantuya from materials compiled by J. Amarsana and O. Batsaikhan (Mongolian Academy of Sciences, 2009). You can find this online at Stanford Libraries.

112   "My eggs" and related: Andrews, "The Eggs and I."

113   "They couldn't know": Andrews, *Under a Lucky Star*.

113   "friendly relations, especially for trade": Addleton, *Mongolia and the United States*.

113   "lesson was obvious enough": Ibid.

113   "worth less than that of a sheep": Andrews, *The New Conquest of Central Asia*.

113   "We rode at full speed": Andrews, *On the Trail of Ancient Man*.

113   "Except for the modern weapons": Ibid.

113   "dashing horsemen," "strange costumes," and "Russians": Andrews, *The New Conquest in Central Asia*.

114   "In short, one was treated as a spy": Ibid.

114   "high-powered binoculars": Morgan and Lucas, *Walter Granger*.

114   "Reynolds": Ibid. In *Dragon Hunter*, Gallenkamp wrote that Andrews was "one of many civilian informants, operating under various guises, whose job it was to gather data on a wide range of subjects for use in formulating American policies in eastern Asia. Andrews, for example, filed reports dealing with, among other things, political conditions in China, communication and rail facilities, troop movements, industrial output, armaments, shipping and ports, and evidence of foreign intervention—particularly Japanese—in China and Manchuria."

114   "salient lines": Morgan and Lucas, *Walter Granger*. Gallenkamp also covers the short-lived "Reynolds" period in *Dragon Hunter*.

114   "the murder of white residents": Roy Chapman Andrews, "Further Adventures of the American Men of the Dragon Bones," *Natural History* 29 (March–April 1929).

114   "packed to the rails": Ibid.

114   "Murder and sudden death": Andrews, *The New Conquest of Central Asia*.

115   Purges: The purges of the 1930s are Mongolia's most terrible secret and legacy. Even after all this time citizens are just now learning the truth of what happened during this period. By some estimates as many as 100,000 (in a population of 700,000 at the time) were killed or went missing. Mass graves filled with monks have turned up. See Kathy Chenault, "Mongolians Seek Truth of Purges," *Los Angeles Times*, August 16, 1992.

115   "Is it surprising that I was filled with regret": Andrews, *On the Trail of Ancient Man*.

CHAPTER 12: MARKET CONDITIONS

116 Ambassador Joseph Lake: A good resource for understanding the first U.S. mission to Mongolia in the early 1990s is an oral history with Lake conducted by the Association for Diplomatic Studies and Training Foreign Affairs Oral History Project. The organization operates out of the State Department's George P. Schulz National Foreign Affairs Training Center, in Arlington, Virginia. Initial interview conducted by Charles Stuart Kennedy on September 5, 1994. Published by ADST, 1998. adst.org.

116 "It is obvious": For excerpts of Gorbachev's address to the UN General Assembly, see "The Gorbachev Visit; Excerpts from Speech to U.N. on Major Soviet Military Cuts," *New York Times*, December 8, 1988. For additional context, see "Gorbachev's Approach to the United Nations: Image Building at US Expense?" (declassified), Office of Global Issues, Central Intelligence Agency, September 1989.

116 "rule of law" and related: Ibid. Gorbachev address to the UN General Assembly.

117 "unwittingly sown the seeds": Kaplonski, *Truth, History and Politics in Mongolia*.

117 Tiananmen Square: On February 26, 1989, in a sit-down in Beijing with President George H. W. Bush, Deng Xiaoping, the leader of Communist China, complained that Russia ceded far too much of Mongolia after World War II. China had lost territory and still wasn't happy about it. Deng had been the leader of Communist China since 1978, assuming power after Mao Zedong. The Nixon administration had eased the way for diplomatic relations by softening trade and travel restrictions, but it was the 1972 visit by Nixon and White House national security adviser Henry Kissinger that became the "historic breakthrough" leading to improved economic ties. Deng had deepened those ties during a historic visit to the United States in 1979, when he said China could "leapfrog the years in which the world had passed it by, but only with American support." "He was ready to cooperate on containing the Soviet Union, even agreeing to the installation of secret American intelligence listening posts along the Chinese border, to track Soviet missiles, wrote Richard Holbrooke, former U.S. ambassador to the United Nations." (See Richard Holbrooke, "The Day the Door to China Opened Wide," *Washington Post*, December 15, 2008.) In the years since, Deng had come to symbolize the "Chinese aspiration to move beyond the ideological extremism that had marked the Maoist era, and reclaim for the Chinese a long-denied prosperity." The *New York Times* noted, "Where Mao had preached 'Communes are good,' Mr. Deng simply preached 'Markets are good.'" Deng stood five feet tall (Kissinger once referred to him as a "nasty little man") and rarely appeared in public. Observers noticed that he showed no emotion, wit, or fleetness of mind. While he participated in groundbreaking talks he rejected democracy and showed a willingness to crush dissent. For him, noted the *Times*, "China's economic reform could only occur under the authoritarian rule of the Communist Party." At the Beijing conference Bush was expected to broach the subject of human rights but never did. Four months later, the Chinese government declared martial law, killing demonstrators at Tiananmen Square. See Patrick E. Tyler, "Deng Xiaoping: A Political Wizard Who Put China on the Capitalist Road," *New York Times*, February 20, 1997. While Deng participated in groundbreaking talks he rejected democracy and showed a willingness to crush dissent. For him, noted the *Times*, "China's economic reform could only occur under the authoritarian rule of the Communist Party."

114 Tiananmen Square: The exact number of dead remains unclear. "No public discussion of the tragedy is possible in China," the *New York Times* reported in 2012. See Didi Kirsten Tatlow, "Tragedy of Tiananmen Still Unfolds," *New York Times*, June 13, 2012.

118 Soviet dissident Garry Kasparov: "What I am fighting against is the same all over the world, wherever it happens to be: It is called evil," he told the interviewer. When the interviewer asked Kasparov about future business projects he said, "I have a friend who has a computer business in Moscow, a cooperative company. When he meets new partners from the West, his first question to them is, 'Have you read *Alice in Wonderland?*' They will answer yes, then he says, 'Imagine you are in Wonderland, and we will start our discussions from there.... I just had a thought the other day: Why don't we sell the Kuril Islands to the Japanese? Frankly speaking, I'm not sure that these islands belong to us, and the Japanese, who claim them, would give us billions and billions of dollars for them! That would keep us going for maybe five or ten years. Then we could sell Mongolia to China and get a few more years that way. But the best deal would be to sell East Germany to West Germany..." See Louis Blair, "Playboy Interview: Garry Kasparov," *Playboy*, November 1989.

118 "the perfect warrior": It's been said that "Genghis Khan" translates to "ruler of all" though "perfect warrior" is often used to describe Chinggis. See Mark Fineman, "Mongolia Reform Group Marches to Rock Anthem," *Los Angeles Times*, January 24, 1990.

118 Politburo's resignation: The old regime never intended a full democratic transformation but rather seemed interested in exploring expanded freedoms under socialism, analysts later said. One leader of a reform faction of the ruling party had become known as "Mongolia's Yeltsin," for instance, "after proposing in a Party magazine that the nation renounce Marxism and initiate a market economy." He said he looked forward to "the emergence of 'a national capitalist—a richest man in our country.'" See Fred C. Shapiro, "Starting from Scratch," *New Yorker*, January 20, 1992.

118 "An isolated and little-known country": Fineman, "Mongolia Reform Group Marches to Rock Anthem." Fineman reported that the hottest rock song in Ulaanbaatar during the democratic revolution was called "Genghis Khan."

119 "21st Century": See Karl Malakunas, "Genghis Khan, a 21st Century Marketing Phenomenon," Agence-France Presse, October 8, 2006.

119 "greatness and complexity": Michael Novacek's richly detailed *Dinosaurs of the Flaming Cliffs* informs much of this chapter. *Dinosaurs* is a first-person narrative account of the American Museum of Natural History's return to the Gobi Desert in 1990, for the first time in the better part of a century.

120 Demberelyin "Dash" Dashzeveg, Altangerel Perle, and Rinchen Barsbold: Descriptions and information come from Novacek's account and interviews with various paleontologists and crew members.

120 "small enough": Ibid.

120 "biological empires": Ibid.

121 "It seemed an absurd predicament": Novacek wrote, "No one had any gas, so it was impossible to get food from the country to the city. Conversely, the country villages and towns could not get basic stores of flour, sugar, salt, or canned goods."

121 "Big Gobi circuit": East from Ulaanbaatar then roughly a circle down nearly to the Chinese border and back up again. "By the second day of travel, the unbroken rolling hills of rusty grass became oppressively boring," Novacek wrote, then quoted the travel writer Paul Theroux, who was referring to northwestern Turkey when he wrote: "Featurelessness is the steppes' single attribute, and having said that and assigned it a shade of brown, there is nothing more to say." See Paul Theroux, *The Great Railway Bazaar: By Train through Asia* (Boston: Houghton Mifflin, 1975).

121 "broiling isolated depression": Novacek, *Dinosaurs of the Flaming Cliffs*.

121 "audaciously penetrated": Ibid.

121 Grand canyon and related: Ibid.

122 "Giant Carnivorous Dinosaurs": See E. A. Maleev, "Giant Carnivorous Dinosaurs of Mongolia," trans. F. J. Alcock, *Doklady, USSR Academy of Sciences* 104, 4 (1955). A more recent translation can be found in the monograph "Giant Carnosaurs of the Family Tyrannosauridae," translated separately by Catherine Siskron and S. P. Welles, and Jisuo Jin, and edited into one volume by Matthew Carrano. Maleev was still working on the descriptions when he died unexpectedly in 1966. The translations can be found at paleoglot.org.

122 "Everything about it": Novacek, *Dinosaurs of the Flaming Cliffs*.

122 "forgotten corner": Ibid. Also see John Noble Wilford, "For Fossil Hunters, Gobi Is No Desert," *New York Times*, September 13, 2005.

123 Xanadu: The dinosaurs included six ankylosaur skeletons, "some with perfectly preserved tails and tail spikes," Novacek reported, plus over a dozen skeletons of small theropods and an "excellent" skull and one of the few known skeletons of *Oviraptor*. Altogether the team had found "the most diverse assemblage of theropods from any single location." See "Major Dinosaur Find," *Washington Post*, April 6, 1994.

123 "I found something" and related: Novacek, *Dinosaurs of the Flaming Cliffs*.

123 "clues": Ibid.

123 "historical, cultural": Mongolian Constitution. See "The constitutions of Mongolia: 1924, 1940, 1960, 1992," compiled by J. Amarsana, O. Batsaikhan; edited by B. Chimid, Ts. Sarantuya; Mongolian Academy of Sciences, 2009. Via Stanford Libraries.

124 "non-bourgeois": See Ichinkhorloo Lkhagvasuren, "The Current Status of Mongolia's Museums: Changes Taking Place in the Practical Activities of Museums Since the 1990s," in *New Horizons for Asian Museums and Museology*, edited by Naoko Sonoda. This is a valuable overview of what was happening with Mongolian museums before and after the fall of Communism in 1990.

124 "made of soft material": Ibid.

125 Market for everything: "The nearest thing to a precedent for what is taking place in this impoverished, sparsely populated nation landlocked between Russia and China may be the Oklahoma land rush of 1889, when a starter's gun sent fifty thousand would-be settlers racing to stake claims on some two million acres of former Indian territory," Fred Shapiro reported in the *New Yorker*. Shapiro acknowledged that the analogy isn't "precise" but the point stands. See Shapiro, "Starting from Scratch."

125 "keep their passports": Kaplonski, *Truth, History and Politics in Mongolia*.

125   "Imagine in the 1930s": Addleton, *Mongolia and the United States.*
125   "almost twice as severe": Ibid.
125   "lesser assets of the state": Shapiro, "Starting from Scratch."
126   Mongolian stock exchange: Ibid. The stock exchange was housed in a pink building on Sukhbaatar Square once belonging to a children's theater. "Economists in your country, if they study and work hard, maybe someday will get to the top and administer your banking system," Naidansurengiin Zoljargal, the exchange's architect and chairman, told *The New Yorker.* "Me, I'm creating an economic system from nothing. I am actually *doing* economics."
126   "trip to Antarctica": Novacek, *Dinosaurs of the Flaming Cliffs.*
126   "epic in scale": Website of Bob Burnham, a senior research computing associate and adjunct assistant professor of business administration at Dartmouth College's Tuck School of Business. Burnham also worked for the U.S. Agency for International Development (USAID) and the State Department. See http://www.dart mouth.edu/~bburnham/mongolia/
126   "a BBC television crew": Shapiro, "Starting from Scratch."
127   Homelessness: For an overview see Stephanie Hoo, "For Children in Mongolia, a Life on the Streets," *Los Angeles Times,* July 17, 2005. Also see Javier C. Hernández, "'We Don't Exist:' Life inside Mongolia's Swelling Slums," *New York Times,* October 2, 2017. Related: pollution. The people who live in the *ger* districts on the fringes of Ulaanbaatar burn tires and other trash, contributing to the city's horrific pollution problem. There's a "huge human cost of mortality, where climate not only has caused the mortality of livestock but also contributed to the loss of livelihoods and culture," researchers reported in *Environment Science Letters.* See et al. "Dzuds, droughts, and livestock mortality in Mongolia," *Environmental Research Letters* 10, 7 (July 17, 2015).
127   "fantastically wealthy": Kaplonski, *Truth, History and Politics in Mongolia.* Also see Thomas Crampton, "A Mongolian Shopping Spree Fizzles," *International Herald Tribune,* June 25, 1998. Collected in *In Their Own Words: Selected Writings by Journalists on Mongolia, 1997–1999,* ed. David South (DSConsulting, 2015).
127   "Under no circumstances": Andrews, *The New Conquest of Central Asia.* According to Shapiro, Buddhist scripture forbade disturbing "the earth's blessed sleep." See Shapiro, "Starting from Scratch."
127   "no encampment is safe": See Frans August Larson, *Larson—Duke of Mongolia* (Worcestershire, UK: Read Books Ltd., 2013).
128   Dinosaur tourism: Most recently, an Odyssey Traveller tour was scheduled for May 2018 "under the auspices of the…Mongolian Academy of Sciences." For $12,363 per person, guests could buy a "unique experience to contribute to the world's understanding of Mongolian dinosaurs through paleontology." Archaeological Institute of America tours long featured guides "who represent the rock stars of the paleontological world," *USA Today* reported in 2001. The trips also called the public's attention to the importance of conservation, and could serve as "a thank-you to private supporters of research or an inducement to provide more support." As one archaeologist put it, "Why should only archaeologists have the right to see things?"
128   "As long as people are interested": Dan Vergano, "Tourists Dig Expeditions Sponsored by Scientists," *USA Today,* August 23, 2001.

128 Watching them through binoculars: In November 2008 the *Cleveland Plain Dealer* reported that Michael Ryan, a curator of vertebrate paleontology at the Cleveland Museum of Natural History, had recently returned to the Gobi to dig out a *T. bataar* found in 2005, only to find the skull, hands, and feet missing. Crude tools had been left behind, including a "chisel fashioned from a sharpened engine rod" and "a hammer made of a rock duct-taped to a stick." Sixty percent of the creature remained but Ryan's travel companions, "adventure tourists," had never excavated fossils so Ryan left the bones in the ground. He couldn't help feeling sorry for the Mongolian poachers, saying, "They see us driving these big fancy trucks and taking the bones away. As rich Europeans and North Americans coming in there, it's hard to say, '*Thou* shalt not do these things,' because that's what it appears we're doing." See John Mangels, "Dinosaur Fossil Poachers Apparently Victimize Cleveland Museum of Natural History," *Cleveland Plain Dealer*, November 30, 2008.

129 "annual influx": Andrew Lawler, "Science Hopes to Rebound in Post–Cold War Era," *Science*, January 22, 1999.

129 "to observe U.S. spy satellites": Ibid.

129 "whatever is left over": In July 1991, representatives from several U.S. government technical agencies visited Mongolia to meet with counterpart agencies and to explore the potential for bilateral projects in science and technology. The team reported that the Mongolian scientists were eager to display their expertise but realistic about their lack of financial support, transportation, and communications. "Though in many fields they desperately need an introduction to modern scientific equipment and facilities, Mongolian scientists are competent, eager to learn and contribute," the team reported in a cable to the State Department. Smithsonian representatives noted the similarities of cultural losses to those of the American Indian. The team reported that Mongolia was still using "Soviet methods of teaching" and that faculty members had not "supported government efforts to stimulate change. It is hoped that future educators will be exposed to Western influences and prompt changes within the country." The Mongolian scientists "possess and are eager to share some impressive historical environmental data that would be of interest to workers in climate change problems," the cable went on. State Department cable C05938924, dated July 1991, obtained through the Freedom of Information Act.

130 "government paid the poor Gobi people": Email interviews with Hollis Butts. The tugrik had "proved resilient in holding its value against most international currency" but fell by some 40 percent against the dollar in the global economic crisis of late 2008 and early 2009, according to "2013 Mongolia Investment Climate Statement," Economic and Commercial Section, U.S. embassy, Ulaanbaatar, Mongolia, January 15, 2013.

131 "We need more scientists": Interview with Chultem "Otgo" Otgonjargal, August 2015, Ulaanbaatar. All of Otgo's quotes in this passage are from this interview.

CHAPTER 13: "GO GOBI"

133 Tuvshin: Email interviews with Hollis Butts, and extensive interviews with Eric Prokopi and other fossil dealers. Butts refused to be interviewed in person, despite my offer to come to Japan. He declined to speak by phone or Skype.

When we were fact-checking the original magazine story that led to this book he refused to speak directly with the checkers, and communicated only by email. Butts told me Tuvshin often brought Mongolian fossils to trade shows in his luggage. "I assumed that as long as Mongolian, Chinese, Japanese, and probably European & American customs controls had no problem with what Tuvshin carried by air, and we heard no objections from Mongolia, and Mongolian material appeared in American auction catalogs (as well as everywhere else), and Tuvshin assured us that there were no legal problems, and the Mongolian museum gift shop offered bone and teeth for sale, and the US government gave no hint of possible problems... Mongolian material was acceptable in the US." Mongolian government officials have said that dinosaur fossils were never sold in the Natural History Museum gift shop, but others have disputed that claim. Butts added, "It seemed that was how things were done in Mongolia." In another email he wrote, "When I asked Tuvshin about Mongolia, he answered that the Stalinist past was being rejected, a new Constitution written, and that laws from that tragic past and everything associated was in change. That seemed reasonable to me at the time."

133　"like a traveling salesman": Email interview with Hollis Butts.

133　"he was trying to get international people involved": Ibid.

133　"very secretive": Interview with a dealer.

134　Location of Ulaanbaatar: The landscape of Mongolia is the country's greatest tourism asset. The lack of infrastructure keeps the terrain unspoiled, yet makes the stunning scenery difficult to reach. Outsiders tend to think of Mongolia as all Gobi, when, in fact the ecosystems include desert, desert steppe, and grassy steppe, which is similar to the American plains; lakes in the north, mountains in the east, and really high mountains in the west, the region of reindeer tribes and eagle hunters. Given the nation's vastness, it would take months to tour the whole country. Oyungerel Tsedevdamba told me that once, when her brother-in-law visited, she and her husband, Jeff Falt, took him to the rolling steppe, and he asked, "But where are the mountains?" They took their book editor to the forested north and the editor asked, "But where are the camels?" Oyuna had been alive for fifty years and had lived in Mongolia for most of them and still hadn't seen the flat steppe of the east.

134　"one of the great empty spaces": Novacek, *Dinosaurs of the Flaming Cliffs*.

134　Traffic: In the summer of 2015 in Ulaanbaatar, I met a Utah geologist, Cari Johnson, who had been working summers in Mongolia since the 1990s. At Millie's, an expat cafe, she told me she loves working in the Gobi because "Mongolia is the most complicated stratigraphy there is." Using a notebook and pen she sketched a crude map, lining off the country not as geographical regions, like provinces or states, but rather as rock regions—"Siberian craton," "plateau," "Cretaceous (old)," "pretty old," and "really really old." The western Gobi is the youngest, the eastern Gobi the oldest, she explained. The Flaming Cliffs and the Nemegt Basin, in the southern Gobi, where *T. bataar* was found, were somewhere in the middle. Much of the country had yet to be mapped. "It's totally different from working in the western U.S., where you have a hundred and fifty years of good, solid geological data," she told me. The public was only beginning to realize that Ulaanbaatar, a city of nearly two million people living and working in a hodgepodge of structures, sits atop active geologic faults. A large earthquake would be "catastrophic," Johnson said. She had been working in the Gobi for so long she was "almost a lizard." She answered her local cell phone "*Baina*"—"I'm here." By now, she knew that in Ulaanbaatar, savvy pedestrians plan their day around which street they refused to cross. In the early days of capitalism the city had no crosswalks because the citizens had no cars. Now that the streets were choked Johnson looked to the locals: if they started to walk, she hustled in among them, trying not to be the last to cross, or she looked for a distinguished-looking older Mongolian who theoretically commanded more respect and stood less of a chance of being struck. When Bolor Minjin heard about this strategy she said, "This is what Mongolians face every day and every hour. It has become part of their life. They found a way to adjust to things that didn't used to be a problem." One day on Peace Avenue, UB's Broadway, Johnson and I ran into an American, a Florida man who now lived in the city. Watching all the vehicles stream past he said, with obvious glee, "Look at all this traffic. That's capitalism!" Another afternoon, Johnson and I left the Blue Sky and crossed a side street, Jamyan, where a new building was going up. New buildings were always going up in UB; it wasn't unusual for the developers to run out of money and leave a project windowless and unfinished, plastic sheeting flapping in the wind. That summer, construction cranes stood like sculpture throughout the city, moving not an inch. When work did resume it was often shoddy, and safety standards were often ignored. The savvy pedestrian had to look up a lot and down a lot because tools fell from the sky and manhole covers went missing, removed by those who took shelter underground in the winter.

135 Mint-green building: On one of the days when I visited, in August 2015, a family was selling vegetables from the trunk of their car in the gravel parking lot behind Tuvshin's building. In the basement, which had been turned into a hip-hop studio, teenagers waited to participate. Upstairs, on the fourth floor, Bobo, Tuvshin's widow, was sitting on an office sofa, thumbing out a message on her cell phone. After her husband died, she had maintained the family travel business. She had on sky-blue pants, a striped sweater, and tan sneakers. Her lipstick was coral and her earrings resembled blue rhinestones. Her eyeglasses were pushed on top of her head. Behind her, in tall glass cabinets, stood enormous purple geodes. We spoke through a translator. The look that came over Bobo's face when I asked about Tuvshin's dinosaur business can be described only as panic. She and a relative had already been detained by the police for weeks. The case was still open.

136 Natural History Museum: Kirk Johnson, head of the Smithsonian National Museum of Natural History, once told me the Ulaanbaatar museum's decrepitude is painful but not surprising: "Mongolians have so many challenges with basic education, access to food, poverty—a museum is legitimately lower on the profile, because you want to take care of your people first. It's hard to do best practices when you have no resources at all."

136 "no substitute for aptitude": On its website, the AMNH told prospective preppers, "Preparation is not one skill but many. It begins when the fossil is excavated from the ground, continues in the laboratory, and never really finishes, because the specimen will require care throughout its life in the collections. It can involve careful digging with small and large tools in the field; use of hand and power tools in the lab; carpentry and metalworking in the construction of

mounts for storage and display; use of a microscope for preparation of very small fossils; and awareness of different materials and their characteristics in consolidation, adhesion, and molding and casting." See http://preparation.paleo.amnh .org/60/training-to-become-a-preparator.

137   "bona fide" and "Really, what an armature is doing": See Maureen Byko, "Steel and Science Bring Dinosaurs into the 21st Century at the Carnegie Museum of Natural History," *Journal of the Minerals, Metals and Materials Society* (June 2008). The preparator Phil Fraley said that when working a new project in an old museum like the AMNH or the Carnegie he always tried to use vintage materials that remained viable: "In some ways it was our way of recognizing and appreciating the work that all these men a hundred years ago had done."

137   tugrik: The Mongolian currency may also be written as *tugrug*. Genghis Khan introduced gold and silver coins during his reign of the Mongol Empire and later was said to have introduced the world's first bank notes; but by the mid-1920s, Mongolia had official money, first written as tögrög. The value was roughly equal to one Soviet ruble.

138   *Ger*: A *ger* may be covered with canvas, skins, felt, or a combination of all three. Felt, which is made by pressing wool, had many uses: clothes, blankets, rugs. During the reign of the Mongol Empire warriors sent felt puppets into battle on horses, "duping the enemy into believing that they had a much larger force than their actual number," Morris Rossabi wrote in *The Mongols: A Very Short Introduction*. You can read about the brilliant history and craftsmanship of the Mongolian *ger* in many places, but a good starting point is the UNESCO website. The customs are fascinating. Outside one family's Gobi *ger*, I committed taboo by stepping over one of the long wooden poles that herders use to lasso wild horses. This horrified two of the little girls who lived there. I don't know what misfortune I brought upon myself but judging from their aghast expressions, it could not have been good.

139   The man of this *ger*: Interviews with Eric Prokopi.

141   "reptiles": Interviews with Eric Prokopi.

141   "I mean how cute" and related: videos, photos, and papers of Eric and Amanda Prokopi.

143   Criminal "mastermind": See John Romano, "Paleontologist May Lose Freedom for Living His Dream," *Tampa Bay Times*, December 30, 2012. Doris told the paper, "This is crazy. He was doing what he loved.... This was like arresting Indiana Jones." Dr. Peter Harries, then of the University of South Florida's geology department, told the paper, "It's a delicate balance in paleontology. You need people out there looking for things because they may eventually be ruined by erosion. And there is a long history of amateurs being involved in the discovery of these bones. But once profit becomes the dominant motive it can change things. I don't think Indiana Jones wanted to sell the Ark of the Covenant for $1 million."

143   "It was more like *Harold and Kumar*": Interviews with Amanda Prokopi.

144   "Big Gobi circuit": According to the road map Prokopi followed, they went east from Ulaanbaatar, south at Bayandelger, down through Choyr, following the old rail line, and way down southeast beyond Erdene, very nearly to the Chinese border. Tuvshin told Eric he'd arranged with the military for special permission to be there. Then they drove west to Zuubayan, Manlay, Dalanzadgad,

Gurvantes. Then northwest to Bugiin Tsav, cutting northeast to Bayanlig and Nariynteel and Arvayaheer, before going on up and over to Ulaanbaatar.

144   Gobi finds: Prokopi collected surface fossils in the desert. He lacked the time, materials, and jacketing experience—much less the fortified transportation—to take out large, heavy dinosaurs. It's possible that diggers went back for any found skeletons and sent them out through Tuvshin, but who knows. The passages about the Gobi trip come from interviews with Eric Prokopi and Tony Perez, and their photos, plus the lengthy interview with Otgo.

CHAPTER 14: THE GHOST OF MARY ANNING

147   Details on early Lyme can be found in *The Great Domesday Book*, published in 1086, a copy of which is available at the U.K. National Archives in Kew, west of London. A good history of this foundational document can be found on the BBC's website: http://www.bbc.co.uk/history/british/normans/doomsday_01.shtml.

147   This chapter is also informed by my visit to Lyme Regis, Charmouth, and London in May 2015. It was remarkable to walk the beaches where Mary Anning did her work and to see her spectacular fossil finds in the Natural History Museum. One drizzly Saturday afternoon in Lyme, some friends and I took a Mary Anning tour, led throughout the hilly town by a woman in a long frock and bonnet. We started in Cockmoile Square; across the way was the Lyme Fossil Shop, which sold fossil-hunting books and geologic hammers. We saw the Congregationalist church (aka the Dissenters' chapel) that Mary attended before switching to the Church of England; it was now Dinosaurland Fossil Museum, a private company packed with curiosities. We passed cottages with names: "Weavers" had a turquoise door, "Lym House" had lavender. After passing Sherborne Lane, a street dating to Saxon times, we came to the site where Mary last lived and Anning Fossil Depot once stood, on the main street through town. Across the way was Ammonite Fine Foods, which had one of the few remaining thatched roofs in town.

147   "verrie daungerous": John Fowles, *A Short History of Lyme Regis*.

147   "wool out, wine in": Ibid. Merchandise came to include honey, oil, tar, figs, tallow, saffron, iron . . .

148   "the beauties of 'wild' nature": This phrase can be found in many texts regarding Romanticism, including "The English Landscape Garden," a chapter in Tom Turner's *Garden History Reference Encyclopedia* (London: Gardenvisit.com, 2002).

148   Mary Anning's biography: An excellent, if short, resource is *Mary Anning of Lyme Regis*, by Sir Crispin Tickell, a descendant of the Anning family. A former warden of Green College at Oxford, Tickell, regarded as a leading expert on climate change, served as president of the Royal Geographical Society, British representative to the United Nations, and the United Kingdom's ambassador to Mexico. Also see the work of Hugh S. Torrens, emeritus professor of history of science and technology at Keele University, in Staffordshire, England, perhaps the greatest Mary Anning expert. See "Anning, Mary," *Oxford Dictionary of National Biography* (Oxford, UK: Oxford University Press, 2004) and "Presidential Address: Mary Anning (1799-1847) of Lyme; 'The Greatest Fossilist the World Ever knew,'" *British Journal for the History of Science* (Cambridge, UK: Cambridge University Press, 1995). Another excellent resource is *The Fossil Hunter*, by Shelley Emling. For your fictional reading pleasure, Tracy Chevalier's *Remarkable Creatures* (New York:

Dutton, 2010) is a deeply researched novel on the life of Mary Anning, informed by Chevalier's immersive studies in Lyme Regis. Also, see Patricia Pierce, *Jurassic Mary: Mary Anning and the Primeval Monsters* (Stroud, UK: Sutton Publishing Ltd., 2006).

149 Jane Austen: In her novel *Persuasion*, Austen wrote, "The young people were all wild to see Lyme."

149 "beyond the value": This often-repeated anecdote can be found in books including Emling's *The Fossil Hunter*.

150 Rockfalls and mudflows: Author's observation. In May 2015, when the spring fields of England were neon yellow with rape, several friends and I walked the beach at Lyme and at Black Ven, whose stratigraphy was striped in blacks, browns, and golds. Wildflowers had forced their way to the surface. The weather was rainy and chilly, but dozens of people were out fossil hunting, their geologic hammers clinking against rock. I saw one man hold a large stone high over his head and bring it down hard on a sharper, bigger rock, trying (and failing) to break it. The waves were relentless; the black shale cliff wore fog. It's a stony and forbidding beach, littered with broad, thick seaweed clumped together like tangled film; you wouldn't necessarily want to sunbathe there. But it's bleakness is gorgeous. The fossil walk put on by the Lyme Regis Museum is excellent, too. It starts at Gun Cliff and is led by geologists, paleontologists, and/or veteran fossil hunters. Our tour day was sunny and blue—"no good," said Chris Andrew, one of the guides. "The dreadful sunshine," said another guide, Paddy Howe, who had been hunting fossils there for the past forty-four years and whose workshop was in the basement of Alice's Bear Shop, a teddy bear and doll hospital. The best time to hunt is after storms, when the finds were like "fossils on a conveyor belt."

151 "found almost all the fine things": See, among many other sources, Dennis R. Dean, *Gideon Mantell and the Discovery of Dinosaurs*. Whenever Mary found an ichthyosaur skeleton, she noticed odd conical stones near the abdominal area. Cracking them open with her rock hammer yielded bones and scales. Mary reasoned that these "bezoar stones" must be fossil feces. Her friend William Buckland eventually agreed with her, formally naming the fossils coprolites and publicly recognizing Mary's discovery. Buckland made a table out of coprolites, which today can be seen in the Lyme Regis Museum.

151 Georges Cuvier: Combining scientific urgency and literary beauty, Elizabeth Kolbert's *The Sixth Extinction* is the finest of all the encapsulations of the work of Georges Cuvier and his contemporaries, including Charles Lyell and Charles Darwin. In 1851, Cuvier wrote, "Why has not anyone seen that fossils alone gave birth to a theory about the formation of the earth, that without them, no one would have ever dreamed that there were successive epochs in the formation of the globe."

152 "persevering female": Also widely quoted, from the *Bristol Mirror*. See Prothero, *The Story of Life in Twenty-Five Fossils*.

152 "The extraordinary thing": This is another widely quoted tidbit about Mary. For this and more information see "Mary Anning (1799–1847)," on the website of the Geological Society of London, https://www.geolsoc.org.uk/Library-and -Information-Services/Exhibitions/Women-and-Geology/Mary-Anning.

153 "men of learning": Emling, *The Fossil Hunter*. Pinney also wrote that Mary "glories in being afraid of no one, and in saying anything she pleases."

153 "Her history shows": See Charles Dickens, "Mary Anning, the Fossil Finder," *All the Year Round*, February 11, 1865.

154 *Duria Antiquior*: You can see *Duria Antiquior* at http://ocean.si.edu/ocean-photos/duria-antiquior-%E2%80%93-more-ancient-dorset-1830. The website notes that an ichthyosaur and plesiosaur "would have likely never battled" but that the Henry De La Beche lithograph "inspired author Jules Verne to pen a similar scene in his book *Journey to the Center of the Earth*."

154 "reduced to straitened circumstances": This was published in the *Dorset County Chronicle* in 1836, where Lyme Regis Museum staffer and researcher Jo Draper found it and wrote about it in June 2010. See Jo Draper, "Mary Anning and Me," *Dorset Life Magazine*, June 2010. Draper wrote that one of the great Mary Anning puzzles was "how someone as sensible as she ended up completely broke towards the end of her life."

154 "many contributions": Various people wrote that Mary Anning made "many contributions" to science, among them Hugh Torrens, who authored the entry on Anning in the *Oxford Dictionary of National Biography*.

155 "I am well known": As quoted in G. Y. Craig and E. J. Jones, *A Geological Miscellany* (Princeton, NJ: University Press, 2016). For an audio slideshow of Mary Anning's discoveries showing Anning's sketches of her finds, see "Jurassic Woman," narrated by Tracy Chevalier, BBC News, October 21, 2010, www.bbc.com/news/science-environment-11590505.

155 FOSSIL WARDEN: Met one at Black Ven. His name was Stuart Godman, and he said his job was primarily to protect people, not fossils. He said he found a fossil every thirty or so stones and that "the really *good* stuff is hard to find."

156 "delightful shop": Some of the biographical information on Chris Moore comes from the various news accounts published or aired about him in the Lyme Regis area and elsewhere in the United Kingdom. In September 2009, *Shoreline*, a local newsletter, ran a package headlined "Paleontology in Charmouth" to mark the bicentennial of Charles Darwin's birth. A delightful package. The editor signed her note to readers, "Here's to mists and mellow fruitfulness," from the John Keats poem "To Autumn."

156 "I always wanted": *Shoreline*, "Paleontology in Charmouth."

157 Befferlands Farms list: The Befferlands and local descriptions are from my visit to Charmouth and the coin-toss list is from a photo courtesy of Eric Prokopi.

158 "A home is more": Interviews with Amanda Prokopi. See also Allison Clark, "A Second Chance for Serenola," *Gainesville Sun*, January 24, 2010.

158 "manufacturer": Entry number UPS-3162168-6, U.S. Department of Homeland Security, Bureau of Customs and Border Protection, "Entry/Immediate Delivery" form. The importation and shipping documents can be found in U.S. District Court case files 12-CIV-4760 and 12-CR-00981, Southern District of New York.

159 "playing both sides" and other details on the conversations between Hollis Butts and Eric Prokopi: Interviews with Prokopi and Butts, backed up by emails obtained by the author.

160 Fossil cases: See Lynn Hollinger, "Two Amazing Seizures: CBP, ICE Seize 100 Million Year Old Fossils, Pre-historic Cultural Artifacts Repatriated to China," *Frontline*, U.S. Customs and Border Patrol, Winter 2010.

161 "Do you have more photos?": Papers of Eric Prokopi.

CHAPTER 15: THE LAST DINOSAUR

162 "I feel very sorry" and related: Papers of Eric Prokopi.

162 Details about the Eric Prokopi/Chris Moore visit to Mongolia are from interviews with Prokopi, the paleontologist Chultem "Otgo" Otgonjargal, and photographs provided by Prokopi.

163 Omnibus Public Land Management Act of 2009, containing the Paleontological Resources Preservation Act (PRPA): The Act passed the Senate on January 15, 2009, and was signed into law on March 30, as Public Law 111-11, Title VI, Subtitle D; 16 U.S.C. §§ 470aaa-470aaa-11. (See congress.gov.) This followed a May 2000 report, for the secretary of the interior, "Fossils on Federal & Indian Lands," which can be found at blm.gov. See also "Obama Signs the Omnibus Public Lands Management Act of 2009," *New York Times*; March 30, 2009. For more information, see the National Park Service's guidelines at nps.gov. Before the 2003 hearing on the PRPA the public had a chance to give opinions on the legislation. A summary of their comments can be found in "Fossils on Federal & Indian Lands." See also "Federal Register Notice 64 FR 27803-27804, May 21, 1999, and Public Meeting, June 21, 1999."

164 "scientific principles and expertise": The full text of Bennitt's open letter to Congress, dated January 15, 2009, can be found on the AAPS website, aaps.net.

165 "dangit": Interviews with Tyler Guynn.

166 "Here is our Tyrannosaurus": Everything Earth Facebook page.

166 "up to and including skulls": Interviews with Michael Triebold, president of Triebold Paleontology Inc., which is in Woodland Park, Colorado. I first interviewed him at the Denver show in 2012.

166 "a big, whole *T. rex*-y skeleton": Ibid.

166 "Luxury Market": See "Luxury Market for Dinosaur Remains Thrives," *Huffington Post*, November 10, 2011, updated on January 9, 2012.

166 The details of prepping and mounting the *T. bataar* come partly from interviews with Eric Prokopi and with paleontologists and other preppers.

166 "Tarbofootpathological": Papers of Eric Prokopi.

167 Basketball: This detail comes from Peter Larson's memoir, *Rex Appeal*.

168 "Buyer X": See case file CV07-00695, Mineralienzentrum/Andreas Guhr v. GeoDecor Inc./Thomas Lindgren, et al., U.S. District Court, Central District of California. The suit was settled in June 2007. According to court documents, Isadore Chait, for I.M. Chait, was the broker.

168 "fighting pair": Heritage Auctions still keeps information about this sale on its website, ha.com. See also Marice Richter, "Dinosaur Auction Features Fighting Pair of Skeletons," Reuters, June 12, 2011; and "Skeletons of fighting dinosaurs sell for $2.5m," *Telegraph*, June 13, 2011. For a profile of Heritage, see Steve Pate, "Rich Heritage," *D* magazine, November 2011. The magazine noted, "For years the top currency and coin house in the United States, Heritage has sold a number of coins for more than $1 million apiece—the most recent ($1.3 million) at an August coin auction in Chicago that brought in a total of $31.5 million. Last year, the company also became No. 1 in the world in sports memorabilia, with almost $13 million in sales. It does $30 million a year in comics and comic art alone. The house has sold $25 million in illustration and fine art just this year." CEO Steve Ivy said, "My partners and I all came up as collectors. We're all seat-of-the-pants entrepreneurs."

168 "impeccably preserved": See Wynne Parry, "Rare Tyrannosaurus Skeleton to Be Auctioned," Live Science, May 14, 2012.

169 "After discussing": Papers of Eric Prokopi.

169 You can see details about the "Fighting Dinosaurs: New Discoveries from Mongolia" exhibit at amnh.org. Also see John Noble Wilford, "Presenting a Fight to the Death from 80,000,000 B.C.," New York Times, May 19, 2000. The spectacular fossil has been repeatedly referenced. For instance, see R. Barsbold, "'The Fighting Dinosaurs': The Position of Their Bodies before and after Death," Paleontological Journal 50 (December 2016); Barsbold writes that the dinosaurs "resemble two small children playing on sand."

169 "More original, brand new": See "Archives of the DINOSAUR Mailing List," hosted at the Cleveland Museum of Natural History, http://dml.cmnh.org/about.html.

170 "Dinosaurs are the one thing": See Ann Marie Gardner, "Digging for Dinos in the Land of Genghis Khan," Discover, August 16, 2007.

170 "Smugglers have gotten so bold": Ibid.

170 "There's no control": Ibid.

171 "You have to take 'dinosaur' out of your name": Interviews with Bolor Minjin.

171 "She married an American": From Bolor and others in Mongolia and the United States.

171 "If we leave": Interviews with Bolor Minjin.

172 "like a service company" and related: Ibid.

172 nationalistic: Mongolia's politics appeared to be increasingly nationalistic. "Media reports and observer reports suggest a rising anti-foreigner sentiment among the public, mostly based on the idea of wanting Mongolian resources developed by Mongolians for the benefit of Mongolians," read one State Department report obtained through the Freedom of Information Act.

172 Oyungerel Tsedevdamba: Oyuna's books include the Mongolian bestsellers Nomadic Dialogues and Notes on My Study in America. In March 2008, she published an award-winning historical novel, The Green-Eyed Lama, coauthored with her husband, Jeffrey L. Falt, about the Mongolian purges of the 1930s. "The book lists the names of 614 victims from Khuvsgul province who were sentenced to death or 10-years imprisonment on just two days on the same burial location where the main characters of The Green Eyed Lama were purged," Oyuna writes on her website. "During the book tour…buyers of the book opened its last few pages first and searched the names of their own relatives. Often they found one. In Renchinlkhumbe soum, a reader found three family members killed on the same day." In a blurb, former President Elbegdorj said, "I hope that the world will see this book some day. Maybe then, the people will understand Mongolia's pains and struggles, and the Mongolians will be better understood by the world community."

173 "her well-earned success": See Georgia de Chamberet, "Dreaming of Outer Mongolia (1), An Editor's Odyssey," BookBlast Diary (blog), bookblast.com.

173 "Sister Oyuna": Interviews with Oyuna Tsedevdamba and Bolor Minjin.

173 "We are losing our heritage": Ibid.

173 "Give me something to read": Ibid. Oyuna later supplemented her reading by studying at Dinosaur National Monument in Utah.

173 "preaching": Interviews with Oyuna Tsedevdamba.

173 "Dinosaur's Dream": The story ran in Mongolian in *Ardchilal.* Oyuna kindly provided me with an English translation.

174 "Dinosaurs!": Interviews with Oyuna Tsedevdamba and Jeff Falt.

174 "You're talking to me about *dinosaurs?*": This account comes from Oyuna.

174 "urgent appeal": "President of Mongolia Is Concerned That T-Rex Skeleton May Belong to Mongolia," Office of the President of Mongolia, Public Relations & Communications Division, May 18, 2012.

175 "held the title": See Wynne Parry, "Tyrannosaur Skeleton for Sale, but Ownership Is Questioned," Live Science, May 18, 2012.

175 "No one knows": Ibid.

175 "That's like saying": Interviews with Jeff Falt.

CHAPTER 16: THE PRESIDENT'S PREDICAMENT

176 "proudly": See James Brooke, "The Saturday Profile; a Mongolian and His Nation, Evolving Together," *New York Times,* December 25, 2004.

176 Lviv, Ukraine: The city lay on the far west side of Ukraine, at the Polish border. The former Mongol conquest was now a "Little Paris." Historians were only just discovering the activities of the secret police in Lviv during World War II, when scores of Jews and Poles were deported, and worse. After the war, Lviv remained a part of the Soviet Union and became a sister city of Winnipeg, Canada, the immigration destination of the family Prokopi.

176 "Mr. Gorbachev": See the nine-second YouTube clip of Ronald Reagan saying the famous line at https://www.youtube.com/watch?v=7NjNL4Nsa4Q.

177 "this incredible clicking": See "Opening an Embassy in the Land of Genghis Khan," an oral history with Ambassador Joseph Lake, Association for Diplomatic Studies and Training, adst.org.

177 "truly progressed": Ibid.

177 "My perspective": Ibid.

177 "third neighbor": For more on what Baasanjav Ganbold, Mongolia's ambassador to the Republic of Korea, called an initial "rhetorical gesture to support Mongolia's first move toward democracy," see http://asiasociety.org/korea/mongolias-third-neighbor-foreign-policy.

177 "Mongolians seem remarkably free": Shapiro, "Starting from Scratch."

177 "teach Christianity": The South Dakotans were following the mandate of Matthew 28:19 (New International Version): "Therefore go and make disciples of all nations, baptizing them in the name of the Father and of the Son and of the Holy Spirit."

178 "Well, do you think this Jesus": Tom Terry, *Like an Eagle: 10 Years in Mongolia* (self- pub., 2014).

178 "lifted verbatim": Shapiro, "Starting from Scratch." As they worked to draft the constitution there "wasn't a single computer" to be found, one participant later said. Tsagaan, the future chief of staff for President Elbegdorj, had to use a typewriter for drafts and edits, later saying, "I think I read the draft constitution more than any other person in the country." Tsagaan went on to serve in various ministry posts including finance and education. Before becoming the president's chief of staff he worked in the private sector on renewable energy options for Mongolian nomads. See Meloney Lindberg, "A Conversation with Tsagaan Puntsag,

Chief of Staff of the President of Mongolia," Asia Foundation, May 21, 2014, https://asiafoundation.org/2014/05/21/a-conversation-with-tsagaan-puntsag -chief-of-staff-of-the-president-of-mongolia/.
178 "We, the people": See *The Constitutions of Mongolia: 1924, 1940, 1960, 1992.* You can find this online at Stanford Libraries. And 1992 is the one you want. Also consti tutionproject.org.
178 "His motive": Terry, *Like an Eagle.*
178 "further a better understanding": For more information on the AMONG Foundation, see the organization's website at amongfoundation.org.
178 "guaranteeing religious freedom": Terry, *Like an Eagle.* The result was a shaky facsimile of the U.S. Constitution, which reads, with unwavering clarity, "Congress shall make no law respecting an establishment of religion, or prohibiting the free exercise thereof; or abridging the freedom of speech, or of the press; or the right of the people peaceably to assemble, and to petition the Government for a redress of grievances."
178 Scourge: Christians had been trying to convert Mongolians for ages. As Rossabi wrote in *The Mongols*, in 1245, Pope Innocent the IV sent an emissary, John of Plano Carpini, to ask the Mongols to stop attacking Christians, and to convert. "John's mission did not produce the desired results," Rossabi wrote. John wrote a "condescending" report saying the Mongols didn't use tablecloths or napkins, "and did not wash their dishes, which he found uncouth and barbaric. He wrote that his Mongol hosts were quick to execute their own people or foreigners who poured out or wasted milk or food, or who killed young birds."
178 Details of the South Dakotans' activities in Mongolia come from Tom Terry's book *Like an Eagle*, but I also drew from English-language newspapers of the time.
179 "epidemic": Terry, *Like an Eagle.*
179 IRI: For an overview of the International Republican Institute's work in Mongolia see iri.org.
179 "Contract with America": By signing the "Contract," candidates promised, among other things, to build more prisons and deny welfare benefits to minors. The *New York Times* advised caution, calling the overall plan "not only reckless but deceptive" because the wealthy would win, the poor lose. Manifested, the "Contract" amounted to a "throwback" that "promised tax cuts and a balanced budget but led to huge deficits and financial scandals." The policies were "likely to usher in an era of greater political turbulence characterized by empty sloganeering, mean-spirited campaigning and the growth of local and national third parties." See John B. Judis, "The New Era of Instability," *New York Times*, November 10, 1994.
179 "conservative revolution": "The changes being debated in America now can provide useful lessons and insights to democratic societies throughout the world...," the conservative Heritage Foundation declared at the time.
179 "Contract with the Mongolian Voter": See Nate Thayer, "In Mongolia, a GOP-Style Revolutionary Movement," *Washington Post*, April 6, 1997.
179 "I read the Contract": Ibid.
180 "The merger, illegal and corrupt": See Michael Kohn, *Dateline Mongolia: An American Journalist in Nomad's Land* (Bandon, OR: RDR Books, 2006).
180 "probably the most important economist": See Peter Passell, "Dr. Jeffrey Sachs, Shock Therapist," *New York Times*, June 27, 1993.

180 "mess of every kind" and "one could go": PBS interviewed Sachs about Russia in the early 1990s on June 15, 2000. You can find the transcript at http://www.pbs .org/wgbh/commandingheights/shared/minitext/int_jeffreysachs.html.

181 "waiting for a major crisis": Naomi Klein, *The Shock Doctrine: The Rise of Disaster Capitalism*. Klein saw this happening around the world, in places like Sri Lanka and New Orleans, where the aftermath of Hurricane Katrina attracted a "nexus of Republican politicians, think tanks, and land developers" talking about "clean sheets and exciting opportunities."

181 "powerful ruling alliance": Ibid.

181 "Communists turned capitalists": Ibid.

181 "unfettered free markets": Ibid.

181 "conservative think tank school": See Jason DeParle, "Right-of-Center Guru Goes Wide with the Gospel of Small Government," *New York Times*, November 17, 2006. Elbegdorj befriended the Mackinac Center president Lawrence Reed, once called the "Johnny Appleseed" behind an extensive network of a "jagged, creative-destructive" brand of capitalism. Reed ran "a national back office of sorts, that allows even policy novices to produce abundant, salable fare," the *Times* reported. Mackinac had fostered the creation of countless state policy groups; Kentucky's largest newspaper called one such group a "conservative propaganda mill," saying its founder had attended Mackinac as a "sales executive with no public-policy background" and "left with access to everything from off-the-shelf speeches and papers to management software." At one point when Elbegdorj won a new seat in government, the Mackinac organization declared its former pupil's political success "another great victory for all of us who believe in limited government and a free society."

181 Robert Painter: Some of the early biographical information comes from interviews with Robert Painter.

181 "sweeping college campuses": See *Rush to Us: Americans Hail Rush Limbaugh*, by D. Howard King and Geoffrey Morris. (New York: Pinnacle, 1994).

182 "Best Western College" and "most bigoted school": See "Editorial: Princeton Says Baylor Is Bigoted," *Baylor Lariat*, February 1, 1996.

182 "There is seemingly": Ibid.

182 "Communist-sympathizer" and "AIDS Quilt": These pieces ran in January 1999, in vol. 1, no. 4, of the paper, archived in the Baylor library.

182 "*Libertas* looks like a newspaper": See "'Libertas' a newspaper in form, not function," unsigned editorial, *Baylor Lariat*, September 16, 1998. Baylor library archive.

182 "something admirable": See Alyson Ward, "'Libertas' abuses journalistic standards, rules," *Baylor Lariat*, February 2, 1999. Baylor library archive.

183 "conservative donors across the nation": Brittney Partridge, "'Libertas' expands distribution," *Baylor Lariat*, January 29, 1999. Painter told the *Lariat* that another group underwriting *Libertas* was the Intercollegiate Studies Institute, a Delaware nonprofit whose founders included William F. Buckley Jr., the conservative author and founder of *National Review* magazine. It was around that time that Baylor's College Republicans chapter held an officers election. Painter, who chaired the voting-credentials committee, angered about forty would-be voters by turning them away at the polls, saying they failed to meet membership requirements because they hadn't filled out the proper forms or paid their

one-dollar dues. The students were so upset that campus security and two administrators had to come out and monitor the election. One of the rebuffed students complained that "the people guarding the entrance believe we won't vote for the people they want." A freshman who observed the chaos said, "I am seriously concerned that a student organization whose basis is the political system cannot hold organized and fair elections."

183 Fulbright & Jaworski: The firm is now called Norton Rose Fulbright.

183 "selling politics to the average person": As the Bellwether members discussed what to teach, Jay Weimer, the future assistant U.S. attorney, said, "Your next question is 'how do we make sure we don't train the next Geraldine Ferraro?' . . . Same way we wound up with all conservatives in our Politics class—we advertise to conservatives, our speakers are all Republicans, word of mouth is that we are a conservative organization." Bellwether Forum's new website is bellwetherforum. com, and while the original site, bellwetherforum.org, is defunct, many of the old posts have been available via https://web.archive.org.

184 "What's *your* story?": Interviews with Robert Painter.

184 "You know, even though Mongolia is seven thousand miles away": This comes from a reprint of Elbegdorj's speech, preserved via the Internet Archive on an early iteration of the Bellwether Forum website. See Tsakhia Elbegdorj, "Mongolian Prime Minister's Remarks: The Power of Ideas and the Power of Hope," September 4, 2004. The speech was reprinted just after Elbegdorj became prime minister.

184 "could not be achieved in Asia": Ibid.

184 "Mongolians' continuing commitment": See U.S. Senate, "Mongolia and Burma," *Congressional Record* 150, 124 (Tuesday, October 5, 2014). McConnell read Elbegdorj's letter aloud, first saying, "While some may not pay much attention to Mongolia—it is literally half a world away—it deserves America's thanks and praise. That country serves to remind us that the fundamental pillars upon which our democracy is constructed—individual rights, freedom of the press, and religious tolerance—are not Western ideals but universal rights."

184 Eagle TV/Donald Rumsfeld: The Department of Defense by now contributed $18 million a year to Mongolia, most of it earmarked for upgrades to the army's equipment. Six U.S. Marines were advising the military, which had shrunk from 70,000 Soviet-era troops to roughly 11,000. Rumsfeld was there to rally support for the war in Iraq and Afghanistan, to which Mongolia had sent troops. When a journalist asked whether the United States planned to establish a military base or a listening post in Mongolia, Rumsfeld said, "We've had no discussions along that line, and I know of no intention to do that." The U.S. ambassador at the time, Pamela Slutz, later said Eagle TV was "a good partner of the U.S. embassy and its programs and policies in Mongolia. This was particularly true of its coverage of the global war against terrorism."

184 "poster child for democracy in Eurasia": "Mongolia's Democratic Identity," by John Tkacik, first appeared in the now defunct *Far Eastern Economic Review*, and was reprinted by the Heritage Foundation on June 21, 2005.

185 "never dared think of themselves as anything but real estate": Ibid.

185 "Americans and Mongolians": For the full speech, see "Remarks in Ulaanbaatar, Mongolia," November 21, 2005. Online by Gerhard Peters and John T. Woolley, *The American Presidency Project*, http://www.presidency.ucsb.edu/

ws/?pid=73752. News coverage included David E. Sanger, "In Mongolia, Bush Grateful for Iraq Help," *New York Times*, November 21, 2005. While the ongoing financial commitment confirmed close U.S.–Mongolian ties and underscored America's decades-long investment in nurturing democracy in the region, it was clear that Mongolia's national security primarily depended on "achieving a sustainable socioeconomic development with guarantee of human rights and freedom to civilians," one Mongolian scholar, R. Bold, wrote for the International Institute of Strategic Studies, a British think tank. "Security" was no longer defined by a country's military but rather by "its capacity to survive and compete."

185 "the godfathers of corruption" and "Mongolia's Thomas Jefferson": See "Mongolia's Former Communist Party MPRP Pulled the Rug under Elbegdorj's Government," prweb.com, January 13, 2006.

185 "It's kind of a coup d'état": Lulu Zhou, "Mongolian PM Out of Office," *Harvard Crimson*, January 20, 2006.

185 Bellwether training: Robert Painter used to tweet about Mongolian politics and Elbegdorj, though those tweets have since been deleted. His archived feed shows that deleted tweets include "Organizing a Bellwether Forum group of election observers for the Mongolian presidential election next month" (April 22, 2009) and "Congratulations to Mongolian President Tsakhia Elbegdorj on his reelection today" (June 27, 2013).

186 "voter ID laws": See "Embarrassed by Bad Laws," editorial, *New York Times*, April 16, 2012.

186 Koch brothers: The definitive source on the political activities of Charles and David Koch is *Dark Money: The Hidden History of the Billionaires Behind the Rise of the Radical Right*, by Jane Mayer.

186 Suspected assassination attempt: Bellwether's Mongolian training was offered in May 2007, in the run-up to the next year's parliamentary election, wherein Elbegdorj was expected to regain a seat and reassume the prime ministership. By now he was an English-speaking, Harvard-trained politician, heavily connected in the United States, with clear ambitions for the presidency. After the training, Elbegdorj treated the Americans to a trip to the Gobi. Not long after that, he was being driven to Kharkhorin, some two hundred miles west of Ulaanbaatar. Mongolian laws were always changing, depending on who was in power, and parliament recently had shortened the amount of time that a former prime minister could retain a government-paid security detail. That night in July, Elbegdorj was traveling with only his driver. A motorcycle appeared out of the darkness, forcing his SUV to swerve. The vehicle flipped, fatally ejecting Elbegdorj's driver and leaving Elbegdorj dangling upside down by his seatbelt, with a head injury. Elbegdorj had instructed his staff to call two people if anything ever happened to him. One was Robert Painter. When Painter's phone rang in Houston, his first thought upon hearing the news about Elbegdorj was of the day Reagan had been shot; the president had wisecracked to his surgical team, "Please tell me you're all Republicans." (Reagan has been quoted as saying, "Please tell me you're a Republican," and "I hope you're a Republican," and "I hope you're all Republicans." I'm quoting the version used in the widely acclaimed biography *President Reagan: The Role of a Lifetime*, by Lou Cannon (New York: Touchstone/Simon & Schuster, 1991). Cannon was quoting *The Master of the Game:*

Paul Nitze and the Nuclear Peace, by Strobe Talbott (New York: Knopf, September 1988).) Painter's second thought was that Elbegdorj was in danger of being poisoned or killed in the MPRP-controlled hospital. Working the phones, he arranged for a medical evacuation to Seoul, where Elbegdorj quickly recovered and was soon en route to Virginia to keynote a libertarian conference. Painter met him in Washington, DC, and they spent a few days together, eating at the Four Seasons and watching rented hotel movies, Painter marveling at the intensity with which Elbegdorj watched *The Bourne Supremacy*. Painter told me they also discussed Elbegdorj running for president.

186   Riots: Elbegdorj won a new seat in parliament on June 29, 2008, but his party, the Democrats, lost. International election observers called the election clean, but Elbegdorj and others argued that it had been rigged, methods of which in Mongolia were known to include bribery, vote-buying, fake IDs, multiple voter registration, and vote-counting shenanigans, exacerbated by a recently passed law that disadvantaged smaller, poorer political parties. Ever since the democratic transition, all parties had alleged election fraud, so the charges weren't unusual; yet as bad as it was, the corruption was actually getting worse, the State Department found. A riot unfolded in Sukhbaatar Square on July 1. One witness later told Amnesty International, "The people's anger had been accumulating for many years, and it just exploded." Eagle TV broadcast live. Tom Terry, the station director, would always ask himself whether he'd contributed to the mayhem by broadcasting the riot; one local newspaper outright faulted the station: "Eagle TV fully accomplished its goal to urge the public for violence...There is evidence that there are Black Powers (foreign investment) who are interfering in our country's political life." Journalists were forced to hand over their footage— then, during the declared state of emergency, the National Police Agency played the footage on state television, overlaying narration that falsely attributed the violence to certain people in the crowd, including reporters, reported Amnesty International. At least four people were shot and killed. The mother of one said, "The Mongolian news said that those shot were instigators and criminals but they were just onlookers at the wrong place at the wrong time." See "'Where Should I Go from Here?': The Legacy of the 1 July 2008 Riot in Mongolia," Amnesty International, March 30, 2009.

186   "system whereby a small elite": See Stephen Noerper, "Mongolia Matters," Brookings Institution, October 8, 2007.

186   Millennium Challenge Corporation: Poor countries were eligible for millions of dollars in grants if they met certain requirements, involving, for instance, government transparency. President Enkhbayar traveled to Washington to sign the compact as well as another agreement, on preventing trafficking in nuclear technology.

187   "tens of billions": "Mongolian Parliament Passes Long Awaited, Absolutely Essential Mining Deal," a cable from the U.S. embassy in Ulaanbaatar to the State Department in Washington, August 26, 2009, obtained through the Freedom of Information Act. A day earlier, parliament had amended laws, allowing the government to execute the much-anticipated agreement for Oyu Tolgoi, the "world class," "titanic" copper and gold mine in the Gobi. The deal was made despite "long delays stemming from fears of the loss of sovereignty and resource nationalism," the embassy noted, and it represented "a success for U.S. commercial

diplomacy and has cemented the U.S.'s position among the key players in crafting these sorts of agreements in Mongolia. Other donors are generally positive but lingering concerns remain that the GOM [Government of Mongolia] gave away too much in some aspects of the deal." Over the next forty years, Oyu Tolgoi was "conservatively" expected to yield "44 million tons of copper and 1,800 tons of gold. At current market prices that equals $264 billion and $55 billion respectively." In a partly redacted section titled "Company perspectives on the deal," the cable noted, "The deal is not perfect, but we share the opinion of most observers that it balances state and private interest in a way that allows all sides to claim victory. One of the key accomplishments is that the deal will set a pattern for all sides to follow in similar projects, the next one being the Tavan Tolgoi coal project, in which we have both commercial and policy interests." The cable in which this information was found was labeled "unclassified but sensitive."

187   "a stable, middle-class society": See Jonathan Kaiman, "Mongolia's New Wealth and Rising Corruption Is Tearing the Nation Apart," *Guardian*, June 27, 2012. A cable from the U.S. embassy in Ulaanbaatar to the State Department, obtained through the Freedom of Information Act, noted, "Firms remain highly critical of the Government of Mongolia's tendency, in recent years, to sometimes be arbitrary, un-transparent, or inconsistent with respect to the implementation of its mining laws."

187   "where an oil boom": Kaiman, "Mongolia's New Wealth." Mongolia had leverage. One Beijing-based mining executive told the *New York Times*, "Mongolia is without a doubt getting more respect from world powers, because it's got something everybody needs." But then China had leverage, too, as did Russia, as did the United States. One Mongolian businessman told the press, "If China closed the borders, we would starve to death." Also see Alicia Campi, "Mongolia's Quest to Balance Human Development in its Booming Mineral-Based Economy," Brookings Institution, January 10, 2012.

187   "abruptly pulled the plug": See Katusa, *The Colder War*. Katusa wrote, "The ostensible reason, as he described it, was that the nuclear waste of other countries is a 'snake grown up in another body.... Receiving back the nuclear waste after exploiting and exporting uranium must not be, as I think this is, a pressure from foreign superpowers...' Perhaps Elbegdorj's concerns were genuine. But many would see the hand of Vladimir Putin once more at work."

187   "Maybe if we caused problems": See Andrew Higgins, "In Mongolia, Lessons for Obama from Genghis Khan," *Washington Post*, June 15, 2011.

188   "Bellwether Mongolia invests in mining": bellwethermongolia.com.

188   Peabody Energy: On November 2, 2007, officials of the U.S. embassy in Ulaanbaatar accompanied Peabody Energy geologists on a helicopter trip deep into the Gobi to "reconnoiter the Little Tavan Tolgoi (LTT) coal mine, a rudimentary facility that... is part of the larger Tavan Tolgoi deposit, which Peabody hopes to develop in partnership with the Mongolian Government," the embassy reported in a "sensitive but unclassified" cable to Washington, obtained through the Freedom of Information Act. "Peabody's specialists noted that everything from roads to power, and water to food, would have to be built from scratch or transported in—an expensive and time-consuming proposition." LTT was owned by the provincial government and a private consortium led by Bat-Erdene, a

Democratic member of parliament, the cable reported, saying, "The managers declined to explain how a contract with a government entity was awarded to one of the private holders of LTT." Tavan Tolgoi was the one commercial project in Mongolia "in which the U.S. has a dog in the hunt," one source told the *New York Times*. "What the Americans are saying is, 'We can't be your best friend and primary third neighbor if all the goodies go to China.'" See Dan Levin, "In Mongolia, a New, Penned-in Wealth," *New York Times*, June 26, 2012.

188   "the latest step": Levin, "In Mongolia, a New, Penned-in Wealth."

188   "What's next?" and related: Interviews with Robert Painter.

188   Voting machines: For general information, see Jill Levoy, "The Computer Scientist Who Prefers Paper," *Atlantic*; December 2017. (The computer scientist Barbara Simons believes the only truly safe voting technology is paper.) Bev Harris's website, where she sought to document electronic-voting malfunctions, is blackboxvoting.org. There, you can download, for free, her book *Black Box Voting: Ballot-Tampering in the 21st Century*. "We are not talking about a few minor glitches," she wrote. "These are real miscounts by voting machines, which took place in real elections. Almost all of them were caused by incorrect programming, whether by accident or by design. And if you run into anyone who thinks we are hallucinating these problems, hand them the footnote section, so they can examine sources and look them up themselves." The issue calls to mind former Louisiana governor Earl Long, who in the 1950s infamously said, "Gimme five [electoral] commissioners and I'll make them voting machines sing 'Home Sweet Home.'"

189   "Our entire governing system": Harris, *Black Box Voting*.

189   "political or quasi-political capacity": The Foreign Agents Registration Act of 1938 requires agents representing the governmental and quasi-governmental interests of foreign powers to make themselves and their activities known to the federal government and the American people. The State Department administered the law before passing it on, in 1942, to the counterespionage unit of the National Security Division of the U.S. Department of Justice. See fara.gov.

189   "I thought it would be good for the country": Interviews with Robert Painter.

190   "naked attempt" and "vanquish a political foe": Dan Levin, "Ex-Leader's Detention Tests Mongolia's Budding Democracy," *New York Times*, May 13, 2012. By 2012 the MPRP had gone back to its 1921 name, the Mongolian People's Party; in typical Mongolian confusion, Enkhbayar had started a splinter party called the MPRP.

190   "Mr. Enkhbayar had released internal government documents" and related: See "Steppe in an Ugly Direction," *Economist*, April 28, 2012. The magazine reported, "The raid has aroused two reactions in the public. One is surprise at the high degree of force deployed in the raid and the low degree of decorum offered to a former leader...The other is the belief that the case has more to do with rabidly partisan politics than corruption."

190   "rather picayune": Kaiman, "Mongolia's New Wealth." An independent observer, Jim Hodes, a former war crimes prosecutor based in Atlanta, agreed that the charges seemed overblown, at one point telling the press, "What I see are huge violations, fundamental violations of his right to a fair trial."

190   "Mongolia...has been widely lauded": Levin, "Ex-Leader's Detention Tests Mongolia's Budding Democracy."

190 "Everyone had hoped": Ibid. Illegal detainments and unfair trials were far too common in Mongolia, "including those that used confessions extracted through torture as evidence," Amnesty International reported.

190 Bat Khurts: In the long-unsolved murder of Zorig, the government charged Enkhbat Damiran, an ex-con in his early forties with small eyes and bushy eyebrows. Enkhbat had served prison time for theft, assault, and fraud, and had been released early for bad health shortly before the murder. By 2003, he was thought to be living in a hotel in Caen, Normandy, where he had applied for asylum under an assumed name. One May afternoon, he went up the coast to Le Havre, where he planned to meet a fellow Mongolian, a female dissident, at McDonald's. Four men jumped him in the parking lot, dragged him into a car, smuggled him into Belgium in a Mongolian embassy vehicle, then drove him to Berlin, where he was tortured. Then he was drugged and loaded by wheelchair onto a MIAT flight to Mongolia; his captors explained his unconsciousness by describing him as a government minister who had "gotten into a fight in Brussels and urgently needed to be brought back home."

Enkhbat's torture continued at a secret location in Ulaanbaatar. Enkhbat later said his interrogators tried to bribe him to confess. He did not confess when they shined bright lights in his eyes, or liver-punched him with a pistol, or cocked what appeared to be a loaded gun, held it to his head, and pulled the trigger. Later, once he was in prison, his lawyer managed to sneak a video recorder inside and tape Enkhbat describing the way he'd been treated; after the lawyer paid a TV station to air the interview, both the lawyer and his client were charged with revealing state secrets. Both went to prison, Enkhbat at a "strict regime" facility northwest of Ulaanbaatar. Manfred Nowak, the UN special rapporteur on torture, visited him there in early June 2005. By then, the murder charges had been dropped "as they obviously had been fabricated," Nowak wrote, but now Enkhbat was suffering untreated health problems. Everything that had happened to him had violated international human rights law, Nowak reported.

The "mastermind" of his kidnapping was said to be Bat Khurts, a GIA operative in his early forties. The son of a prominent architect, he was married with three children and a fourth child on the way. His precise position within Mongolian intelligence at the time of Enkhbat's kidnapping never quite became clear—he was variously identified as a GIA agent, the first secretary of the Mongolian embassy in Budapest, the chief executive of Mongolia's National Security Council, and "chief spymaster." After a German investigation determined that Khurts drove the car that carried Enkhbat across Europe, the Germans charged Khurts with kidnapping and unlawful imprisonment. Investigators issued a European warrant for Khurts's arrest, which allowed law enforcement in any country of the European Union to pick him up if they found him within their borders. Khurts reportedly never knew about the warrant. In 2010, as Mongolia and Britain talked about supporting each other in anti-terrorism efforts, the British ambassador in Ulaanbaatar, William Dickson, arranged a London meeting between British national security officials and Khurts, who by that time headed the GIA. Khurts arrived on September 17, expecting to participate in talks. "He believed that he would meet senior Downing Street figures as part of preparations for the Mongolian President's visit to London to meet David

Cameron," the *Times of London* reported. When Khurts deplaned at Heathrow Airport, he found Scotland Yard detectives waiting with handcuffs. They took him to Her Majesty's Prison Wandsworth, in south London, the largest men's prison in the United Kingdom, where he remained for ten months as the courts argued over whether he could be extradited to Germany.

Khurts's defense against extradition consisted primarily of two arguments. First, he'd been set up by the Brits, who had surely "lured" him to London in order to arrest him under the pretext of high-level national security talks on "a new era of intelligence cooperation relating to Muslim fundamentalism." Although the British government insisted that Khurts hadn't been baited, the practice—*mala captus bene detentus*—"the court once in possession of the accused has jurisdiction over the person and all that is required is a fair trial"—was known to happen. "At times governments use deceit, fraud, and tricks to lure individuals from the country of their residence to a location where there is jurisdiction to arrest them," Bruce Zagaris wrote in *International White Collar Crime: Cases and Materials.* "Unlike in an abduction by force, weapons are not used to get the suspect to the location where the arrest will occur. Whether a trick can be just as coercive as a gun remains controversial." Khurts's second defense was that as a government agent running a "secret mission" for his country he had diplomatic immunity. In one court appearance, wearing a gray sweatshirt and jeans, and blowing kisses to relatives, he spoke in translated Russian when he called the case against him an "insult"—he'd been invited to London only to be betrayed.

The Brits argued that the Foreign Office never set up "formal" meetings with Khurts and that he'd been arrested merely because law enforcement authorities had learned of the German warrant and were obligated to detain him. It was true that it made no sense for Great Britain to antagonize Mongolia. The two had been Cold War adversaries but now were allies in Afghanistan, and Britain reportedly wanted a piece of Mongolia's promising mining sector as well as the contract to build and run its new stock exchange. Khurts wasn't high ranking enough to qualify for diplomatic immunity—he was a civil servant, not a minister or head of state, a German representative told the court when arguing for extradition. As everyone awaited the court's decision, the Mongolian government described Khurts's target, Enkhbat, as a "dangerous career criminal who organized gangs of women shoplifters in France, Belgium, the Netherlands, Germany, and Eastern Europe" and claimed he'd controlled the women with "rape and torture." Enkhbat's family insisted he was simply a refugee. Ambassador Dickson, meanwhile, was recalled from Mongolia to London, for "operational reasons," and soon retired.

Khurts was ordered extradited. He appealed and lost. His Mongolian supporters demonstrated outside the British embassy in Ulaanbaatar, demanding that "London stop its intervention in the internal affairs of other countries." But Mongolia ultimately acknowledged the illegality of the kidnapping in Germany and apologized to that country, France, and England. Khurts was extradited to Germany on August 19, 2011; afterward, his relatives complained that British law enforcement had failed to return some of his personal belongings, including a "Vacheron Constantin watch worth 28,000 Euros, his Versace glasses, and Montblanc pens."

The Germans scheduled Khurts's trial for October 24, 2011. The Mongolian government had hired two German lawyers to represent him as they figured out how to navigate a precarious diplomatic situation. Britain and Germany were both interested in Mongolia for its strategic location and its untapped mineral wealth and rare earths; Mongolia wanted good relations with friendly superpowers as well as help developing its natural resources. In fact, German Chancellor Angela Merkel was scheduled to make an historic trip to Mongolia in early October. She would be the first German government leader to make an official trip there and the first G7 leader to visit since George Bush in 2005.

Just ahead of Merkel's state visit, Germany released Khurts without explanation. Nothing more was said on the matter. Fifteen days after he returned home, Merkel was welcomed to Ulaanbaatar, where she signed a trade agreement with Mongolia, whose president, by now, was Elbegdorj.

Elbegdorj made Khurts vice chairman of the new Independent Authority Against Corruption, an agency described as having "very broad powers to question, investigate, and formulate charges." And just after former President Enkhbayar's sentencing on corruption charges, Khurts was awarded the Sukhbaatar Medal, one of Mongolia's highest honors.

Before long, Khurts was working on a mining case involving an Australian lawyer, Sarah Armstrong, chief counsel of a Rio Tinto subsidiary, who was stopped from boarding a flight to Hong Kong and banned from leaving the country as the IAAC questioned her. One source told the *Sydney Morning Herald* that Armstrong was questioned in "retaliation" for accusing government officials of corruption. Friends told *The Australian* they feared she was being used as a "political football to gain leverage over the company in the nation's notoriously corrupt business environment and to appease rising xenophobia." The lawyer, the press noted, faced "one of Mongolia's most feared investigators"—Khurts. That was in 2012, the year the *T. bataar* skeleton went to auction in New York and the Mongolian government started wondering if they could have Eric Prokopi brought back to Ulaanbaatar for questioning. For more information, see *Federal Court of Justice Germany v. Bat Khurts*, Oxford Public International Law, http://opil.ouplaw.com/view/10.1093/law:ildc/1779uk11 .case.1/law-ildc-1779uk11. The case was widely covered in newspapers including the *Financial Times*.

190 Zorig's murder: Sanjaasuren Zorig, by trade a political science professor who had studied in Moscow, was the best known of the democratic revolutionaries, and was considered a national hero. A moderate, he had won successive seats in parliament alongside Elbegdorj, and in the autumn of 1998 was the compromise choice for prime minister. Inflation was finally down and the economy was expected to grow, with steady global demand for exports like cashmere and gold. Chinese traders were selling cheap goods all over the place, and the Chinese government was buying half of Mongolia's copper. Datacom had become the nation's first internet service provider, allowing for the proliferation of internet cafés. Those with money were buying their first personal computers. Tourism looked promising—international "jet setters" liked to hunt gazelles and wild boar, while others preferred theme camps like "Be Genghis Khan for a Day," the *Asian Wall Street Journal* reported. Mongolia's

airline, MIAT, served meat-and-butter sandwiches packaged with a personal bottle of Mongolian vodka. The pace of free-market reforms worried Zorig, who feared that moving too fast would create a wealth gap that would be hard to walk back, with many Mongolians pushed below the poverty line. Zorig was thirty-six, with thick dark hair, dimples, and round eyeglasses. His recent ministry work had involved vetting multimillion-dollar construction contracts. The British ambassador called him "Mr. Clean." As prime minister, Zorig was expected to spearhead the efforts against public corruption. The announcement of his new position was scheduled for Monday, October 5. Three days earlier, he stayed late at his Government House office, playing chess with a friend. He was driven home around ten, to the apartment where he lived with his wife, Bulgan, down the street from the Russian embassy. Waiting for him inside his fourth-floor apartment were two assailants armed with knives. The details of Zorig's murder were everywhere and nowhere and all over the place, and it's still hard to know what happened; by one account, the assailants were a man and a woman in masks who had bound and gagged Bulgan on the bathroom floor and then waited in the dark for Zorig to get home. They stabbed him sixteen times, three through the heart. They stole a pair of gold earrings and, bizarrely, a bottle of soy sauce and a bottle of vinegar, though other reports said they also took three rings and five silver cups. Almost everyone attributed Zorig's death to a contract killing. Politics, not robbery, was the motive. Theories included "bloodyminded communists, corrupt democrats, the Russian mafia...and fascist extremists," wrote Michael Kohn, an American journalist living in Ulaanbaatar who reported on the aftermath. Mourners packed Sukhbaatar Square for vigils, spelling out Z O R I G in candlelight. In the press his death was compared to the assassination of John F. Kennedy. Some of the anger was turned at members of Zorig's own party. "Newspapers were brimming with allegations about the Zorig murder and corruption within the Democratic ranks," Kohn wrote. "It was easy to see why the Democrats were so loathed. They lived luxurious lifestyles; driving around in expensive SUVs, chatting on mobile phones (a privilege at the time) and throwing away thousands of dollars at the country's only gaming place, Casinos Mongolia. No one believed they were spending their own salary, which amounted to just $100 per month." See also Erik Eckholm, "A Gentle Hero Dies, and Mongolia's Innocence, Too," *New York Times*, October 25, 1998.

190   "demise of democracy": Addleton, *Mongolia and the United States*.

190   "Their task was to make me a criminal": Dan Levin, "Mongolian Ex-President Denounces Timing of His Graft Trial," *New York Times*, May 23, 2012.

191   "pawn in a global game": Ibid. Addleton, the former ambassador, also wrote that a good many Mongolians wanted outsiders to stay out of it: "It was precisely this unhealthy mix of wealth, economic interests, connections, access, and political power that, increasingly, seemed to be undermining democracy in Mongolia." See Addleton, *Mongolia and the United States*.

191   "This is really a case": Levin, "Mongolian" Ex-President.

191   "It has been deeply troubling": Feinstein's full statement was published in "Sen. Feinstein Warns of Political Crackdown in Mongolia ahead of June Elections," by Julian Pecquet, *The Hill*, May 14, 2012.

191 "Win *T. bataar*": Interviews with Oyuna; she also told the story in the Mongolian media.

191 "Assess the big picture": Interviews with Robert Painter.

191 Temporary restraining order application: Mongolia provided Robert Painter with an English translation of Mongolian laws. See "Plaintiff's Original Petition and Application for Injunctive Relief," Elbegdorj Tsakhia, as President of Mongolia, v. Heritage Auctions Inc., DC-12-05591-H, District Court of Dallas County, Texas, May 18, 2012. The application for a temporary restraining order included legal translations including, "Any territory and underground (cq) containing items of historical, cultural, or scientific value shall be protected by the State and any findings are the property of the State." The document also cited Mongolian criminal law describing the smuggling of "prohibited or restricted goods" as punishable by a fine equal to 51 to 151 times the minimum salary; 251 to 500 hours of a "forced job;" or up to six months in jail. A second offense could bring up to eight years in prison.

192 "found in Mongolia": "Waiting to Be Snapped up, T-Rex's Cousin: Near Perfect Dinosaur Skeleton to Go Under the Hammer," *Daily Mail (U.K.)*, May 16, 2012.

192 Carlos Cortez: The judge left public office in January 2015. A *Dallas Morning News* investigation had found that he had "long been plagued by legal crises of his own making," including a drunk-driving charge and allegations that he "used cocaine, paid for sex, choked a woman who accused him of fathering her son, and sexually assaulted a little girl," the *Morning News* reported. Police were investigating allegations that he had raped a woman at his condo, and the state ethics commission was looking into his practices. The newspaper investigation had found that Cortez had "used campaign contributions to repay himself for about $15,000 in unsubstantiated travel expenses" to a Canadian ski resort. Cortez had also been arrested "at least three times," the paper reported, noting that the cases had been dismissed. For more information, see Brooks Egerton, "Former Dallas Judge Carlos Cortez Could Face Financial Misconduct Prosecution," *Dallas Morning News*, March 2015. See also Brooks Egerton and Matthew Watkins, "Files Detail Allegations That Dallas Judge Used Coke, Bought Sex, Raped Young Girl," *Dallas Morning News*, April 2014.

192 "The sale of this next lot": The details of the moment of the *T. bataar* sale come largely from an evidentiary video shot by Painter Law Firm staffer Andrew King. The exclusive footage was provided by Robert Painter.

CHAPTER 17: *UNITED STATES OF AMERICA V. ONE TYRANNOSAURUS BATAAR SKELETON*

194 "the English crown": See Sarah Stillman, "Taken," *New Yorker*; August 12 & 19, 2013. In that passage Stillman cites the scholars Eric Blumenson and Eva Nilsen.

194 "unreasonable searches and seizures": See "The Bill of Rights: A Transcription," National Archives, archives.gov.

194 "Congress soon authorized": Stillman, "Taken."

194 Asset forfeiture: See "Audit of the Assets Forfeiture Fund and Seized Asset Deposit Fund Annual Financial Statements," Office of the Inspector General, U.S. Department of Justice; this audit is done every fiscal year. See also Peter

Lattman, "For U.S. Attorney's Office, Forfeiture from Crimes Pays (Sometimes in Dinosaur Bones)," *New York Times*, January 1, 2013.

194 *United States of America v. Approximately 64,695 Pounds of Shark Fins*: U.S. District Court case 03-CV-0594, U.S. District Court, Southern District of California, January 19, 2005.

195 "taking the profit out of crime": "U.S. Restrains 1909 Pablo Picasso Painting Valued at $11.5 Million," U.S. Department of Justice, June 24, 2013.

195 "Asset forfeiture is an important part of the culture here": Lattman, "For U.S. Attorney's Office."

196 *Portrait of Wally*: For more information, see Judith H. Dobrzynski, "The Zealous Collector—A Special Report," *New York Times*, December 24, 1997. Also "Strategy in Schiele Art Case Questioned," also by Dobrzynski, *New York Times*, October 12, 1999. Also Nicolas Rapold, "The Multidimensional Fate of a 1912 Schiele Portrait," *New York Times*, May 10, 2012.

196 "rocked": John Anderson, "Portrait of Wally," *Variety*, April 28, 2012.

196 "correct a historical injustice": See Ben Protess, "Sharon Levin to Leave U.S. Attorney's Office for WilmerHale, Joining Other Ex-Prosecutors," *New York Times*, April 26, 2015.

196 "most accomplished female attorneys": "The National Law Journal's Outstanding Women Lawyers," *National Law Journal*, May 4, 2015.

196 "Want some help?": Interviews with Robert Painter and Sharon Cohen Levin.

196 "This is a victory": Elbegdorj's press release has since been removed from his former office's website, president.mn, but an Internet Archive copy can be found. See "President of Mongolia Is Concerned That T-Rex Skeleton May Belong to Mongolia," May 18, 2012.

196 "legal action": The letter from Elbegdorj to Sharon Cohen Levin and other case documents can be found in 12-CIV-4760, U.S. District Court, Southern District of New York. Heritage sent investigators documents including "Entry/Immediate Delivery," U.S. Department of Homeland Security, Bureau of Customs and Border Protection, showing the importation of "fossil specimens" by Prokopi, from Museum Imports Co. Ltd., in Saitama, Japan, dated March 15, 2007; an invoice from Chris Moore Fossils, dated March 12, 2010, for three crates of unprepared "fossil reptile heads" and other materials, along with import documents; and the Heritage Auctions "Consignment Agreement: Natural History," for "Complete Tyrannosaurid, Res $850,000," March 22, 2012. David Herskowitz signed for the item, writing out a note that said Heritage would temporarily "insure this tyrannosaurs [*sic*] skeleton" and that "50% of the proceeds...will be made out to Chris Moore" and the rest to Eric Prokopi.

197 "Steven Spielberg's dinosaur advisor" and related: These details come from an interview with Don Lessem and emails obtained by the author.

198 "Although I am sure": Various Eric Prokopi emails can be found in 12CRIM981, U.S. District Court, Southern District of New York.

198 "We do have documents": Eric Prokopi email, dated May 28, 2012.

200 "fairly light": The descriptions and written reports of Philip Currie, Bolor Minjin, Mark Norell, and Tsogtbaatar can be found in federal case files including 12-CIV-4760, U.S. District Court, Southern District of New York. *USA v. One Tyrannosaurus Bataar Skeleton* cited the 1924 Mongolian Rules to Protect the

Antiquities as covering "remnants of ancient plants and animals"; the 2001 Mongolian Protection of Cultural Heritage Law, which covered the "territory and land bowels where historically, culturally, and scientifically significant objects exist;" and the 2002 Criminal Code of Mongolia, which established punishment for smuggling "valuable findings of ancient animals and plants, archeological and paleontological findings and artifacts." These would be used by U.S. prosecutors to establish a criminal case for trafficking in materials known to have been taken illegally from another country.

200 "several misstatements": See 12-CIV-4760, U.S. District Court, Southern District of New York.

200 "2 large rough (unprepared) fossil reptile heads": Ibid.

200 "You couldn't be in better hands": Email from David Herskowitz to Eric Prokopi, May 26, 2012.

201 "white knight": Interviews with Eric Prokopi, Peter Tompa, and Michael McCullough.

201 A year of his life: See "Claim of Interest in the Defendant *in Rem*," *USA v. One Tyrannosaurus Bataar Skeleton*," 12-CIV-4760, U.S. District Court, Southern District of New York. "A year" wasn't true; Prokopi had *owned* the bones for some time, but he had spent several months at most working feverishly on the skeleton. As the case lawyers came to realize the skeleton had been crafted with parts from different specimens, Sharon Levin asked Michael McCullough, one of Prokopi's lawyers, whether it contained any *T. rex* parts. McCullough told her correctly that it didn't.

201 "In her mind" and related: Interview with Greg Rohan. Rohan worked for Heritage founders and collectors (and former archrivals) Steve Ivy and James Halperin, a Massachusetts polymath who had dropped out of Harvard in the early 1970s to deal full time in rare stamps and coins. Like a lot of collectors, Rohan also had started young. (Someone once asked Ivey, "What were you doing before you were in the coin business?" Ivey replied, "I was in the third grade.") Heritage had expanded its business the way collectors often broaden their interests. Ivey wasn't just a coin collector, he also collected Texas historical documents. In addition to stamps, Halperin collected million-dollar comic books and Thomas Moran paintings. Rohan collected not just coins but also American Indian textiles.

202 "for sure" and related: Eric Prokopi press statement, via Peter Tompa and Michael McCullough.

202 "fossil collecting is well established": See "Statement of Facts," 12-CIV-4760, U.S. District Court, Southern District of New York.

202 Thomas Jefferson was an avid fossil collector, as was, to a lesser extent, George Washington. In 1780, when Jefferson was governor of Virginia, the French sent a questionnaire to each of the American colonies, whose independence it was backing. Each was to provide all kind of data, from "limits & boundaries" and "A notice of the best seaports" to "The particular customs & manners" of the people. Jefferson was thirty-eight, five years away from succeeding Benjamin Franklin as the U.S. minister to France and twenty years away from the presidency. In the past five years, his daughter, his son, and his mother all had died, and he was soon to lose two more daughters and his wife. The American Revolution was ongoing. He'd had a bad fall off a horse, and had been all but run

out of Richmond and Monticello. The questionnaire may have been a welcome distraction, and the task perfectly dovetailed with his detail-oriented tendencies. In addition to Virginia's geographic boundaries, seaports, and mannerisms, he looked at its mountains, caves, climate, soils, springs, military, colleges, religion, and commerce.

He also took on the French naturalist Georges-Louis Leclerc, count de Buffon, a keeper of the royal cabinet, who had gotten everyone's attention with a theory of "American degeneracy." Buffon had never set foot in the New World, but after reading the accounts of early explorers he had decided the continent's life forms and natural prospects were deeply inferior to those of his own. In the New World, "animated Nature is weaker, less active, and more circumscribed in the variety of her productions," he had written in 1761. The reptiles and insects were okay, but overall the number of species was fewer, and the species that did exist were much smaller than those found elsewhere. No New World animal could compete with the elephant, the rhinoceros, the hippopotamus, the dromedary, the lion, the tiger. Buffon had spared not even the humans, writing, "In the savage, the organs of generation are small and feeble. He has no hair, no beard, no ardour for the female. Though nimbler than the European, because more accustomed to running, his strength is not so great. His sensations are less acute; and yet he is more timid and cowardly. He has no vivacity, no activity of mind." If you were thinking of moving to America, you'd be wise to know that you couldn't possibly last in an environment too frigid and damp to produce or sustain robust life. Too bad America was not Senegal, a place of "perfectly scorching" sun; or Peru, where "an agreeable temperature prevails." North America—so uncultivated, so wild, so full of all those super-wet rivers and lakes—was too saturated and tangled to be of substance. "As the earth is every where covered with trees, shrubs, and gross herbage, it never dries," Buffon had written, saying, "The transpiration of so many vegetables, pressed close together, produce immense quantities of noise and noxious exhalations. In these melancholy regions, Nature remains concealed under her old garments, and never exhibits herself in fresh attire; being neither cherished nor cultivated by man, she never opens her fruitful and beneficent womb."

If Buffon had had no audience, his diss might have passed without consequence; but he happened to be a bestselling author whose readers comprised the upper classes of Europe. The negative assessment of the New World, whose colonial leaders would soon go to war for their independence, was a sort of thrown glove. In her soiled undergarments, amid chilly drear, America noxiously exhaled an eventual response, from no less than the eventual author of the Declaration of Independence: Jefferson answered Buffon's claims one by one. To the question of America's embarrassing poet gap, he wrote, "When we shall have existed as a people as long as the Greeks did before they produced a Homer" (or the English a Shakespeare, or the French a Voltaire)—well, *then* they could talk. Re: the observation about the absence of "one able mathematician, one man of genius in a single art or a single science," Jefferson noted that in war, America had produced a Washington, and in astronomy a Rittenhouse, and in physics a Franklin, adding, of his burgeoning nation, "We...have reason to believe she can produce her full quota of genius." To the question of physical robustness, he moved to weights and measures, making a size chart of animals common to both continents, right down to the shrew mouse. It was ludicrous to suggest that Nature favored one side of the globe over the other—"as if both sides were not warmed by the same genial sun…" Jefferson's report became *Notes on the State of Virginia,* his only book. In a much-repeated anecdote, recounted in a letter by Jefferson dated December 4, 1818, Ben Franklin was said to have taken a more direct approach to rebuffing the baron: at a dinner party in Paris where half the guests were American and half were French, the writer Guillaume-Thomas Raynal "got on his favourite theory of the degeneracy of animals, and even of man, in America." Franklin, noticing the "accidental stature" of each side of the table—the French happened to be seated opposite the Americans—said, "Let both parties rise, and we will see on which side nature has degenerated." *Voilà qui est réglé*—that settled that.

202 "openly sold": In *USA v. One Tyrannosaurus Bataar Skeleton,* Prokopi made various arguments for being awarded the dinosaur: The government's actions were unfairly based upon "technical violations" he hadn't been aware of and "upon foreign laws that are largely unavailable to American citizens and are unclear in both their meaning and application"; Mongolia's laws on private property were murky or contradictory. Mongolia's cabinet minister for culture and science had the authority to approve "the sell [*sic*] or temporary exchange of historical and cultural [*sic*] valuable object." But this fact remained: it was illegal to sell or transfer ownership rights of such objects to foreigners. The Mongolian legal adviser to President Elbegdorj who researched the relevant laws was Bayartsetseg Jigmiddash, who had been a visiting scholar at Columbia Law School and, in 2008, earned her law degree at Harvard.

202 "I stand to be educated" and related: I covered the hearing and obtained a transcript from Southern District Reporters, the court reporting agency that took dictation for the Southern District of New York. The paleontologists' letters can be found in 12-CIV-4760, U.S. District Court, Southern District of New York.

203 "I'm just a guy": Eric Prokopi statement to the press.

203 "Frankenstein": Hearing transcript, 12-CIV-4760, U.S. District Court, Southern District of New York.

CHAPTER 18: RAID!

207 eBay items: I documented sales items/comments from the Florida Fossils account as the *T. bataar* legal matter unfolded in 2012 and 2013.

207 "It's not like we murdered someone!": Interviews with Amanda Prokopi.

207 "People might think": Ibid.

208 "Like, America just got *involved*": Ibid.

208 "This couldn't come at a worse time": Papers of Eric Prokopi.

209 "Dear Butts": Hollis Butts told me by email, "I owed no money."

210 Otgo: These details came from my extended interview with Otgo in Ulaanbaatar in August 2015, and via other sources.

212 "Isn't it because Mongolian paleontologists": B. Odontuva, "Ts. Oyungerel: When Eric Prokopi Was Arrested, Delivery of Another Dinosaur Skeleton Arrived at His Home," *UB Post,* October 26, 2012.

212 "I *also* had the fear": Ibid.

212 "Financing Geology, Paleontology, and Natural Science Study": I had this contract translated from Mongolian to English by someone unrelated to the *T. bataar* case.

212 Libel laws and media transparency: The Mongolian government generally respected press freedoms and did not interfere with internet access, as
happened in other countries, but "many journalists and independent publications practice[d] a degree of self-censorship to avoid legal action...," Freedom House reported in its annual human rights report in 2012. The burden
of proof fell, unusually, upon the defendant. Journalists had been sued by
members of parliament, businesspeople, and private organizations, and had
been unfairly subjected to tax audits. In 2015, one blogger would be sentenced to a hundred days in prison for badmouthing a government minister on
Twitter, "the first time Mongolia's libel laws were extended to comments made
on social media." Reports can be found at freedomhouse.org. Also see Lisa
Gardner, "Mongolia's Media Laws Threaten Press Freedom," *UB Post*, May 4,
2014.

213 "Is Eric Koproki here?": Interviews with Amanda Prokopi.

213 "These nice men": Ibid.

214 "Why are there TV trucks": Ibid.

214 "do a bail" and "always been angels": Ibid.

214 "It's not just *stuff*": Ibid.

214 Told Amanda not to open that box: Interviews with Amanda Prokopi and
Georges Lederman.

214 "This fresh conduct": AUSA Martin Bell's request and other documents can be
found in *United States of America v. Eric Prokopi*, 12-CRIM-981, U.S. District
Court, Southern District of New York.

215 "We want to make this illegal business practice extinct": United States Attorney's
Office, Southern District of New York, "Manhattan U.S. Attorney Announces
Charges Against Florida Man for Illegally Importing Dinosaur Fossils," press
release, October 17, 2012.

215 "merely the tip of the iceberg": Ibid.

215 "like he was some bad guy" and related: Interviews with Amanda Prokopi.

215 "treasure hunters": Interviews with Tyler Guynn and Eric Prokopi.

215 "This is going to sound": Interviews with Tyler Guynn and Eric Prokopi.

216 "I didn't *mean* to kill you": Interviews with Amanda Prokopi. Later, she said,
"Maybe it's my stupidity, but I kind of get it." People do crazy things when
they're stressed. Not that she was *completely* chill about it. At one point, Tyler
texted Amanda to try to smooth things over because they were all about to
attend the same black-tie event in Gainesville. "I'm sorry that my decisions have
affected you—I mean I'm a class act," Tyler told me she told Amanda. "She sent
me this long thing: 'I don't know what to tell ya. You stole the love of my life,
and it wasn't nice.'"

216 "I can't go through my mourning": Interviews with Amanda Prokopi. The
children "love Tyler and she's really good with them, and as a mom I can't ask
for more," Amanda later said. "And she's taken care of Eric. I love Eric enough
that I just want him to be happy, and if being with her makes him happy, that's
fine."

216 Guilt: Interviews with Amanda Prokopi and Tyler Guynn.

216 "Don't worry": Interviews with Amanda Prokopi.

216 October 22 court hearing: I covered this hearing.

216 Count 1: Prokopi had ordered a *Microraptor* skeleton from a dealer in China. When UPS called to ask for an accurate description and value, Prokopi had emailed the dealer, "What did you write on the shipment? I need to know what to tell them." The dealer replied, "Sample of craft rock" with a value of thirty dollars. After days passed without the shipment clearing customs, Prokopi had emailed the dealer, "I am worried that they will take the item because you declared the value too low." The dealer had asked what he should do with the "other 8 ones," and what value to declare. Prokopi had responded, "Geological specimen for collection. Value $1000." The dealer had told Prokopi, "Selling fossils like this is not allowed in China." Homeland Security had intercepted the *Microraptor*. Prokopi had protested the seizure by letter, complaining that he couldn't afford to lose his thousand-dollar investment. He had faulted the Chinese supplier for the "mixup," saying he "did not speak English well." See 12-CRIM-981, U.S. District Court, Southern District of New York.

216 "multiple containers": See 12-CRIM-981, U.S. District Court, Southern District of New York.

216 "taken from Mongolia": Ibid.

217 "about half a million" and related: For more information, see 12-CRIM-981, U.S. District Court, Southern District of New York.

217 stricter bond agreement: The prosecution's evidence that day included the long-ago letter from David Webb, of the Florida Museum of Natural History, who had written to Prokopi, banning him from the collection and quarries. The judge noted that the Brooksville incident had happened eighteen years earlier. Prokopi's criminal defense attorney, Georges Lederman, argued that the land-owner had once given Eric permission to hunt at the Brooksville site, and that Prokopi was the person who had discovered it. He argued against the higher bond amount, saying the case had left Prokopi with no way to make a living. "I mean is there really a risk he'll leave the country?" the judge asked AUSA Martin Bell, who replied, "He's a party in the black market.... These are real concerns." Meanwhile, the attorneys in the *in rem* case were on their way out but filed one last attempt to have the *T. bataar* action dismissed, writing, "If Government investigators honestly believe that Prokopi engaged in criminal conduct cumulating [*sic*] in consigning the Tarbosaurus Bataar Display Piece...for sale in *public auction*, why not treat the matter as a criminal one from the start?" See "Reply Memorandum of Law in Support of Claimant Eric Prokopi's Renewed Motion to Dismiss and Request for Expedited Consideration," 12-CIV-4760, October 23, 2012, U.S. District Court, Southern District of New York.

217 "was like rubbing the tarbo": I obtained emails confirming this conversation.

218 "Tell me what you did": The federal magistrate was Ronald L. Ellis. I obtained a transcript of the hearing, which is part of *USA v. Eric Prokopi*, 12-MG-2634, U.S. District Court, Southern District of New York.

CHAPTER 19: VERDICT

219 Divorce: Details can be found in the Prokopi divorce case file, 2013- DR-002572, a public document that I viewed in person via the Alachua County Clerk of Court, in Gainesville.

220 "This is the least of what's depressing": Interviews with Eric Prokopi.

221 "treasure hunting": Interviews with Eric Prokopi and Tyler Guynn.

222 "Ultimately we're the house": See "Author's Note."

222 "It did take some work...speaking with him": Hearing transcript.

222 "Is that moral?": *The Orchid Thief,* by Susan Orlean (Random House, 1998).

223 "He's such a difficult person for the average person to believe": See Author's Note on sourcing.

223 Mongolian detainment: One State Department report quoted Mongolian law as stipulating "there need be no actual arrest warrant or any sort of official determination that charges are warranted: mere complaint by an aggrieved party is sufficient to deny exit." See U.S. embassy, "2013 Mongolia Investment Climate Statement," Ulaanbaatar, January 15, 2013. For information about detainments of foreign businesspeople see Hannah Beech, "The Jailing of Foreigners in Mongolia Is Unnerving the Business Community," *Time,* February 5, 2015. Also see Julie Makinen, "American, 2 Others Pardoned in Mongolia Tax-Evasion Case," *Los Angeles Times,* February 26, 2015.

223 "There the unfortunate prisoners": Roy Chapman Andrews made this comment at a New York City lecture in front of nearly two thousand people, his first public address after returning from the Gobi. See "Gobi Desert Life Seen by Andrews," *New York Times,* November 25, 1923. This and other hard-copy clippings otherwise unavailable via the *Times*'s online database were found in the American Museum of Natural History's research library.

223 "The most fiendish tortures": Ibid.

223 "Torture persists, particularly in police stations": See "Civil and Political Rights, Including: The Questions of Torture and Detention," by Manfred Nowak, UN Special Rapporteur on torture and other cruel, inhuman, or degrading treatment or punishment. United Nations Commission on Human Rights, December 20, 2005.

224 "no one shall be subjected to torture": Mongolian Constitution.

224 "flying to space": Nowak, "Civil and Political Rights." The term calls to mind descriptions of torture by one old Mongolian who participated in the purges of the 1930s: "They were placed on tall stools next to a hot stove for interrogations that sometimes lasted days....Soviet advisers had told the Mongolians to use such methods." See Kathy Chenault, "Mongolians Seek Truth of Purges," *Los Angeles Times,* August 16, 1992.

224 "Mere complaint by an aggrieved party": U.S. Embassy, "2013 Mongolia Investment Climate Statement," January 15, 2013.

224 "state secret": Nowak, "Civil and Political Rights."

224 "among the most restrictive and punitive": See "State Secrets Law: An Invitation to Corruption and a Blight on Mongolia's Human Rights Record," a "sensitive but unclassified" cable from the U.S. embassy in Ulaanbaatar to the State Department, dated March 31, 2006. The state-secrets law extended the definition to "not only national security interests but also to maps finer than a 1:200,000 scale, to statistics on the number of prisoners, to basic economic and census data, to the identity of shareholders in private companies, to audits of state owned companies, to access by citizens to state archives." Mongolia had not only a 1995 Law on State Secrets but also a 2004 List of State Secrets, which altogether "set up such far reaching restrictions on access to government records in Mongolia

as to make it possible for virtually anything to be classified as 'secret' and hidden from the public view for an indefinite period." This was the kind of secrecy that bred "irresponsibility" among government officials, the embassy pointed out, saying, "The lack of transparency leads to corruption."

224 "Authorities remain fearful": See "Assessment of Corruption in Mongolia," a 2005 report by USAID.

224 "complete isolation": Nowak, "Civil and Political Rights."

225 "I feel that he did something wrong": This conversation took place in Land O' Lakes in August 2014, a few weeks before Prokopi went to prison.

226 "I am doing thomthing": Papers of Eric Prokopi.

226 "law enforcement renaissance": Sentencing transcript, *USA v. Eric Prokopi*, 12-CRIM-981, U.S. District Court, Southern District of New York.

227 "He described his childhood": "Presentence Investigation Report," *USA v. Eric Prokopi*, 12-CRIM-981, U.S. District Court, Southern District of New York. Obtained through the Freedom of Information Act.

228 "My concern": Sentencing transcript, *USA v. Eric Prokopi*, 12-CRIM-981, U.S. District Court, Southern District of New York.

228 "Everyone averts his eyes" and related: Ibid.

228 "I have to feel": Judge Hellerstein said, in court, "No one in this business takes things or buys things without looking at the provenance of an item. You are not buying a fake in a store. You are not buying a copy of a piece of art. You are buying something because of its quality as a fossil. Your client was an expert at that. So much so that he had the ability to create an exact replica of a dinosaur from these fossils, and he did [so] to great value." Sentencing transcript, *USA v. Eric Prokopi*, 12-CRIM-981, U.S. District Court, Southern District of New York.

228 "in recognition and consideration": Martin Bell wrote what's called a "501 letter" to the court, essentially recommending leniency at sentencing. Bell's comments in court are found in the sentencing transcript.

229 "size up Mr. Prokopi": Sentencing transcript.

230 "in a place where there was a climate": See Rachel Abrams and Peter Lattman, "Ex-Credit Suisse Executive Sentenced in Mortgage Bond Case," *New York Times*, November 22, 2013.

230 "This is a deepening mystery": Ibid. Of course nobody had the answer. Susan Orlean, in *The Orchid Thief*, might've come close in her assessment of the poacher Laroche: "I never thought very many people in the world were very much like John Laroche, but I realized more and more that he was only an extreme, not an aberration—that most people in some way or another do strive for something exceptional, something to pursue, even at their peril, rather than abide an ordinary life." Laroche told Orlean, "It's not really about collecting the thing itself. It's about getting immersed in something, and learning about it, and having it become part of your life. It's a kind of direction."

230 "Mr. Prokopi is an unusual person" and related: Sentencing transcript, 12-CRIM-981, U.S. District Court, Southern District of New York.

230 "Okay, we can deal with this": Interviews with Amanda Prokopi.

231 "Just a word of prison advice": Prokopi used his phone to record the monologue and post it to Facebook. Also see "Late Night with Seth Meyers Monologue Highlights—6/4," June 5, 2014, https://www.broadwayworld.com/bwwtv/

article/LATE-NIGHT-WITH-SETH-MEYERS-Monologue-Highlights -64-20140605.

231 "Well, at least you're famous" and related: Eric Prokopi Facebook page.

CHAPTER 20: TARBOMANIA

232 "You're in the media": Interviews with Oyuna Tsedevdamba.

232 "Thank you, dinosaur!": Ibid.

233 "something of a (literal) black box to voters": See Julian Dierkes, "Monitoring the Election," *Mongolia Focus*, June 27, 2012. Dierkes is an associate professor at the University of British Columbia's Institute of Asian Research.

233 "Despite their alleged": See Bumochir Dulam and Rebecca Empson, "The Black Box of Presidential Politics," UCL Emerging Subjects Blog, June 17, 2017, blogs.ucl.ac.uk. Bumochir and Empson sought not to prove or disprove wrongdoing but rather to capture the *"form that politics is currently taking in Mongolia"* (italics in the original).

233 "significant promotion": Oyuna said this to a Mongolian tourism-centric website that's no longer active, but you can find a good overview of her thoughts on Mongolia's enormous potential and obstacles in "Widening Appeal: OBG Talks to Ts. Oyungerel, Minister of Culture, Sports, and Tourism of Mongolia," *The Report: Mongolia 2013* (Dubai: Oxford Business Group, 2013), available at oxford businessgroup.com.

234 "retrieval operation": B. Khash-Erdene, "Economic Forum 2013 Aims to Establish Mongolian Brand," *UB Post*, March 7, 2013.

234 "Mongolia's been criticized": Al Jazeera video, http://www.yousubtitles.com/ Mongolian-dinosaur-returns-to-heros-welcome-id-1291147.

234 "Roy Chapman Andrews's expedition": Interviews with Oyuna Tsedevdamba.

234 "If they decided": Tania Branigan, "It's Goodbye Lenin, Hello Dinosaur as Fossils Head to Mongolia Museum," *Guardian*, January 27, 2013.

234 "We have a wonderful dinosaur heritage": Ibid.

235 "In this business": Ralph Blumenthal, "Dinosaur Skeleton to Be Returned to Mongolia," *New York Times*, May 5, 2013.

235 "We simply cannot allow the greed": Video of the *T. bataar* repatriation ceremony can be found on YouTube.

235 "every Mongolian's pride and idol": See "Another Name Change for Ulaanbaatar's Main Square," News.MN, September 16, 2016.

235 Auspicious days for weddings: In the summer of 2015 I was walking near the square when I looked up and saw a couple of newlyweds on the open ledge of an unfinished building, hundreds of vertiginous feet above the pavement, posing for a camera like a couple of costumed daredevils. Other details about the current state of the square are my observations from my time in Ulaanbaatar; but the fin/sail/etc. characterizations are mentioned anytime a reporter describes the shape of that distinctive hotel.

236 "She's mostly castrated": Interviews with Oyuna Tsedevdamba.

237 "dying on the square": Ibid.

237 "It's not bone" and related: Ibid.

237 "made up": Interviews with Oyuna Tsedevdamba, Jeff Falt, and Robert Painter.

237 "Okay, you guys" and related: Interviews with Oyuna Tsedevdamba. Various videos are available on YouTube of the *T. bataar*'s reemergence in Mongolia. One

short spot shows the skeleton at the Sukhbaatar Square pop-up exhibit before the LED screens and display cabinets went in. See https://www.youtube.com/watch?v=RwI4wgmlhd4.

238 "Those expeditions": One day at the city museum the curator, B. Tungalag, was kind enough to show me the original Roy Chapman Andrews signature on a contract from the 1920s, "leader" just barely showing anymore, in blue ink.

239 "enriched the true nature": The statement has been removed from the president of Mongolia's website but an Internet Archive copy is available. See "Remarks by Mr. Puntsagiin Tsagaan Chief of Staff of the President of Mongolia at the Tarbosaurus Bataar Return Ceremony," May 18, 2013.

239 Rumors about vote-rigging: Bumochir and Empson, "The Black Box of Presidential Politics."

239 "overall lack of transparency": See "Mongolia Presidential Election, Election Observation Mission Final Report," Office for Democratic Institutions and Human Rights, Organization for Security and Co-operation in Europe, June 26, 2013. (The OSCE is an intergovernmental partnership among fifty-seven states that monitors arms control, elections, human rights, and press freedoms.) Many related reports are available online through OSCE and other organizations such as Freedom House and Transparency International.

239 "programming error," "correctly calculating," "essential," "would not affect": Ibid.

239 "directly or indirectly owned": Ibid.

239 "black PR": Ibid.

239 "A lack of transparency": Ibid. The OSCE report notes, "Freedom of information, including the right to access information held by public authorities, is a core element of the guarantee of freedom of expression. This principle allows only exceptional limitations that must be previously established by law in case of a real and imminent threat to national security. Allowing for other secrecy provisions to override the right to information fails to respect these principles." The struggle for transparency is ongoing. See Michael Kohn, "Nothing to See Here: Mongolia Media Goes Dark to Protest Curbs," Bloomberg, April 27, 2017.

239 Hillary Clinton: Clinton first visited Mongolia in 1995, as First Lady. "It was important for the United States to show support for the Mongolian people and their elected leadership, and a visit from the First Lady to one of the most remote capitals of the world was one way to do it," she wrote in her memoir *Living History* (New York: Simon & Schuster, 2003). Deputy Secretary of State Winston Lord, who had been the first to say Mongolia should be "held as an example for anyone doubting democracy's ability to take root in unlikely places," told the *New York Times,* "A visit like this can put a spotlight on a country like Mongolia, which Americans pay little attention to." In diplomatic terms, the First Lady outranks even the secretary of state, so Clinton was the highest-ranking American to visit Mongolia since a fleeting stop by Vice President Henry Wallace in 1944. Her ten-car motorcade set out for the *gers* of a herder named Zanabaatar, his wife, Haliun, and their six children, roughly an hour's drive outside the city. Wearing black suede boots, Clinton carried a handmade saddle as a gift, which she presented to Zanabaatar as she explained that her husband too "came from a region with horses and cattle." She learned how to milk a horse. She exulted at

the sight of a yak. In a speech at Mongolian National University, she praised the Mongolian people's courage and urged them to "continue their struggle toward democracy." Then she told a story she'd heard, about a lesson Genghis Khan's mom had tried to teach her children: "To make her point, she showed them how easily one arrow could be broken, but how hard it is to break five arrows when they are held together." See Seth Faison, "After China, Hillary Clinton Finds Mongolia a Gentler Place," *New York Times*, September 8, 1995.

240 "continue down the democratic path": See Hillary Clinton, *Hard Choices: A Memoir* (New York: Simon & Schuster, 2014).

240 "perfectly at home": Addleton, *Mongolia and the United States*.

240 "hope that the U.S. pivot to Asia": This is from a *Washington Post* editorial, "A Proper Pivot toward Asia," July 14, 2012.

240 Peabody Energy: I obtained details of the meetings involving Clinton and Elbegdorj from the State Department through the Freedom of Information Act. Much of the file, including large amounts of the material involving Peabody, was redacted.

240 "our Oval Office": Addleton, *Mongolia and the United States*.

240 "information": This, detail, too, was obtained through the Freedom of Information Act. U.S. Department of State to the Department of Homeland Security, *Report of Investigation No. 006*, May 19, 2012. The State Department's mission is to "shape and sustain a peaceful, prosperous, just, and democratic world and foster conditions for stability and progress for the benefit of the American people and people everywhere," according to its mission. More and more, this cannot help but involve Russia, China, and North Korea.

241 "dinosaur *bataar*": Elbegdorj's speech is archived on YouTube at https://www .youtube.com/watch?v=qV16q2PMZG4. See "A Public Address by His Excellency Tsakhiagiin Elbegdorj, President of Mongolia," September 21, 2012. Before the speech, Elbegdorj had stopped off at the Coop, at Harvard Square, to buy twenty-five Harvard T-shirts (five for his biological children and the rest for the kids he and his wife had adopted). His old dean and adviser, David Ellwood, introduced him. Former Massachusetts governor William Weld sat on the front row. Former classmates and heads of state from Ecuador and Greece were in attendance. Elbegdorj often speaks the language of inspirational posters: "A man can never ruin what he built with heart and soul" and "No authoritarian government can stand against the collective will of a people determined to be free."

241 "Opportunities for corruption": For more information, see USAID, *Assessment of Corruption in Mongolia, Final Report*, August 31, 2005. Also see Transparency International's corruption-perception index. In 2017, the organization gave Mongolia a score of 36 on a scale of 0 ("highly corrupt") to 100 ("very clean"). Mongolia fell between Côte d'Ivoire and Tanzania, and behind 103 other countries. (The U.S. had a score of 75, tied with Belgium and just ahead of Ireland.) See https://www.transparency.org/country/MNG.

CHAPTER 21: PETERSBURG LOW

242 "God has subpoenaed": I was there and saw it. You can find photos at www .paigewilliams.com.

243 "Get up here" and the rest of this section and the next: Ibid.

244 Fortieth birthday and fish funeral: Ibid.

245 "Aren't you sick": Interviews with Amanda Prokopi.

245 All-night box-truck trip: I was there.

245 "It was the size of a small dog": Ibid.

245 Tony's photos: Ibid.

246 Joe and Charlene's: Ibid.

246 "You loot, we shoot": Ibid.

246 "not Braille blind": Interviews with Amanda Prokopi.

246 "Fuck the government!": Interview with Joe Kutis.

247 "That's Prokopi": I was there.

247 Prison day: I was there for everything but the school dropoff, which the Prokopis told me about later. Other information comes from interviews with Amanda and Eric Prokopi. On the night before he went to prison, Eric helped Amanda load the van with Bizarre Bazaar items, then parked it in Betty's garage, beam to floor, in preparation for the Christmas show. Eric watched *Indiana Jones and the Temple of Doom* with the kids, then made a final trek to the river house. Homeland Security had returned some of his possessions, including the family iMac, so when he got back to Betty's, he checked his email, then posted to Facebook: "On Tuesday I will be surrendering myself to the federal prison in Petersburg, Virginia, to begin the three-month incarceration that my government has deemed necessary to rehabilitate me." It worried Amanda that he felt like such victim. She kept thinking, "There's gonna have to be a point where all the excuses end, and he just moves on." Tyler, on the other hand, had heard Eric acknowledge his mistakes, and at one point told me, "I guess with him it feels like they're taking so *much*, that it's cost so *much*."

250 "Mayday": Eric Prokopi prison diaries and interviews, plus interviews with Amanda Prokopi, as well as photos and documents she had saved.

251 "great apple muffin," "Started playing," "Felt sad": Eric Prokopi prison diary.

251 "fantasizing": Papers of Amanda Prokopi. Amanda planned to visit Eric in prison maybe weekly, but that didn't happen. When he told her he planned to eBay full time when he got out, and to prep for other dealers, she worried that he planned "just to step right back into his old life. Which is gone." At one point, he talked about starting a scrapyard, and she sent him a book on the topic.

251 "I loved that trip": Ibid.

251 "just go": Ibid.

251 "a cozy nest": Interviews with Amanda Prokopi.

CHAPTER 22: THE DINOSAUR BUS

252 "Like, the *hole*": Interviews with Bolor Minjin.

252 Bolor Minjin in Ulaanbaatar: I was reporting in Mongolia at the time Minjin made her Gobi excursion, and accompanied her on part of it.

253 "Nature-Study and Literature": A good resource is the AMNH research library. Another is *Teaching Children Science: Hands-on Nature Study in North America, 1890–1930*, by Sally Gregory Kohlstedt (Chicago: University of Chicago Press, 2010).

253 "started with a suitcase": Jonathan Mandell, "All Aboard for a Museum that Rolls On and On," *New York Times,* August 20, 2000.
254 "Tom Sawyer of paleontology": Gardner, "Digging for Dinos."
254 "gets people excited": Ibid.
254 "one of Silicon Valley's favorite villains": Nick Wingfield, "Debating the Merits of Patent Warfare," *New York Times,* May 30, 2012. See also Jennifer 8. Lee, "An Auction of 'Nature's Sculpture': Rare Dinosaur Skulls," *New York Times,* June 1, 2009.
254 "To be blunt": John Noble Wilford, "Did Dinosaurs Break the Sound Barrier?," *New York Times,* December 2, 1997.
255 PERC: For more information, see perc.org. Archives of *PERC Reports* can be found at https://www.perc.org/report-archive/.
255 Ohrstrom foundations: Bolor Minjin credited Gerry Ohrstrom's Epicurus Fund with the donation that covered the shipment of the dinosaur bus from New York to Mongolia. There are also Ohrstrom family foundations, both listed on Charity Navigator at 665 Fifth Avenue in New York. One is the George L. Ohrstrom Jr. Foundation, which in 2016 reported over $59 million in assets and paid $3.5 million in grants and contributions. The other is Ohrstrom Foundation Inc., which in 2016 reported over $55 million in assets and paid $1.2 million in grants and contributions. Regular beneficiaries include hospitals, libraries, the CIA Officers Memorial Foundation, fire departments, food pantries, and nature and wildlife conservation organizations.
255 "huge disservice": Ibid.
255 "private landowners": Ibid.
255 "Once upon a time": Linda E. Platts, "Fossil Farming Blooms Where Barley Withers," *PERC Reports,* Fall 2007.
257 UAZ Bukhanka: The UAZ is a photogenic torture box ubiquitous in Mongolian tourism. At least that was true of the one I hired. No discernible shock absorbers, no seat belts. Every bump in the road registered in the teeth. I'm no fragile flower but forget it, I was done. If your seat belt–free UAZ happens to catch the jagged edge of the eroding pavement at a high speed, you are finished.
258 "tourism": I was there.
259 Children and horse racing: For a fascinating twenty-five-minute TV segment on this, see Drew Ambrose and Daniel Connell, "Mongolia's Child Jockeys Risk Death to Race," *Al Jazeera,* August 28, 2017, aljazeera.com.
259 "We are going to party": Interviews with Bolor Minjin.
259 "Nomads are really good singers": Ibid.
260 "Someday they can sell it": Interviews with Selenge Yadmaa.
260 "Good for health": I was there.
261 "Maybe I will make an arrest": Ibid.
261 "Thank you very much": Ibid.
262 "drooling with fossils": Interviews in Mongolia and later by Skype with Dr. Rosemary Bush and Dr. Steve Won. Bolor Minjin later recounted the discovery of the eggs to a reporter, saying, "We didn't know what to do. I had a park ranger with me, but she didn't have the skill to dig out that fossil. Even if she had the skill, she has no place to store that fossil. If she even had a place to store it, she doesn't have the lab to prepare it. It's one problem after another that we were

facing." See Jacqueline Ronson, "How to Shut Down Black Markets in Velociraptor Country," inverse.com, February 7, 2017.

262 "Those are *eggs*": Interviews with Rosemary Bush and Steve Won.

262 "they would wind up": Ibid.

EPILOGUE

263 "unfailing gift": "Dr. Andrews Resigns," *New York Times*, November 11, 1941.

263 "not only made the Museum a vital force": Ibid.

263 stolen: "Mongols Accuse Explorer of '25," *New York Times*, September 9, 1956.

263 "perfectly ridiculous": see "$5,000 Dinosaur Egg stolen from Colgate U," the Troy Record, March 26, 1957. The egg was insured for $10,000. On the night of March 22, 1957, the glass case in which the egg was kept in Lathrop Hall was shattered, the egg stolen. The fossil was quickly found beneath the steps of a local church after Dr. Carl Kallgren, dean of the college, received an anonymous phone tip. (Some press reported that the egg was found on the church lawn.)

263 Auctioned egg: The Robert M. Linsley Museum at Colgate, in Hamilton, New York, calls the egg its most famous fossil.

264 "necessarily permanent": Novacek, *Dinosaurs of the Flaming Cliffs*.

264 Philip Currie: See Tristin Hopper, "How a $34M dinosaur museum in northern Alberta went from big awards to bailouts," *National Post*, December 4, 2016. See also Terry Reith and Briar Stewart, "Dinosaur museum in small-town Alberta hopes to hit the big time," CBC News, September 2, 2015. Also Dan Ilika, "Celebrity Dino Dig Brings Stars," *Grand Prairie Daily Herald Tribune*, July 3, 2011.

265 *Once upon a time*: I bought an English version of *A Story of Tarbosaurus Bataar* at the Central Museum of Mongolian Dinosaurs in August 2015.

266 "I fulfilled my duties": Interview with Otgo, Ulaanbaatar, August 2015.

266 Tsogtbaatar: Tsogtbaatar and I talked in his office in Ulaanbaatar in August 2015. He told me about coming to the States for the inspection, naming the team as "Phil Currie, from Canada, and Mark Norell, from MNH; and I'm from Mongolia." He didn't mention Bolor Minjin except to say that after she finished her PhD, he "invited her to join this official institute, and she organized some NGO instead." He said that in the early days, after the fall of Communism, the Mongolian Academy of Sciences always attached paleontologists to foreign expeditions. He said Bataa, or Bat, the digger Eric Prokopi had met through Tuvshin, had once worked in the prep lab. "There are many herds people, especially old people, old herdsmen, they love to protect the illegal diggers' activity," he said, adding that they probably did so to get paid. Herders acted as park rangers but also as scouts. I didn't yet know about the *GEO* article, much less have an English translation. In Tsogtbaatar's office that day, I asked him what the difference would be, between the Natural History Museum (should it ever reopen) and the Central Museum of Mongolian Dinosaurs. By now the United States was repatriating skeleton after skeleton, to be housed in the old Lenin Museum. "The idea [for a dinosaur museum] is very good but it's materialized wrong," he told me. "There isn't any specialty of paleontology. I was trying to help them but we have our own job. By my opinion it's illegal diggers' specimens, never published. It's a shame to show the public."

267 "Are you kidding": See "Author's Note." Dealers tended not to like it when report-
ers like me showed up, asking questions. "You might as well be the FBI," one dealer
told me one year at Tucson. Another said, "You know Tuvshin is dead."
"I know," I said.
"Do you know how he died?" the dealer said.
"How'd he die?"
"Now see, that's the question."
"What's your understanding?" I said.
"I don't have an understanding; I just have a fear."
"Of?"
"Mmm hmm."
"You feel he was killed?"
The dealer just stared at me.

267 "In most cases": See George F. Winters, "International Fossil Laws," Association
of Applied Paleontological Sciences website, aaps-journal.org.

267 "I thought it was okay": Interviews with David Herskowitz.

268 Coleman Burke: The buyer of the *T. bataar* hoped to erect the skeleton in an historic building he had bought in 1983 at West Twenty-Sixth Street and Twelfth
Avenue, along the Hudson River in New York City. This is West Chelsea, and
the property, once home of the Central Stores Terminal Warehouse Company,
held over a million square feet of space. Double railroad lines once ran directly
from the river and through the riverside arch and out the other side, onto Eleventh Avenue. The largest of its kind in the city, the building held various goods
in transit—carpets, furniture, furs, stage props, antiques. By the 1980s West
Chelsea was desolate but coming back; in December 1986, a nightclub called
Tunnel became a tenant. It occupied the cavernous ground-floor space that still
hinted of trains. Tunnel would last, famously, until 2001, after hosting perfor-
mances by artists like Prince and being burdened, infamously, by drugs and violence. There wasn't anything quite like it in New York.

By the time the *T. bataar* went to auction, Burke's real estate firm, Water-
front New York Realty Corp., operated on the seventh and top floor, and Burke
kept a private crow's nest of an office just above that, at the top of a spiral staircase. Homey and compact, the office seemed like less a work space than a living
scrapbook of Burke's many interests, including the piano (he played), antique
maps, and the Explorers Club, where he was a member. Picture windows overlooked the Hudson—Burke could watch harbor cruisers, sailboats, and water
taxis plow the river. On a small observation deck, he kept a large signal light and
nautical flags, and enjoyed using them to communicate with passing ships. A
cast *Giganotosaurus carolinii* skull occupied half the eastern wall.

One of Burke's chief interests was Charles Darwin. When Burke was about
fifty-five, he was walking the badlands of Argentina and came upon a moon-
scape of a boneyard. Argentina is among those countries that ban fossil collecting without a permit. Burke logged the location by GPS and told me he "went to
tell the good guys." He began funding expeditions. In thanks, the paleontologist
Fernando Novas later named a late Cretaceous dinosaur after him, *Orkoraptor
burkei*, one of the most southernmost carnivores ever found in South America. Burke funded fossil expeditions through Yale, and he once backed Nate

Murphy, the Montana dinosaur hunter who went to prison. "People say, 'How's the archaeology going?' and I say, 'That's not my bag.' If it's not a hundred million years old, I don't do it," he told me.

The architect Richard Meier was once interviewed at a party at Tunnel. As he studied "the dungeonlike expanse of brick, granite, and steel girders," he told the *New York Times*: "It's a great space—you can't make it better than this." But Burke thought of a way. In the spring of 2012, he heard about the Heritage auction and went to see the *T. bataar* at the preview, where he got so excited he camouflaged his interest by pretending to browse the other items. He had only twice bought anything at auction: an antique coffee grinder and, on Nantucket, a set of candlesticks on behalf of his wife, who was busy playing golf. The dinosaur was a lark, but a big one. Burke won the auction. But as the legal entanglements set in, he withdrew his offer.

268 Chait: For more information on Joseph Chait's criminal case see the Department of Justice release from June 22, 2016, "Senior Auction Official at Beverly Hills Auction House Sentenced to Prison for Wildlife Trafficking." Also see Brendan Pierson, "Auctioneer Gets Year in U.S. Prison for Ivory, Rhino Horn Smuggling," Reuters, June 22, 2016.

268 "the Babe Ruth of forfeiture": See Robert Lenzner, "The Babe Ruth of Forfeiture is After Your Ill-Gotten Gains," *Forbes*, August 14, 2012.

269 Mickelson: See Bob Verdi, "Q&A with Phil Mickelson," *Golf Digest*, April 23, 2009.

269 Hollis Butts: He emailed me a couple of times, asking for information (and not getting it) about "what Prokopi & Tuvshin and others were up to" during a certain period. "Most of the Mongolian fossils in the market would not otherwise have been collected however some possibly would have been collected and that loss needs to stop," he told me. "The great majority of what was offered were duplications of already common fossils, protoceratops [sic] for example, and I was actually pleased that these specimens had been saved from the fate of all uncollected fossils, but occasionally I would see or hear about a specimen that should have been carefully excavated for study...Tuvshin assured us that export was permitted and that, additionally, nobody actually cared." Butts claimed to have received emails from an American prepper, only to learn that the account was being used by a Homeland Security Investigations agent in disguise. Anyway, he had a lot of thoughts on the whole matter. At one point he noted how "distant these events revolving around Prokope [sic] are from the actual work done by the average full-time or part-time person in the fossil trade who will be negatively impacted by the politics of these events." He also said there are two kinds of fossil buyers, "those who love fossils and those who want an interesting decoration, but mostly it was love."

269 Documentary: *Attenborough and the Sea Dragon*. The one-hour documentary is available on *BBC One* though not, apparently, in the United States.

269 Mary Anning: The Mary Anning wing opened in July 2017, overlooking the beach that Mary once combed. For more information see the Lyme Regis Museum's website, http://www.lymeregismuseum.co.uk/about-us/mary-anning-wing/.

269 Case outcome: The dealer Tom Lindgren told me the case led to more self-monitoring in the trade. "If anything came out of this that's good it's that everyone is afraid to not do the homework that they should do. Everyone is policing everyone now. It's going

to be much more difficult in the future for people to do illegal business." Lindgren said he sold a *T. bataar* skull with Guhr because at the time "you couldn't find a published law. Even if you went on the internet you couldn't find a lot. It was on the books in Mongolia, but in Mongolia you could go to customs to ship something, give them some money, and it's gone, no problem. No one was policing it. What good is a law if it's on the books for seventy years and no one's enforcing it? How do people then know that the law really has teeth? Or that it even exists?" Mongolian law enforcement had, in fact, arrested a few people in fossil-trafficking cases, as the government relayed to U.S. authorities as the Prokopi cases developed, but that news was difficult to find.

270 "When I die": See Jay Cridlin, "The Bone Collector," *St. Petersburg Times*, March 21, 2003.

270 "buy the very same [voting] machines": See Bumochir and Empson, "The Black Box of Presidential Politics."

271 "too far away": See Andrew Higgins, "In Mongolia, Lessons for Obama from Genghis Khan," *Washington Post*, June 15, 2011.

271 "The president feels bad for him": Interviews with Robert Painter.

272 "sort of holy pilgrimage": mongoliandinosaurs.org.

273 "All the things": Ibid.

274 "Hey, guys!": I was there.

274 "Aren't you anxious": Ibid.

274 "Where's the moon": Ibid.

275 "That's called a mold": Ibid.

275 "We declared our love": Interviews with Tyler Guynn.

276 "fate unknown": See "Noka (YNT-22) ex Noka (YN-54) (1941)," navsource.org.

276 "The tiled floor": Sales ad, archived on eBay.com.

277 "Why can't you just be normal": Interviews with Eric Prokopi.

INDEX

Page numbers followed by "n" *indicate notes.*